Agrarian Transformation in Western India

This book examines the economic gains and social costs of agrarian transformation in India. The author looks at three phases of agrarian transformation: colonial, post-colonial, and neoliberal. This work combines macro and micro economic data, economic and non-economic phenomena, and quantitative and qualitative aspects while exploring the context of historical and contemporary changes with special reference to Maharashtra in western India. It discusses regional disparities in agricultural development, issues of modernisation and social inequality, land owning among scheduled castes and tribes, women in agriculture, pattern of labour migration and farmer's suicides, and documents the experiences and conditions of the rural poor and socially weaker sections to provide a comprehensive understanding of the significant changes in agrarian rural economy of western India. It also discusses contemporary development policy and practices and their consequences.

Lucid and topical, this volume will be useful to scholars and researchers of agrarian studies, rural sociology, social history, agricultural economics, development studies, political economy, political studies, and public policy, as well as planning and policy experts.

B. B. Mohanty is Professor of Sociology at Pondicherry University in Puducherry, India. His areas of interest include agrarian transition and crisis in India. He is the Editor of *Agrarian Change and Mobilisation* (2012) and *Critical Perspectives on Agrarian Transition: India in the Global Debate* (2016).

Agrarian Transformation in Western India

Economic Gains and Social Costs

B. B. Mohanty

Routledge
Taylor & Francis Group

LONDON AND NEW YORK

First published 2019
by Routledge
2 Park Square, Milton Park, Abingdon, Oxon OX14 4RN

and by Routledge
605 Third Avenue, New York, NY 10017

First issued in paperback 2020

Routledge is an imprint of the Taylor & Francis Group,
an informa business

British Library Cataloguing-in-Publication Data
A catalogue record for this book is available from the
British Library

Library of Congress Cataloging-in-Publication Data
A catalog record has been requested for this book

ISBN 13: 978-0-367-73300-1 (pbk)
ISBN 13: 978-1-138-32428-2 (hbk)

Typeset in Sabon
by Apex CoVantage, LLC

Dedicated to the memory of
Professor D. N. Dhanagare and My *Bou*

Contents

Figures

Tables

Maps

Preface and acknowledgements

With a special focus on Maharashtra, the heartland of western India, which experienced rapid change both in industrial and agricultural sectors, the book makes an attempt to explain how and why the process of agrarian transformation, which was intended to enhance agricultural productivity and growth, generated undesirable social consequences like inequality, distress migration, poverty, and farmer suicides.

Though I began working on this book in the mid-2000s due to a variety of professional as well as personal reasons, the writing of the book was somehow slow and came to a standstill by 2010. Unabated agrarian crisis, continuous media reporting on farmers' distress and suicides, and inspiring advice from two senior most Indian sociologists, late professors B. S. Baviskar and D. N. Dhanagare (who did pioneering work on rural/agrarian politics and change in Maharashtra) to extend my writings on agrarian issues of Maharashtra, motivated me to revive this book project. Sujata Patel also advised to prioritise the project. I resumed writing towards the end of 2014. Though I completed the main chapters by early 2016, the introduction and conclusion chapters and the revision of the three published papers took longer time than expected. For about one-and-half years, I went through very difficult time. My mother-in law, who was a great source of support for my family and admirer of all my academic endeavours, left us. Few months later, D. N. Dhanagare, my academic mentor and long-time friend passed away following a cardiac problem in Pondicherry. Thereafter, I met with a car accident, and my mother developed serious health problems and remained bedridden. Writing this book somehow helped me to recover gradually from these shocks. I lost my mother when the book was in the process of publication. I dedicate this book to her.

While developing the proposal and outline of the book, I was benefitted from my regular consultations with D. N. Dhanagare. He meticulously went through earlier draft of the first and the second chapters despite his preoccupation with other pressing commitments and made critical suggestions. Unfortunately, he could not wait to see the book in print. Till a few days before his last breath, he was keenly enquiring about the progress of publication of the book. I pay tribute to his creative and critical thinking. I also thankfully acknowledge the helpful comments and suggestions made by Gaurang Ranjan Shahay, Thanuja Mummudi, N. Benjamin, and Kanchan Dhar on a later version of the manuscript.

Though arguments and interpretations put forward are entirely mine, my greatest debt goes to those people of rural Maharashtra on whose activities, experiences, thoughts, and voices this book is largely drawn.

Earlier versions of Chapters 2 and 3 were published in *Economic and Political Weekly*. Chapter 8 is a revised and updated version of an article of the same title that appeared in *The Journal of Peasant Studies*. The original publication details of these articles are cited in the respective chapters. I am indebted to the editors and the publishers of both the journals for their permission to reproduce these articles. Chapter 7 is the revised version of the presentation made at the Indian Statistical Institute (ISI), Kolkata as a part of Ramakrishna Mukherji commemorative lecture series. I am grateful to P. N. Mukherji for inviting me to ISI for the lecture series and thankful to the participants, especially T. K. Oommen, Staffan Lindberg, D. N. Dhanagare, and Swapan Bhattachraya for their insightful comments. The earlier draft of Chapter 5 was presented at the department of sociology, Pune University. My thanks to Shruti Tambe, who invited me to the department, and the participants for their useful comments.

I am glad to acknowledge the authorities of Pondicherry University for granting me sabbatical leave for the 2015–16 academic year, which enabled me to complete a major part of the manuscript.

I owe thanks to the anonymous reviewers of the publisher, who read carefully the previous version of the manuscript and suggested important changes. The encouraging cooperation and support extended by the Commissioning Manager of Routledge India, Shoma Choudhury, is admirable. The other members of the Routledge team, especially Apala Bhowmick, and Rimina Mohapatra, were helpful and supportive. I profusely thank all of them. My special thanks to Kate Fornadel at Apex CoVantage for taking care of copy-editing and typesetting of the manuscript. Thanks are due to Joan P. Mencher, Barbara Harriss-White, and Surinder S. Jodhka for endorsing the book.

I happily thank my colleagues Sudha Sitharaman, K. Gulam Dasthagir, C. Aruna, Pradeep Kumar Parida, M. Mancy, and Imtirenla Longkumer for their cooperation and encouragement. I owe special thanks to Papesh Kumar Lenka, my research scholar, who helped me greatly in processing some of the macro data, arranging references, and handling much of the computer related work. He read enthusiastically many chapters and raised interesting queries. I also thank my other research scholars Saubhagayalaxmi Singh, Shasana Yomso, Subhashree Mohanty and Sunil Turuk for correcting as well as updating many references. Hemalatha Bhatt, Debadatta Pradhan, K. Shafique and Atmaja Acharya also helped me in many ways. Some of the maps and images included in the book were prepared by Chandrakanta Ojha. I thank him amply. Thanks are also due to Rohan and Rasika who translated the illustrated history of *Shetkari Sanghatan* from *Marathi* to English. I acknowledge the timely help of Aprameya Mohanty, who collected a much needed set of materials from the library of Gokhale Institute of Politics and Economics, Pune while she was on a personal visit to Pune. The help provided by the staff of the library of the Gokhale Institute of Politics and Economic, especially P. N. Rath and N. A. Choudhari was greatly beneficial. ICRISAT and its research staff, especially Surijit V., Ranjit Kumar, Anupam G., and Uttamkumar, gave me access to their village-level data on Maharashtra. I am grateful to all of them.

Without the firm support and constant encouragement of my wife, Pritinanda, and son, Samarpit (Gudul), this book would never have been completed. As always, I record my debt to my father for his unlimited love and blessings, which inspired and guided me at each stage of this long journey and can never be repaid.

Needless to say, any errors and omissions are solely mine.

B. B. Mohanty
Puducherry
July 14, 2018

Abbreviations

AEZ	Agri Export Zones
AOA	Agreement on Agriculture
APEDA	Agricultural and Processed Food Products Export Development Authority
APMCs	Agricultural Produce Marketing Committees
BJP	Bharatiya Janata Party
BKU	Bharat Kishan Union
CACP	Commission for Agricultural Costs and Prices
CPI	Communist Party of India
CPI(M)	Communist Party of India (Marxist)
DFID	Department for International Development
EGS	Employment Guarantee Scheme
FAO	Food and Agricultural Organization
FFC	Fact Finding Committee
GATT	General Agreement on Tariffs and Trade
GDP	Gross Domestic Product
GoI	Government of India
GoM	Government of Maharashtra
GSDP	Gross State Domestic Product
HYVP	High Yielding Varieties Programme
IBC	Indicators and Backlog Committee
ICAR	Indian Council of Agricultural Research
ICRISAT	The International Crops Research Institute for the Semi-Arid Tropics
IMF	International Monetary Fund
IRDP	Integrated Rural Development Programme
ITC	India Tobacco Company
KRRS	Karnataka Rajya Raitha Sangha
LFPR	Labour Force Participation Rate

MAHYCO	Maharashtra Hybrid Seeds Co.
MAIDC	Maharashtra Agro – Industrial Developmental Corporation
MGNREGA	Mahatma Gandhi National Rural Employment Guarantee Act
MHDR	Maharashtra Human Development Report
MIDC	Maharashtra Industrial Development Corporation
MLAR	Market-Led Agrarian Reform
MPCE	Monthly Per Capita Consumer Expenditure
MSAMB	Maharashtra State Agricultural Marketing Board
MSHMPB	Maharashtra State Horticulture and Medicinal Plants Board
MSSC	Maharashtra State Seed Corporation
NAP	National Agricultural Policy
NCDC	National Cooperative Development Corporation
NDPLR	National Draft Policy on Land Reform
NGO	Non-Governmental Organisation
NHB	National Horticultural Board
NMPB	National Medicinal Plant Board
NPF	National Policy for Farmers
NREGA	National Rural Employment Guarantee Act
NREP	National Rural Employment Programme
NSC	National Seeds Corporation
NSDP	Net State Domestic Product
NSSO	National Sample Survey Organisation
OBC	Other Backward Classes
OGL	Open General License
PACS	Primary Agricultural Cooperative Society
PWP	Peasants and Workers Party
RBI	Reserve Bank of India
RDRC	Report of the Deccan Riots Commission
SAP	Structural Adjustment Programme
SBP	Swatantra Bharat Pakshya
SC	Scheduled Caste
SEZ	Special Economic Zone
SS	Subsidiary Status
ST	Scheduled Tribe
TE	Triennium Ending
TRYSEM	Training of Rural Youth for Self-Employment
UK	United Kingdom
UNDP	United Nations Development Programme

UPS	Usual Principal Status
US	United States
USA	United States of America
USAID	United States Agency for International Development
VCD	Video Compact Disc
VHS	Video Home System
WDR	World Development Report
WM	Western Maharashtra
WPR	Worker Population Ratio
WTO	World Trade Organisation

Glossary

Adivasi a member of any of the aboriginal tribal peoples living in India.

Bajra millet food grain.

Balutedar village artisan or provider of village services.

Beedi country cigarette.

Benami anonymous.

Bhatias a community mainly from Punjab, Rajasthan, and Gujarat.

Bighas a unit of area, approximately ½ acre.

Chitpawan Brahmin a Hindu Brahmin community from Konkan, Maharashtra.

Deosthan inam these are lands granted to religious bodies for maintenance of temples and mosques or similar institutions.

Deshmukh a Marathi term for a landed aristocrat.

Dhangers a pastoral community of Maharashtra.

Dharna the mode of compelling payment or compliance, by sitting at the debtor's or offender's door until the demand is complied with.

Guntas a unit of area, approximately 10 guntas = ¼ acre of land.

Gur unrefined brown sugar made from sugarcane (known as jaggery).

Inam gift or a grant of land in perpetuity (generally tax free).

Izardar an izardar is, for most purposes, almost exactly the same position as a jagirdar.

Jagirdar holder of land grant, usually for services rendered to a king/emperor.

Jowar millet food grain.

Khatedar land owner.

Khot holder of office and superior proprietary rights in the Konkan.

Komtis small traders dealing in grain, cotton, sugar, and other products.

Kunbis the caste of cultivators in Maharashtra.

Mahalwari where village bodies jointly owned land and were responsible for the land revenue.

Mahar ex-untouchable caste.

Malis a caste group belonging to OBC.

Mamlatdar a revenue officer who was in charge of a taluka under the Marathas. It also refers to a native officer responsible for the collection of tax in a taluka under the British government.

Mandal board.

Maratha a member of the princely and military castes of the former Hindu kingdom of Maharashtra in central India.

Mirasidars a cultivator who holds land on the meeras tenure.

Mirasis a form of tenure which confers 'proprietary' rights upon a cultivator.

Mohosu a servant whose maintenance had to be paid daily by the debtor.

Morchas an organised march or rally.

Mukadams labour recruiting agents appointed by sugar factory management.

Panchayat village council.

Patil the headman of a village.

Patta registration of landholding.

Rajput descent from the ancient Kshatriya or warrior and are said to have come from upper India.

Rastaroka road block.

Ryot a peasant.

Ryotwari a form of settlement in which the peasants pay their land tax directly to the state.

Sabha meeting.

Samaj society.

Samyukta united.

Sanghatan association or collective body.

Sarpanch elected head of the panchayat.

Satyagraha Gandhian protest campaign.

Sowkar moneylender.

Takavi loan made by government for agricultural purposes.

Talathis village official.

Taluka a revenue sub-division of a district.

Thalkaris collective owners of the village agricultural lands.

Tolis labourers in groups.

Uchal advance payment.

Upari newcomer with no proprietary claim to the soil.

Vani a village moneylender.

Vanjari an agriculturist community of Maharashtra.

Vatandari revenue-free inam land is called vata, and the holder was the vatandar.

Watan a grant in land or a right in office.

Zamindar lord of the land or proprietor-owner of a landed estate.

Introduction

The current phase of agrarian transformation in India induced by market integration and neoliberal globalism in the process of setting agriculture on a high growth and profit mode has dynamically and significantly altered both scale and relations of production to a far greater extent than previously. Stretching over almost the last three decades, such a transformation has brought a kind of turbulence in rural landscape in terms of income, employment, livelihood patterns, labour relations and mobility, and distribution of and access to productive resources, which have affected not only those directly involved in agricultural production, but also the entire rural-based population. Although there is no clarity about economic gains of this transformation, its social costs are more apparent. Peasant farmer suicides have become an everyday phenomenon, and a large tract of the countryside, which was previously known as an agriculturally 'progressive and productive zone' and had witnessed peasant rebellions during the colonial times and farmers' movements in the post-independent period, is now identified as 'farmer suicide belt'. Apart from this, disparities across regions and social groups, incidences of rural poverty and distress migration often figure in public discourse, and hit the headlines of media reports.

Although volumes have been written on the recent changes in agrarian economy and society, rarely an attempt is made to examine how and why the process of agrarian transformation intended to bring economic gains by enhancing productivity and growth turned counter-productive, leading to a variety of undesirable social consequences. Though some of the recent books (Reddy and Mishra 2009; Deshpande and Arora 2010; Breman 2010; Vasavi 2012; Mohanty 2012, 2016; Singh *et al.* 2016; Kumar 2016) have touched upon the social implications of agrarian change to some extent, they largely focus only on selected aspects. Some of the books which focus on core issues like agrarian distress and farmer suicides ignore the central question – why do peasant farmers, who had earlier participated in protests during the colonial and post-colonial periods, now commit suicides. The experience of the socially weaker sections such as scheduled castes (SCs) and tribes (STs) and women and regional responses to development disparities, which are core to understanding the nature and extent of the social costs of transformation, have, by and large, remained unattended. Moreover, an in-depth and larger study linking economic gains of agrarian transformation with its social costs on a state like Maharashtra (the heartland of western India of which Gujarat was a part till 1960), which experienced rapid changes in the industrial as well as agricultural sectors, accompanied by striking social costs,[1] was almost non-existent. This book attempts to fill in this critical gap.

It addresses five major questions, which are as follows:

1   What are the significant changes in agrarian/rural economy of Maharashtra under the influence of colonialism, early phases of development planning and neoliberalism? To what extent, and in what ways, are new processes at work either a continuation or reversal of the past trends?
2   How have different regions of Maharashtra, being at different stages of development, responded to this transformation, and how are they reflected in the politics of regional development?
3   Does the transformation help improve the levels of income of different categories of rural population and especially the vulnerable sections, and what is the emerging pattern of social inequality today?
4   How are the rural labouring poor coping up with demands of new agriculture, and how do the rural women and the farming communities belonging to SCs and STs respond to these changes?
5   Why did the farmers who had shown rebellious spirit earlier in history resort to killing themselves in the post-reform period? To what extent did the classical sociological paradigm on suicides explain the farmer suicides in the context of neoliberalism?

The book will analyse economic gains[2] and social costs[3] of agrarian transformation in a comprehensive manner combining macro with micro data and analysing them in the context of historical and contemporary changes with reference to Maharashtra.[4] Its purpose is to initiate a discourse on economic gains and social costs of agrarian transformation that would cover a wide range of issues central to contemporary development policy and practices, and their consequences. Though social implications relating to aspects like health and environment are not included, the book offers a new take on the diversity of social costs and addresses core issues, which result from this transformation, afflicting the rural-agrarian society. It also reveals how responses from farming communities to agrarian changes are context-contingent, shaped by local circumstances, the opening and closing of opportunity structures, and historical peculiarities.

Agrarian transformation: the global discourse

Agrarian transformation in the modern sense of the term[5] was first introduced in Europe, especially in France and England in the eighteenth and early nineteenth centuries preceding the Industrial Revolution.[6] Its meaning deepened further following the Industrial Revolution with the growing interaction between industry and agriculture, and the consequent international liberalisation of agricultural trade.[7] It is believed that the dramatic Industrial Revolution would not have been possible without the agricultural transformation that preceded it (Nurkse 1953: 52–53). Arguably, agricultural transformation in England was a necessary prelude for the 'rise of the West' (Moore 1966; Wrigley 1988; Hopcroft 1994). By the end of the eighteenth century, or sometime thereafter, the rest of Western Europe and North America witnessed some sort of agricultural revolution following the stimulating growth of farm exports in response to the British demand for food and farm-based materials. However, the 'spectacular' economic gains of agrarian transformation in England and Western Europe also led to a variety of unwanted social consequences.

Nineteenth-century sociologists like Marx, Weber, and Durkheim, who saw changes in agriculture as reinforcement of capitalist transformation, indicated the social costs of this transformation. Marx dealt with costs of agrarian transition in his discussion on the Corn Laws of England in *Address on Free Trade*, 1848. He also explained social costs of capitalist agriculture in *Capital*. He said,

In agriculture as in manufacture, the transformation of production under the sway of capital . . . becomes the means of enslaving, exploiting, and impoverishing the labourer; the social combination and organization of labour-processes is turned into an organised mode of crushing out the workman's individual vitality, freedom, and independence . . . all progress in capitalistic agriculture . . . not only of robbing the labourer, but of robbing the soil. . . . The more a country starts its development on the foundation of modern industry . . . the more rapid is this process of destruction. Capitalist production, therefore, develops technology, and the combining together of various processes into a social whole, only by sapping the original sources of all wealth – the soil and the labourer.

(Marx 1867/2007: 555–556)

According to Marx, with the development of capitalism, the mass of people were expropriated from land and agriculture. In his words, "it transforms feudal landed property, tribal property, and small peasants' property in mark communes, whatever may be their legal form, into the economic form corresponding to the requirements of capitalism" (1867/2007, III: 723). To him, the parliamentary enclosure movement of the eighteenth and nineteenth centuries resulted in depeasantisation, which led to the emergence of the English proletariat. He recognised that the process of capitalist transformation in agriculture created both peasant dispossession by displacement or enclosure and peasant dispossession by differentiation (Akram-Lodhi 2007; Araghi 2009).

Weber, who dealt more elaborately with agriculture in Germany, in his early writings, narrated the adverse impacts of capitalist transformation of agriculture, especially on the labourers. He indicated the proletarianisation of peasants and the emergence of *Junkers* – a new class of capitalist entrepreneurs – in the second half of the nineteenth century. Weber emphasised that the capitalist transformation of labour relations in eastern Germany had tended to depress the workers' standard of living and illustrated their miserable conditions in the agriculturally advanced region of Silesia as compared to that of the backward province of Mecklenburg (Riesebrodt 1986: 497). He pointed to the frequent employment of women, the barracks-like living quarters of day labourers and their families, and their lack of wage supplements in the form of truck gardening or a few head of cattle in Silesia (Bendix 1978: 43). Although

Durkheim did not write extensively on agriculture, he also deplored the negative impact of industrialisation on what he saw as the organic solidarity of rural society. He indicated the rising number of suicides in agriculture with growing industrialisation. He said, "Agriculture is less prone than industry, but its share in the number of suicides is continually growing" (Durkheim 1984: 193).

More than half a century later, Polyani (1944) expressed similar sentiments in *Great Transformation* while narrating the social devastation caused by the new uncontrolled market system, which had disentangled the rural farm economy from the socio-cultural milieu, into which it was previously embedded. In his words,

> Much of the massive suffering inseparable from a period of transition is already behind us. In the social and economic dislocation of our age, in the tragic vicissitudes of the depression, fluctuations of currency, mass unemployment, shiftings of social status, spectacular destruction of historical states, we have experienced the worst. Unwittingly we have been paying the price of the change.
>
> *(Polyani 1944: 250)*

Many subsequent writers, while highlighting the economic gains of agrarian transformation, also indicated its adverse social consequences in terms of the rising inequality, unemployment, migration, and social tension (Allen 1992; Overton 1996). To quote Allen (1992: 1),

> By nineteenth century, a unique rural society had emerged in England. This new society was characterized by exceptional inequality. English property ownership was unusually concentrated. Rents had risen, while wages stagnated. By the nineteenth century, the landlord's mansion was lavish, the farmer's house modest, the labourer's cottage a hovel.

The technological inventions and policy changes of the nineteenth and twentieth century's US agriculture, which made it the world's 'breadbasket' also gave rise to problems like labour displacement, distressed migration to poorer farms, and decline of sharecroppers and tenants among the blacks, which was the subject of debate among many rural sociologists (Hamilton 1939; Raper 1946; McMillan 1949).

Nevertheless, the development trajectory of the West underscored the importance of transformation in agriculture as a prerequisite for economic development. Many early seminal studies held the view that agricultural growth was a critical step towards economic development, and developing nations aspiring for a higher level of economic development would have to follow the same or similar trajectories of agrarian transformation as these so-called advanced nations did (Clark 1940; Johnston and Mellor 1961; Timmer 1969). After the Second World War, most of the newly emerging nation states, which were at different stages of development,[8] started transforming their agriculture. Though the experience of these developing countries recognised the economic gains of agricultural transformation, its social costs generated scholarly research and policy debates, particularly in the context of recent neoliberal reforms, especially after the end of the Cold War. While one group argues that this transformation in the process of enhancing agricultural growth and productivity has led to unemployment, inequality, social tension, poverty, occupational dislocation, distress migration, and many other health and environmental issues (Balisacan 1993; Razavi 2002; Harvey 2007a; Pauw and Thurlow 2011), the other group views that it has rather increased employment opportunities, reduced poverty, and improved economic conditions of all categories of population (Irz *et al.* 2001; Thirtle *et al.* 2003; Diao and Pratt 2007).

India, however, does not fit neatly into this global debate because the trajectories of its transformation have been quite distinct, indicating the common processes at work in the countryside and the substantive diversity rooted in region-specific trajectories of variations conditioned by varying local settings and broader socio-political and historical contexts. Moreover, it is argued that India's neoliberal reforms, unlike many other countries, have gradually evolved through a process of learning and doing (Ahluwalia 2002). The influence of coalition politics, emergence of regional parties and their demand politics, and grassroots resistance and mobilisation were attributed to this gradualist approach (Kaviraj 2010: 234; Chakrabarti *et al.* 2016: 175). Additionally, the influence of neoliberalism on agriculture is country-specific, as it entails different national, class, ethnic, and gendered dimensions, which are largely determined by the structure of the farming sector, organisation of social forces in the countryside, the degree of integration into the global economy, and the hierarchical position of a given country within the world market (Prügl *et al.* 2013: vii; Marois and Pradella 2015: 4).

Agrarian transformation in India

(a) Colonialism and agrarian change

Agrarian transformation started in India with British colonialism mainly to enhance land revenue and to produce raw materials for Britain's industries following the Industrial Revolution. The British inherited a tax system that served as a baseline in which the burden fell heaviest on the cultivators. While in principle the British followed their predecessors, in practice, the actual amount collected was often higher (Dharma Kumar 1965; Peers 2013). Keeping in view the vast discrepancies across India, new land tenures (*zamindari, ryotwari,* and *mahalwari*)[9] were introduced to standardise the methods to ensure efficient collection of land revenue and to provide greater protection for property rights as a means of encouraging investment in land which could lay the foundation for commercial agriculture. By and large, all cultivable land in British India fell under one of the three alternative systems, which superseded the traditional land rights and compelled the cultivators to become increasingly involved with the markets.[10]

The British used taxation to force locals to grow wheat and commercial crops like sugarcane, cotton, and indigo. By the 1880s, wheat in north-western and central India, cotton in Bombay Presidency, groundnuts in Madras, and jute in Bengal had become major staples of agricultural production. In short, the structure of agriculture had changed, and peasants no longer produced traditional food for local consumption but cash crops or monocrops for the world market instead. Efforts were made between 1840 and 1860 to teach the Indian peasants how to grow better crops by importing American experts, setting up agricultural research stations, and creating a set of inducements recommended by British businessmen (Tomlinson1993).[11] After the 1850s, railway and road transport made possible a huge expansion in cash cropping for national and international markets (McAlpin 1974; Hurd 1975; Tomlinson 1993; Washbrook 1994). The agricultural produce of the deep interior was drawn into the world market. As noted by Thorner and Thorner (1962/2005: 69), "wheat poured out of the Punjab, cotton out of Bombay, and jute out of Bengal. As commercial agriculture and money economy spread, the older practices associated with self-subsisting economy declined". India was thus transformed into a producer and exporter of agricultural products, and the railways facilitated this export. The first railway line was laid out of Bombay in 1853, followed by others from Calcutta (1854) and Madras (1856); then there was a patchwork process of construction culminating in the building of the main trunkline network inland from the major port cities in the 1880s (Tomlinson 1993: 55). The expansion of railways was not only massive but also rapid, with state encouragement of huge British private investments. Between 1865 and 1894, the railways accounted for 77 per cent of total British investments in India (Das 2015: 21). India's rank, in terms of the world's total track mileage, went up from ninth in 1860 to fourth in 1910, only lagging behind the United States, Russia, and Germany. Besides, the opening of the Suez Canal in 1869 made bulk shipment of grain and other produce cheaper.

To expand the area under commercial cropping, attempts were made to bring more of the area under irrigation. British canal-building activity in India commenced in 1817. However, most of the early British efforts were centred on extending indigenous works undertaken previously by the Mughals in various parts of the country[12] (Stone 1984: 13). All irrigation-related works up to 1854 were carried out by the Military Board. It was Dalhousie who abolished the Military Boards and created the Central Public Works Secretariat in Calcutta to oversee them with an aim to attract capital through raising loans from the city of London. In fact, Dalhousie attempted to make India the centre for the British capital in the wake of financial failures in the United States and rising competition for the city in continental Europe (McGinn 2009: 7). However, with Dalhousie's retirement and the 1857 mutiny, these financing proposals were delayed. In the post-mutiny period, between 1858 and 1866, private companies were engaged in irrigation works. Sir Arthur Cotton, an adherent and supporter of Dalhousie, was the driving force behind the Madras Irrigation Company and the East India Irrigation and Canal Company, which started its operations in the coastal plains of Orissa and the adjoining areas to construct a series of canals. It was noted that this period was an important experiment in irrigation development through private companies (Stone 1984: 18–19). However, this venture failed, and on the initiatives of Lord Lawrence, changes[13] were made in the principles and policies governing the execution and financing of irrigation projects in

1866. As consequence, five larger irrigation projects – the Sirhind Canal in Punjab, the Lower Ganga and Agra Canals in the United Provinces, the Lower Swat Canal in the North-West Frontier Province, and the Mutha Canals in Bombay Presidency – were undertaken (Randhawa 1983: 160). After 1880, a number of irrigation projects, including the Lower Chenab and Lower Jhelum Canals (Punjab), the Jamrao and Nira Canals (Bombay), the Son Canals (Bihar), and the Periyar and Kumool-Cuddapah Canals (Madras), were carried out. In addition, the construction of the Godavari and Krishna Canal in South India and the upper Bari Doab in north India were also undertaken. The First Irrigation Commission, which was set up in 1901 for ascertaining the utility or irrigation as a means of protection against famines, expanded the scope for further irrigation works.[14]

Though commercialisation of agriculture, along with the demand for higher land revenue, expanded the need for working funds, there was hardly any institutional arrangements for agricultural credit till the end of the nineteenth century. The needy peasants had to turn to the moneylenders who exploited them by charging exorbitant interest rates on loans, and thus pocketed the vast economic surplus generated in the countryside. The growing discontentment against the exploitative moneylenders (like the Deccan Riots of 1875) compelled the British to consider institutionalisation of rural credit. As a follow-up, in 1904, the Cooperative Societies Act was passed, and cooperatives were introduced to give peasants an alternative to moneylenders. Subsequently, a new act was passed in 1912 giving legal recognition to credit societies, and several committees were set up to improve the rural credit structure.[15] In 1935, the Reserve Bank of India was established, which created an institutional mechanism for agricultural credit. However, the British seemed to have had little impact on the development of rural institutional credit for agriculture (Mohan 2004).

Occasional efforts were made during the nineteenth century to introduce European and North American agricultural machinery, although it was largely unsuccessful. Initially, the British administrators and reformers believed that there was enormous scope for the introduction of 'mechanical contrivance', but the response from the countryside was not very encouraging. As a result, the farms that tried to introduce them 'speedily became museums' for obsolete and rusting implements (Randhawa 1983; Arnold 2000). Nevertheless, a few mechanical implements like cotton gins and maize shellers were adopted, and harrows and seed drills of European patterns were imported. The Indian Industrial Commission in 1918 called for greater mechanisation of agriculture to enhance agricultural productivity and to stimulate industrial output.

(b) National developmentalism and agrarian change

Although the Indian agrarian economy witnessed a series of substantial changes during the British Raj, at the time of independence, the condition of agriculture was miserable; it lacked dynamism and was based on traditional and low productivity technology (Blyn 1966; Moore 1966; Heston 1973; Varshney 1994; Corbridge 2010). The area under irrigation was insignificant, and agricultural production was largely at the mercy of the monsoons. Between 1881 and 1941, while the population of British India rose from 257 to 389 million, the per capita availability of food grains declined from an already low level of 200 kilograms per person per year to a mere 150 kilograms (Guha 2008). The terrifying shortage of basic foods caused by agricultural stagnation led to a series of famines at the end of the nineteenth century, culminating in the occurrence of the Bengal famine (1943–44). Hence, after independence, the task of transforming agriculture was of urgent priority in the country as a whole. However, the intense debate[16] after the Second World War on whether to prioritise industry or agriculture in economic development created a situation for policy ambivalence. Perkins (1997: 157) noted,

> Should India aspire to be an industrialized, urban society? Or should India create a more prosperous but agrarian society based on hundreds of thousands of largely self-sufficient rural villages? Finding an acceptable answer to this question posed far more difficult problems for Indian politics than the central question that had existed for nearly a century: how to get the British to leave.

Notwithstanding this dilemma between an industrialised urban society and a prosperous agrarian society, post-colonial India began its First Five-Year Plan in 1952, which was something of a damp squib, maintaining a kind of neutrality between agriculture and industry sectors. Although Nehru, the architect of modern Indian planning, who would have preferred industrialisation given his vision[17] and weakness towards the soviet model of industrialisation, finally agreed half-heartedly to prioritise agricultural development due to the pressure of internal politics within Congress, particularly the influence of Gandhi on the one hand and Vallabhabhai Patel on the other.[18] Even though industry did not receive high priority during the First Plan period, Nehru's priority for the construction of mega dams and hydroelectric power projects[19] aimed to boost the output of power, which was an essential component for the development of heavy industries, and the establishment of the first fertiliser plant at Sindri reflected his inherent weakness towards industrialisation. The strategy for agricultural transformation in the First Plan included three types of institutional reforms: land reforms (to provide incentives to the actual tiller), formation of cooperatives (to bring in economies of scale), and community development programmes linked with the Panchayati Raj (to enable the poor to avail the benefits of development measures).

In the early 1950s, the emerging intellectual[20] and political[21] conditions inspired Nehru and P. C. Mahalanobis,[22] the architects of the Second Five-Year Plan, to shift the focus of the planning from agriculture to industry. The Industrial Policy Resolution of 1948 was revised in 1956 to accelerate the pace of industrialisation. The budget allocation for industry increased substantially from 8 per cent to 18 per cent between the First Plan and the Second Plan (Sharma 2009: 30). To generate the agrarian surplus required for sustaining the process of industrialisation, provisions were made to ensure mandatory procurement of food grains from the countryside at low prices and to improve foreign exchange reserves. However, the implications of this development strategy necessitated ever-rising food production. To tide over the critical gap between demand and supply of food, subsidised wheat from the United States under its Public Law 480[23] was considered an alternative supplement to domestic production. In 1958–59, the government of India invited an agricultural production team sponsored by the Ford Foundation to study the food problem and make necessary recommendations to increase production on an emergency basis. Consequently, the allocations for agriculture rose marginally (14 per cent) during the Third Plan (1961–66), although it continued to be much lower than that of industry (Panagariya 2008: 41).

By the early 1960s, it was realised that desirable changes in agrarian structure did not take place due to the absence of an effective strategy based on price incentives and profit maximisation in agriculture, accompanied by failures of land reforms, cooperatives, and the Panchayati Raj system. While narrating the agricultural situation of the time, Moore (1966: 397) noted, "Indian agriculture in a word remains today what it was in Akbar's and still was in Curzon's time: a gamble in the rains, where a bad crop means disaster for millions of people". The war with China in 1962, failure of monsoon in 1965–66, along with the war with Pakistan in 1965, which led the United States to cut off funds to India, worsened the situation further. The growing burden of debt and the spiralling inflation forced the Indian government to devalue the rupee under the pressure of the World Bank and the United States, resulting in an economic recession. Amidst these crises, the planning process was stressed to a breaking point, and a 'plan holiday' was declared (from 1966 to 1969). Instead, three annual plans were introduced in which emphasis was laid on agriculture. Meanwhile, major revision in development thinking took place at the global level, following the contributions of scholars like Schultz (1964), who emphasised investment in technological inputs in agriculture to revolutionise agricultural production. These developments ultimately shook the foundation of Nehru-Mahalanobis's 'industry first' policy.

After Nehru's death, Prime Minister Shastri reverted the policy of highest priority to agriculture, as well as the rural sector, and insisted on a 'common man's plan' (Varshney 1994: 61). Realising the importance of agriculture, he coined the slogan *'jai jawan jai kisan'*. Shastri appointed C. Subramaniam the minister of food and agriculture, who under Nehru was the minister of steel, mines, and industry. Subramaniam developed a new model, shifting from the earlier 'institutional model' of Nehru. This new model emphasised price incentive and profit maximisation, and the need for improved technology and required institutional mechanism to bring agricultural revolution. The Indian Council of Agricultural Research (ICAR) was reorganised, and on the recommendations of the L. K. Jha Committee (1965),[24] likewise, the Agricultural Prices Commission and Food Corporation of India were established to carry out the new pricing and marketing policies, and to provide the necessary institutional support. It was during this time that reports of new high yielding varieties (HYVs) of wheat and rice began to show extremely promising results. The US economist Bernard Bell, who visited India in 1964 under the aegis of World Bank to review India's development performance, impressed upon C. Subramaniam to increase the use of improved inputs emphasised earlier by the

Ford and Rockefeller foundations and US Agency for International Development (USAID). Additionally, USAID started to use conditional loans to encourage reform in the agricultural sector. With pressure from within and external forces[25] like the World Bank, the government of the United States, and US-based private agencies (Rockefeller and Ford), the union cabinet and parliament accepted in principle Subramaniam's proposal for agricultural reform, which included the green revolution package in November 1965. Indira Gandhi, who became prime minister in 1966 upon Shastri's death, pushed ahead the Subramaniam model and the green revolution was launched in year 1966–67.[26] Agriculture received priority during the Fourth Plan (1969–74).

As earlier emphasis on either large-scale industrialisation or agriculture development did not solve the problem of widespread poverty and unemployment, the subsequent two national plans (1975–79 and 1980–84) focused on solving these problems. When the Janata Party coalition government came to power in March 1977, it focused on agricultural and rural development to bring parity of prices between agriculture and industry so as to reduce the asymmetrical terms of trade. Moreover, the rise of agricultural interests in parliament,[27] a clear indication of overwhelming response to the agrarianism of Charan Singh, and the emerging non-party peasant mobilisation in the 1980s motivated Indira Gandhi to adopt populist schemes favouring rural and agrarian population. The subsidies for fertilisers, irrigation, electricity, and other farm inputs received priority, and the search for a foreign market for agricultural exports was started aggressively. A series of programmes like Training of Rural Youth for Self-Employment (TRYSEM), National Rural Employment Programme (NREP), and Integrated Rural Development Programme (IRDP) were introduced to address rural unemployment and poverty. However, in the early 1980s, many parts of the country, including Maharashtra, experienced acute famine conditions leading to a striking decline in the agricultural output. To meet this crisis, the subsequent Rajiv Gandhi government focused on food and agricultural productivity during the Seventh Plan (1985–90).

(c) Post-reform changes

Indian economy witnessed dramatic changes in the 1990s. A significant part of the 1990s was dominated by short-duration governments,[28] which affected the process of economic development in general and agriculture in particular. The populist loan waiver scheme of the V. P. Singh government pushed the country into fiscal deficit and balance of payment crisis in 1991. Subsequently, in the year 1991, the Narasimha Rao government went all out for the Structural Adjustment Programme (SAP) and adopted new economic reforms, including liberalisation with the International Monetary Fund (IMF) conditionality. It implied the opening of agricultural produce, particularly with respect to the international trade, without having an extensive market economy or depth of competition comparable to the West (Chakrabarti *et al.* 2016: 176). It is during the 1990s, therefore, that the importance of Marxism declined, and Regan and Thatcher came to power in the United States and the United Kingdom, respectively, and propagated the neoliberal ideology. This had profound effect on Indian economy, which got itself pushed closer to the United States via the World Trade Organisation (WTO). These developments urged the economy to gear up and accelerate the process of liberalisation during the Eighth Plan (1992–1997). The BJP coalition government led by Vajpayee (1998–2004) also endorsed liberalisation-oriented policy reforms. The next three plans (1997–2012) echoed the spirit of liberalisation and targeted towards 'faster economic growth', focusing on the improvement of efficiency and major reforms for the agricultural sector.

Though India introduced a series of neoliberal economic reform measures following the endorsement of the structural adjustment programme in 1991, initially, there was no direct reference to agriculture. Nevertheless, changes in exchange and trade policy, gradual dismantling of the industrial licensing system, and reduction in industrial protection had a direct bearing on agricultural trade and marketing. The Dunkel Draft of the General Trade Agreement was signed in April 1994, and India joined WTO. As a result, the state that was playing a dominant role in controlling the economy had ironically reduced its own importance (Kaviraj 2010: 237). Although the outcome of the Uruguay Round Agreement brought the agricultural sector under the purview of multilateral trading rules, its role in opening up the farm sector to global competition was restricted, and it is only after the Doha negotiations that agricultural liberalisation process started with full spirit (Panagariya 2005: 1277).

Neoliberal policy weakened the pre-existing institutional mechanism in agriculture by repealing land reform laws to promote large-scale private corporate investment in industry; that practically rendered the cooperatives redundant and disrupted the formal rural credit network. The policy on agricultural credit underwent significant changes following the recommendations of RBI's Committee on Financial System known as the Narasimham Committee (1991), which suggested banks to follow commercial lines, keeping profitability as the prime objective, and advised the control of unnecessary expansion of their rural branches. Besides, there were many other changes like placing all Indian product lines in a general system of preferences, reduction of average tariffs on agricultural imports, withdrawal of the provision of minimum export price, removal of quantitative restrictions on the agricultural trade, decanalisation of agricultural trade, further liberalisation of the import of seeds, permission of 100 per cent of foreign equity in the seed industry, reduction of fertiliser subsidies, and reforms in power sectors (Reddy and Mishra 2009; Tripathy 2014). The state absolved itself off the responsibility to produce or procure and distribute major inputs. The National Agricultural Policy (NAP) was formulated in 2000, and it was succeeded by the National Policy for Farmers (NPF) in 2007, along with many other new policies and programmes.[29] The NAP emphasised integrating Indian agriculture into the global order and encouraged corporatisation of agriculture. To put it succinctly, the neoliberal reform policy adopted in 1991 dismantled the so-called welfare capitalism by sidelining the Nehruvian policy of a mixed capitalist economy. In other words, the traditional agrarian structures (i.e. land control and social structures), which were transformed under the impact of colonialism and planning, were drastically altered, paving the way for urban-industrial manufacturing activity promoted by capitalist economy.

Debate on economic gains

The extent of the contribution of this agricultural transformation towards economic growth triggered extensive discussion and debate. The debate on the impact of colonial changes on agricultural growth started with Blyn's classic work published in 1966. Blyn provided a pessimistic account of the British impact on agricultural growth based on analysis of official statistics. He argued that there was a small expansion of the total output of all crops during the 1890s, but it fell off thereafter, especially for the food crops leading to a kind of stagnation. On the other hand, Heston (1973) challenged Blyn's notion of stagnation and questioned the reliability of British official statistics. He viewed that there was substantial growth in food grain as well as in total output during the last four or five decades (1860–1920) of British rule, which was faster than the rate of population growth, though he noted the slower overall growth compared to the earlier period. Subsequently, many scholars contributed to this debate, though it remained largely inconclusive (for instance, Sarkar 1983; Tomlinson 1993; Charlesworth 1982).

The debate on agricultural growth in post-independent India started with the advent of the green revolution and continued for over two decades. One group of studies reported that the adoption of new technology brought a revolution in Indian agriculture, leading to a marked increase in the growth of agricultural output in different parts of India (Dantwala 1970; Frankel 1971; Mellor 1976; Narain 1977; Ahluwalia 1978; Rao 1980; Bhalla and Chadha 1983). The other group, on the contrary, held the view that growth of food production and agricultural production did not experience an acceleration; rather, it led to an overall stagnation in the growth rate (Griffin 1974; Dasgupta 1980; Bardhan 1984). Yet it is argued by another group of scholars that though the green revolution technology was projected as 'scale neutral', it created conditions for the economic gains of large farmers because of preferential access to means of production (Lockwood *et al.* 1971; Byres 1972). Though after a long debate, a consensus emerged that green revolution contributed substantially to the agricultural growth and productivity, the issue of distribution of its gain still remained contentious.

This debate was revisited further by the development economists from 1990s following economic liberalisation. While the proponents of economic liberalisation argued that there were significant signs of growth in agriculture and the effects of neoliberal measures had promoted tradable agriculture (Ahluwalia 1996; Gulati 1998; Gulati and Kelly 1999; Panagariya 2008), the other line of argument was that the growth of agriculture has visibly decelerated in terms of gross product and output during the post-reform period as compared to the pre-reform period (Reddy and Mishra 2009; Patnaik and Moyo 2011; Bhalla and Singh 2012).

Debate on social costs

Notwithstanding extensive debates on the economic gains of such a complex agrarian transformation, its social implications received little attention. However, some of the early village studies in the

1960s and the 1970s reflected upon the adverse impact of agricultural modernisation on rural society, particularly with regard to the structure of social inequality in terms of caste relations (Beteille 1965; Epstein 1973). Subsequent village studies undertaken in the 1980s discussed at length the phenomenon of social inequality in the context of agricultural modernisation (Gough 1981; Harriss 1982). The wide-ranging changes in agriculture following the green revolution also generated an extensive debate on the trends of social inequality among development economists and sociologists (as discussed in Chapter 3).

The changing conditions of agricultural labourers in terms of wages and incomes were subjects of an interesting debate that surfaced in the 1970s among economists, following the introduction of the green revolution. On one hand, while scholars like Bardhan (1973) argued that the real wages of agricultural labourers in agriculturally advanced states like Punjab had declined, the positive impact of growth on economic conditions of agricultural labourers was highlighted by scholars like Lal (1976) on the other. Still, scholars like Jose (1978) were of the opinion that although real wages of labourers increased, their per capita income declined due to loss of employment opportunities. Significantly enough, many sociologists contributed to this debate with their rich data and qualitative analysis (e.g. Breman 1974, 1985). Besides, socio-economic conditions of agricultural labourers, changes and continuity in labour relations, and the emerging pattern of labour mobilisation were analysed by many scholars (Oommen 1975; Alexander 1978).

Similarly, the issue of rural poverty following agrarian transformation in the aftermath of the green revolution was the subject of discourse. While village studies made by some sociologists and anthropologists reported the rising poverty and deprivation among agricultural labourers and marginal peasants (Mencher 1980; Rodgers 1983; Breman 1985), macro level studies based on national statistics by economists indicated a gradual decline in the number of rural poor (Ahluwalia 1978; Rao 1985). The famous 'mode of production' debate in the 1970s and 1980s among scholars of Marxist persuasion on the emerging social relations of production in Indian agriculture touched upon the social costs of agrarian transformation while discussing the growing polarisation and class relations among the peasantry. Of late, there have also been debates on the impact of modernisation of agriculture on women relating to their employment, wages, and earnings (discussed in Chapter 5).

Most of these debates, however, died down by the mid-1990s, and interest in agrarian studies declined largely due to the rise of the critique of village studies, the problems of identifying modes of production, and the move to urban and industrial studies for addressing the pressing problems of liberalisation. Nevertheless, recently, attempts were made to renew the earlier debates on social inequality, labour relations, and farmers' movements, and to initiate new discourses on emerging issues like agrarian crisis, labour migration, survival of peasantry, and gender disparity (Chakravarti 2001; Gupta 2005; Chatterjee 2008; Breman 2010; Vasavi 2012; Agarwal 2016; Dhanagare 2016a). This book intends to extend these debates to Maharashtra, one of the few states in India which experienced rapid commercialisation during the colonial period, relatively successful land reforms, and cooperative movement during the post-planning era, and responded promptly to the demands of the new global order in the post-reform period.

Theoretical issues

The agrarian rural economy of third world countries like India is moving more clearly along the axis of capitalist development than ever before. As noted by Sanyal (2007: 1), the triumph of capitalism is now unquestionable and self-evident, and the entire third world is being integrated into one global capitalist network. The policies of neoliberal globalisation have given a new impetus to the forces of capitalist development. Under the conditions of this 'great transformation', agriculture has come progressively under the domination of the capitalist market, altering radically the economic and social relations of production.

In the debates on agrarian transformation and its economic and social consequences, four main theoretical views have been expressed. One view led by classical Marxists holds that capitalist transformation in agriculture entails economic prosperity for the large farmers while destroying the economic base of the small peasants. To Marx (1867/2007), the rigours of capitalist agriculture were so severe that the peasants, particularly the small peasants, would not be able to survive, and they would be transformed to wage labourers. Engels agreed with Marx's prediction and noted, "The development of the capitalist form of production has caught the life-strings of small production in agriculture; small production is irretrievably going to rack and ruin" (Engels 1950: 382). Lenin, who extended Marx's idea, stated that capitalist transformation would lead to a process of differentiation among peasantry, ultimately resulting in depeasantisation. He held that the old peasantry would be ousted by a new type of rural inhabitants: rural bourgeoisie and the rural proletariat (Lenin 1964). To him, at a certain stage when agriculture is fully organised on capitalist lines, the number of rural proletariats will decline as opposed to the industrial sector. Though Kautsky differed from Lenin on the nature of capitalist development in agriculture and highlighted the speciality of agriculture, he also indicated the ultimate shift of peasants to the industrial sector as wage labourers. He hinted that the condition of the working population would be miserable if they did not move out of the countryside.[30] Stated precisely, the so-called Marx-Kautsky-Lenin laws suggest that capitalist transformation in agriculture destroys peasantry, and the displaced peasants are eventually absorbed in the expanding industrial sector.

This approach, however, is criticised for its taken-for-granted analysis, suborning the roles of state, politics, and culture in favour of a narrow economism grounded in 'capitalist logic' that neglects upstream and downstream links to and from agriculture, and says little about farmer choices, beliefs, and political actions. Moreover, the rapid changes in agricultural production and marketing, which has taken place following neoliberal reforms, have raised a number of issues concerning the socio-cultural and institutional embeddedness of agrarian livelihoods, agricultural development, and commodity systems whose scopes reach well beyond the conceptual domains of this classical approach.

The other view led by the reformists (modernisation school) argues that agricultural transformation is mostly driven by market and technology logic (Hayami and Ruttan 1985; Timmer 1988). The farmers, those who make use of the opportunities offered by new technology, enlarge production, follow the market in a coherent way, and are engaged in fierce mutual competition remain in advantageous position to appropriate the economic gains of the transformation. On the other hand, the peasants who lack required entrepreneurial skills and resources become marginalised, and quit to join the ever-growing ranks of urban workers. Such a view is based on the price, market, and technology responsiveness of the farmers, as guided by economic rationality in conformity with the standard economic theory, as postulated by the neoclassical economists and tells less about the process that leads to the failure of certain category of farmers. It is rightly remarked that positivistic social sciences are quite good at explaining the success or failure of the individual within the system, but the system itself is left out of analysis (Apthorpe 1977).

Both the Marxists and liberal economists argue that that agrarian transformation leads to the demise of small and poor peasants who ultimately migrate to the developed industrial/urban centres. The neoclassical argument based on the Lewisian framework also postulated the inevitable shift of labour from agriculture to the fast-developing modern industrial activities under the capitalist development in the predominantly agricultural countries. Though the developed countries of the West more or less witnessed this trend earlier, such historical experiences are not repeated in countries like India where industrial and urban capitalism has not been expanded to the extent desirable to absorb the displaced mass of peasants. That is why the disenchantment with grand theories of agrarian transformation in India brought about a shift from 'vanishing peasantries' to 'persistence of peasantry' (Mohanty 2016). However, the conditions of this neoliberal global capitalist development with its emphasis on 'accumulation through dispossession', as stated by Harvey (2007a), threatened the basis of the subsistence of peasantry, unleashing mass impoverishment and displacement. Although the question of managing this vast section of population and its 'welfare' impinges upon capitalist production itself, capitalism never ameliorates the condition of the masses, because its functioning at the global level is predicted on the growth of riches at one pole and the promotion of unemployment, impoverishment, and hunger at the other (Patnaik 2016). Though in such a context it is left to the state to take care of this impoverished population, the state under the conditions of neoliberal globalism withdraws itself, shifting responsibility to the market and offers individual-based solutions to collective problems (Brown 2006: 704). As a result, the cost of this transformation ultimately falls heavily on peasants themselves.

Keeping in view the blockage of an exit option, these peasants look to agriculture as their only source of survival. In order to secure their subsistence and reproduce themselves, they either work hard than before or are forced to intensify the cultivation by adopting capitalist methods of production or both. Marx himself acknowledged this when he noted that peasants would survive only if they are capitalistically inclined (Nelson-Richards 1988: 1). The third view adopts this stance, which is close to that of Chayanov and his Organisation and Production School. Though this view also suggests a process of differentiation based on the demographic cycle in terms of consumer-labour balance, it demonstrates strategies peasants adopt to withstand the capitalist forces. According to this view, the peasant economy has a logic of its own that defied the disintegrative forces of modern capitalism. It is premised that small peasant firms are more efficient than large-scale units because of the intensity of labour use upon them or because of the inherent suitability of small-scale units for agricultural production. The argument is that peasants would survive the adverse effects of capitalism through their hard work or self-exploitation. To Chayanov, "the peasants would put in greater effort and work more hours of work or work more intensely, sometimes even both" (Thorner 1966: xiv). In such a context, if the hardworking peasants of Chayanov are eventually unable to reproduce themselves despite all their efforts, what are the other alternatives available to them? Though Chayanov specify a range of possibilities covering external conditions and the peasants' response to them, they represent only so many adaptations derived from the static logic of household reproduction and its mode of calculation (Bernstein 2009: 65).

The fourth view on agrarian transformation addresses this issue by arguing that the peasant economy may interact with commercial markets, but when its negative impacts destroy the economic security, it fosters indignation and resistance (Moore 1966; Wolf 1969; Scott 1976). It is argued that economic exploitation inherent in agrarian capitalism leads to peasant struggles and revolutions. As Wolf (1969: 278) puts it, "Capitalism necessarily produces a revolution of its own". However, the capacity of peasants to mobilise against these threats depends largely on the strength of traditional village institutions, mobilisation of peasants, alliances with other classes, and social organisation of peasantry, etc. (Moore 1966: 475–479; Wolf 1969: 291–292; Scott 1976: 4). As the conditions of neoliberal global capitalism pose formidable challenges to unity and solidarity among the peasants by promoting heterogeneity, values of competition, individualism, and consumerism, strong movements with genuine leadership appear to be a difficult proposition at this juncture. The movements which emerge in such a situation in a country like India are mostly of the populist variety with different class characters mainly seeking to protect the interests of rich farmers and corporates (Dhanagare 2016a). Hence, the peasants remained in a dilemma whether or not to join these movements. Popkin (1979), who considers peasants as rational beings, argues that peasants make cost-benefit analysis before joining movements. To him, peasants join movements only when they are convinced that it will bring economic gains for them. If this point holds, what choices are peasants left with when they realise that they are at a loss either way, whether they join the movements or not?

In a nutshell, under the capitalist method of farming, with intensive cultivation and self-exploitation, if the peasants are unable to secure their subsistence and there is withdrawal of state support, absence of protest movements, and blockage of an exit option, they remain in a state of helplessness in the context of growing individualism. This situation ultimately leads to a process of socio-economic 'estrangement'. This condition may be linked to the Durkheimian explanations of suicide as a means of viewing peasants' behaviour. In his classic theory on suicide, Durkheim (1952) says that increasing individualisation breaks down the basis of organic solidarity and creates an anomic condition that ultimately results in the growth of the 'suicidogenic impulse'.

The book throws some light on this issue, showing under what conditions agrarian transformation operates. Who are the gainers and losers of this process? How do the losers respond to the crisis? Though the aforesaid four approaches provide useful insights to understand the process of agrarian transformation and its social and economic consequences, adherence to any of this particular school/approach may restrict our scope of inquiry, as they do not capture the complexity of the entire process. Therefore, we follow an approach which is close to a structural-historical approach that rejects the old dichotomy between agency and structure and states and markets, and seeks to develop an integrated analysis by combining parsimonious theories, which looks into agency in terms of a conception of rationality, with contextual theories, and analyses structures institutionally and historically. We incorporate into our exploration dimensions which were relatively neglected in classical agrarian studies, such as the dynamics of gender, caste, livelihood's diversity, mobility, rural-urban link, and the like. To be precise, we look into the economic gains and social costs while analysing the trajectory of agrarian change, its different degrees of integration with national and global economy, local cultural and ethnic relationships, role of the state, and attempt to offer an innovative view on a broad spectrum of collective responses ranging from overt to passive resistance and later suicides.

A note on the methodology

The analysis in the book is woven around two central questions: (a) Why does the process of agricultural transformation intended to bring economic gains generate social costs? (b) Who bears these costs? In order to address these questions, we have looked into the texture, nuance, and variations in the peasants' responses to the contours of changes in agrarian economy across time, space, and social categories. Information has been gathered both at micro and macro levels, covering economic and non-economic, as well as quantitative and qualitative aspects, connecting historical with contemporary conditions. In a sense, we have followed J. M. Clark's dictum that "it is unscientific to exclude any evidence relevant to the problem in hand" (cited in Kapp 1950: 284). Actor-oriented methodologies are combined with macro historical perspectives to understand the interaction between economy and society, as emphasised by Hefner (1990). Unlike many other studies on agrarian change where the analysis was entirely based on certain specific selected districts and/or villages, we cover different districts and villages which are considered to be representative of the state of Maharashtra on the concerned issues. The analysis of the impact of agricultural modernisation on social inequality is based on the most agriculturally advanced district of the state (Chapter 3). To analyse land and agriculture of SCs and STs, two different districts having a greater concentration of the population of these categories have been considered (Chapter 4). The pattern of labour migration and the gendered impact of agrarian transformation is assessed both at the state as well as the district levels (Chapter 6 and 5). To study the problem of farmer suicides, districts have been selected based on high incidences of suicides (Chapter 8). In many chapters, macro analysis is supplemented with micro-level data, which were collected through field surveys involving a wide range of data gathering techniques (in-depth interviews, focus groups, observation, etc.). Field surveys were made in several phases spanning over a period of five years (between September 1998 and May 2003). The required macro-level information was collected from a wide variety of data sources covering government as well as non-government agencies. To put it succinctly, the analysis in the book combines local rural ethnography and reflexive theorising in order to bridge the gap between those who consider the village a unit of analysis and those who focus on the determining influence of exogenous and global forces.

Outline

The rest of the book is divided into eight chapters. Chapter 1 traces the trajectory of agricultural transformation in Maharashtra in a historical perspective and brings out linkages with changes in national developments and world economy. This chapter provides a broad base to view the rest of the chapters in the book. Chapters 2 and 3 focus on the disparity across regions and social groups in the context of agricultural transformation. The analysis in Chapter 2 indicates that agricultural development in Maharashtra over the last four decades (from 1970 to 2010) has been unequal across regions, with western Maharashtra (WM) being much ahead of other regions in terms of major development indicators. Chapter 3 examines the nature and extent of social inequality. It shows that social equality remains elusive as agricultural modernisation measures generate further inequalities. The experiences of the weaker sections – SCs, STs, women, and rural poor – are examined in the next three chapters. Chapter 4 analyses changes in landholding and agricultural practices among the SCs and STs. The response of women to agricultural transformation and the consequent gender disparity is discussed in Chapter 5. The emerging pattern of migration among the rural labouring poor is elucidated in Chapter 6. The last two chapters, i.e. Chapters 7 and 8, document the nature of the agrarian crisis and the resultant distress that has led to farmer suicides. Chapter 7 questions why the peasants in Maharashtra, who had shown a rebellious spirit during the second half of the nineteenth century and almost till the aftermath of the green revolution, resorted to killing themselves in the post-reform period. Chapter 8 reinterprets Durkheim's theory of suicide to explain farmer suicides in Maharashtra. The concluding part reflects the contribution of the book to some of the major ongoing debates and indicates how the economic gains and social costs of agrarian transformation are far more complex and nuanced in India than is commonly assumed.

Notes

1 For almost the last two decades, since 1996–97, the state continued to be one of the top-ranking states with very high incidences of farmer suicides in the country. It was time and again a case in point in policy discourse on regional developmental imbalance, specially pertaining to disparity in economic development. It was also one of those few states that experienced peasant rebellions during the colonial period (like the Deccan Riot in 1875) and organised farmers' movements in the post-independence period. In terms of rural labour migration, the state often hits the headlines of media reports. According to the recent NSSO (64th round) data, of the total migrants in cities like Mumbai, Maharashtra accounts for nearly 70 per cent, and most of them come from the rural areas. Despite these signals on diverse social implications, social science researchers have not paid the attention Maharashtra deserves.

2 Though economic gains of agrarian transformation include a variety of direct and indirect economic benefits, we have assessed the gains in terms of only major indicators like agricultural production and farm income.

3 The concept of social cost in economic discourse is largely related to losses to the people or community that were amenable to estimations in monetary terms. While some groups of economists relate the concept to the efficiency of the market, others relate it to the violation of social rights (Franzini 2006). However, neither of the perspectives provide a useful frame to view social cost associated with cultural aspects, politics, and undesirable changes in the social sphere. K. William Kapp, in his seminal work *The Social Cost of Private Enterprise* (1950), noted that economic orthodoxy both in its traditional and amended versions omitted serious treatment of social cost due to its assumption that the economic sphere can be separated from the people it affects. To him, social costs are direct and indirect losses suffered by sections of a community as a consequence of unrestrained business activities (Kapp 1950: 13). Taking his lead, we define the term social cost here as the undesirable social consequences of changes in agricultural productive technology, marketing, and social relations of production. However, the concept is not used to measure the losses caused to the community and groups in economic terms, rather to understand how the process of agrarian transformation generates these costs and how they are socialised.

4 It is noteworthy that, barring some earlier village studies and the subsequent studies on agrarian change impacting rural politics in the 1970s and the 1980s (Carter 1975; Brahme and Upadhyaya 1979; Attwood 1979, 1992; Baviskar 1980; Lele 1981; Walker and Ryan 1990; Dhanagare 1994), a fresh analysis of agrarian economy and the society of Maharashtra is conspicuous by its absence. Though there are a few recent studies (such as Mohanty 2005; Mishra 2006a; Dhanagare

2016a), they cover only limited aspects of the socio-political effects of agrarian transformation like farmers' suicide and movements.

5 Generally, agricultural transformation is understood as the gradual shifts in production concept, economic growth mode, farming structure, leading role of agriculture, and employment structure in rural areas of national economy.

6 Till recently, the timing of the agricultural revolution in England was a subject of interesting debate. While Overton (1996) assigned the revolution to the late eighteenth and early nineteenth centuries, during the period of parliamentary enclosures, the subsequent revisionist historians such as Havinden, Jones, and Kerridge, however, argued that output and productivity had risen significantly in the sixteenth or seventeenth century. Recently, the revisionist view has been reformulated and extended by Allen and by Clark. Allen (1992) offers new evidence to show that there were two agricultural revolutions marked by rising output and productivity. The first preceded the parliamentary enclosures and was accomplished by small-scale farmers in the open fields, while the second occurred in the first half of the nineteenth century. For more details, see Allen (1992).

7 The protectionist Corn Laws in England were moderated in the 1830s and phased out in 1846–49, followed by the liberalisation of agricultural trade policies in other countries, particularly after the French-British treaty in 1860.

8 Timmer (1988: 282) classified the stages of agricultural transformation based on historical and contemporary cross-section perspectives into four phases: getting agriculture moving, agriculture as a contributor to growth, integrating agriculture to macro economy, and agriculture in industrial economies.

9 Permanent settlement was made in Bengal, Bihar, Orissa, the Central Provinces, and some parts of Madras Presidency, giving rights to lands to the *zamindars* in perpetuity, with the objective of giving landlords an incentive to improve their land, increase the rent they could charge, and hence the profit they could make over the fixed land revenue demand. However, as an instrument for agricultural improvement, permanent settlement was a failure, because it accelerated and intensified the trend towards parasitic landlordism (Moore 1966; Dhanagare 1983; Tomlinson 1993). Learning from this experience, and under the influence of nineteenth-century utilitarianism, individual peasant proprietorship known as the *ryotwari* system was adopted in Bombay Presidency along with most parts of Madras Presidency, Berar, and Assam. However, the myth of peasant proprietorship in the *ryotwari* system was exploded as a viable alternative to *zamindari* system, as the landholders in the *ryotwari* areas also began subletting the land. A new variety of land system known as *mahalwari* system was introduced in the North-West Provinces and Panjab where village bodies jointly owned land and were responsible for the land revenue. However, this tenure made no system better than the others, as the landlord spirit was very much dominant.

10 In fact, the relationship between the British land tenure system and agricultural development led to a debate among economic historians. While

scholars like Banerjee and Iyer (2005) held that the non-landlord-based revenue systems were conducive for agricultural development, Iversen *et al.* (2013) did not find any such correlation. On the other hand, Dutt (1904) concluded that the zamindari system improved agriculture and defused prosperity among the people.

11 This happened particularly in the case of cotton crops, as Indian cottons were unsuitable for Lancashire mills due to their short-staple variety. However, this effort was in vain, even though the great boom in Indian cotton export took place during the time of the American Civil War.

12 However, not all pre-British works were built primarily for irrigation purpose. In many cases, they were built to supply water for the Moghul emperor's personal use. For example, the Western Jumna Canal supplied water to the emperor's hunting grounds at Hissar, while an extension of the Western Jumna Canal to Delhi was intended to feed the fountains and gardens of the imperial palace. Similarly, a 12-mile cut linking the Kali Nadi with Meerut was excavated with a view to supplying Meerut's groves and gardens, rather than for irrigation. See Stone (1984: 14).

13 The policy changes include the following: (a) irrigation projects would be constructed by the state through its own agency; (b) these projects would be financed from public loans raised specifically for the purpose; (c) political boundaries would not be considered while utilising water from a river for irrigation purposes. See Randhawa (1983: 160).

14 The introduction of perennial irrigation during the British period generated an intense debate with regard to its impact on the agrarian economy and society. While scholars like Whitcombe (1972) presented a pessimistic view, citing the adverse ecological consequences, such as waterlogging and salinisation, and destruction of traditional well irrigation, which ultimately led to islands of plenty amidst an ocean of depressed peasantry, the optimistic account of Stone (1984) highlighted canal irrigation as a source for economic dynamism and innovations in terms of adoption to new technology, quest for maximisation of output, and ability for rational decision-making. A few other scholars took a different position altogether and pointed to the emergence of the contradictory economic pattern following canal irrigation in which growth was accompanied by underdevelopment (Ali 1987; Gilmartin 2003).

15 While the Maclagan Committee on Cooperation (1915) advocated the establishment of provincial cooperative banks, the Royal Commission on Agriculture examined the programme of rural credit in 1926–27, and Sir Malcolm Darling submitted another report on cooperative credit in 1935.

16 It is well-known that in the 1930s and the 1940s, Indian political leadership started discussing how independent India would be reconstructed. Three major blueprints came into being in the 1940s: the Bombay Plan, the People's Plan, and the Gandhian Plan. While the Bombay Plan, which was prepared by leading industrialists emphasised on capitalist development, the People's Plan advanced by M. N. Roy highlighted a leftist

vision. The Gandhian Plan, developed by Shriman Narayan, stressed on self-sufficient village economy.

17 In 1934, it was resolved by the Congress Working Committee at Banaras not to support the idea of large and organised industries. Nehru was critical of this approach and rallied support to reformulate the resolution with a Congress backing for industrialisation. In his 1936 Faizpur presidential address to the Congress, he strongly favoured heavy and rapid industrialisation.

18 Gandhi emphasised 'village development' and opposed heavy industries, as they would not contribute to human welfare but lead to the degradation of human life. On the other hand, Patel, who was the most powerful figure in the politics of the Congress Party of that time, opposed Nehru's high socialist and modernist vision of development.

19 A number of large multi-purpose river valley projects such as the Bhakra-Nangal, Hirakud, Damodar Valley, Kosi, Chambal, and Riband, regarded as 'temples of modern India', were undertaken by Nehru in the early years of the planned development.

20 In the early 1950s, development economists did not consider agriculture to be an engine of growth because traditional farmers were believed to be unresponsive to economic incentives. Following Clark's analysis (1940) on the empirical link between economic growth and the shift from primary to secondary production, the neoclassical view dominated the development discourse. It considered agriculture as a declining sector and opposed the policy initiative to modernise agriculture as it does not lead to efficient allocation of resources. Neoclassical theorists led by Lewis (1954) argued for relocation of factors of production from an agricultural sector to a highly productive modern industrial sector as the former was characterised by low productivity and the use of traditional technology. Moreover, the Soviet style of industrialisation and the debate that followed had considerable impact on many developing countries. This debate took place between 1924 and 1928 within the communist party among the politically active economic theoreticians over the pattern and speed of industrialisation in Soviet Union. Preobrazhensky argued for an industrialisation strategy which was based on exploitation of peasantry through the fixation of price of industrial goods at a high level while depressing the price of agriculture goods, thereby leading to the extraction of surplus from agriculture to industry. Other leaders like Bukharin had opposed this idea, as it left little incentive for peasants. See Erlich (1960)

21 Nehru emerged as the unchallenged leader of the Congress Party after the death of Sardar Patel in 1950, who was his major rival. Nehru dominated the state apparatus and party organisation.

22 Beside Nehru's tilt towards the soviet model of industrialisation, Mahalanobis was also reminiscent of G. A. Feldman's famous multi-sector growth model put forward in Russia in the 1920s. See Cardoso et al. (2014).

23 Between 1952 and 1961, wheat producers in the United States faced chronic overproduction. In response to this, the United States created export subsidies for food aid through Public Law 480 (popularly

phrased as 'food for peace' law), which was approved in 1954. Through export subsidies in food aid, the United States found an outlet for its agricultural surpluses in the newly independent nations. Under this law, surplus wheat was shipped directly to India, and the payment was accepted in local currency, which was mostly spent by the US embassy inside the local economy.

24 The Jha Committee emphasised the protection of both consumers and producers. It recommended market interventions for procuring food grains as well as for a protective price levels (minimum support price) for farmers.

25 While Varshney (1994) argues that the sources of such policy changes emerged from within and the external agencies only facilitated its implementation through financial support and supply of information, scholars like Perkins (1997) attribute these changes largely to the external agencies.

26 However, the introduction of the green revolution sparked a wider debate. While the proponents of the High Yield Variety Programme (HYVP) argued that it would eliminate acute food scarcity, hunger, and widespread malnutrition in India, the opponents viewed it as a strategy to make farmers dependent on foreign technology, which would benefit the large landholders and disrupt the traditional social relationship in rural India.

27 The percentage share of the members of Lok Sabha with a background of agricultural occupation rose consistently from 23 per cent in 1952 to more than 40 per cent in 1989. For details, see Varshney (1994: 89)

28 Barring P. V. Narasimha Rao's tenure (1991–1996), for many years, there was political instability till the NDA-led A. B. Vaipayee's government took over in 1998 for the second time. While V. P. Singh, Chandra Shekhar, Deve Gowda, and I. K. Gujral ruled for less than a year each, the first Vajpayee-led coalition government continued for only 16 days.

29 The National Seeds Policy came into being in 2002, which is meant to 'strengthen the seed industry' in view of the liberalisation in the farm sector under the WTO. The National Commission on Farmers, which was set up in 2004 to suggest an action plan for farmers and the farm sector, submitted three interim reports between 2004 and 2005. In these reports, the Commission emphasised, among other things, enhancing agricultural productivity and profitability, and strengthening and expanding the horticulture revolution, quality and global competitiveness and prosperity of cotton, agricultural market reforms, and technology mission (*Economic Survey of India 2005–06*: 163). *Vishesh Krishi Upaj Yojana* (Special Agricultural Produce Scheme) was introduced in the year 2004 with the objective to promote mainly the export of fruits, vegetables, and flowers. The National Horticulture Mission (NHM) was launched in May 2005 as a major initiative to bring about diversification in agriculture and to augment the income of farmers through the cultivation of high-value horticultural crops. In 2007–08, the Rashtriya Krishi Vikas Yojana (RKVY) was launched for extending the green revolution to the eastern region of the country to increase crop productivity by intensive cultivation through recommended

agricultural technologies and package of practices (*Economic Survey of India 2010–11*: 201). The Integrated Scheme of Oilseeds, Pulses, Palm Oil, and Maize (ISOPOM) was launched in the year 2010 for promoting crop diversification. The Technology Mission for Integrated Development of Horticulture (TMIDH) was launched in 2001–02 to address issues on production and productivity, post-harvest handling, marketing, and processing of horticultural crops. The government announced the Export and Import Policy (EXIM) in 2001–02 to set up Agri Export Zones for developing and sourcing raw materials and their processing/ packaging leading to final exports.

30 See *Review of Karl Kautsky's Die Agrarfrage*, V. I. *Lenin, Collected Works*, Moscow, Progress Publishers, 1964, vol.4, pp. 94–99.

Chapter 1

From colonialism to neoliberalism

The trajectory of agricultural transformation

Although the trajectory of agrarian transformation in Maharashtra is similar to the rest of India in many respects, it has myriad cleavage and certain peculiarities due to its unique historical features, socio-political conditions, and regional variations, which need to be explained in order to provide a broad analytical background for comprehending its social and economic consequences. Historically a bridge between north and south, Maharashtra (Map 1.1) occupies a substantial portion of the Deccan plateau in the western peninsular region of India. It was formed as a separate ethno-linguistic state in 1960, following the division of the former Bombay state. The newly formed state also included some parts of Marathi-speaking areas of Madhya Pradesh and the princely state of Hyderabad. Although the origin of the present form of agricultural practices in the state lies elsewhere in history, it was largely shaped by colonial, national, and recent neoliberal policies. This chapter presents a critical historical overview of Maharashtra's agricultural sector, linking it with changes in global and national economy, and explores the policies in three different time periods (colonial, national, and neoliberal), underlying ideological approaches and their apparent connections.

The presentation takes the following form. It begins by explaining the agricultural setting under colonial policy and then goes to demonstrate the changes brought about by national developmentalism. The subsequent section highlights the broad features of the agricultural policies under the neoliberal regime. The last section makes the concluding remarks.

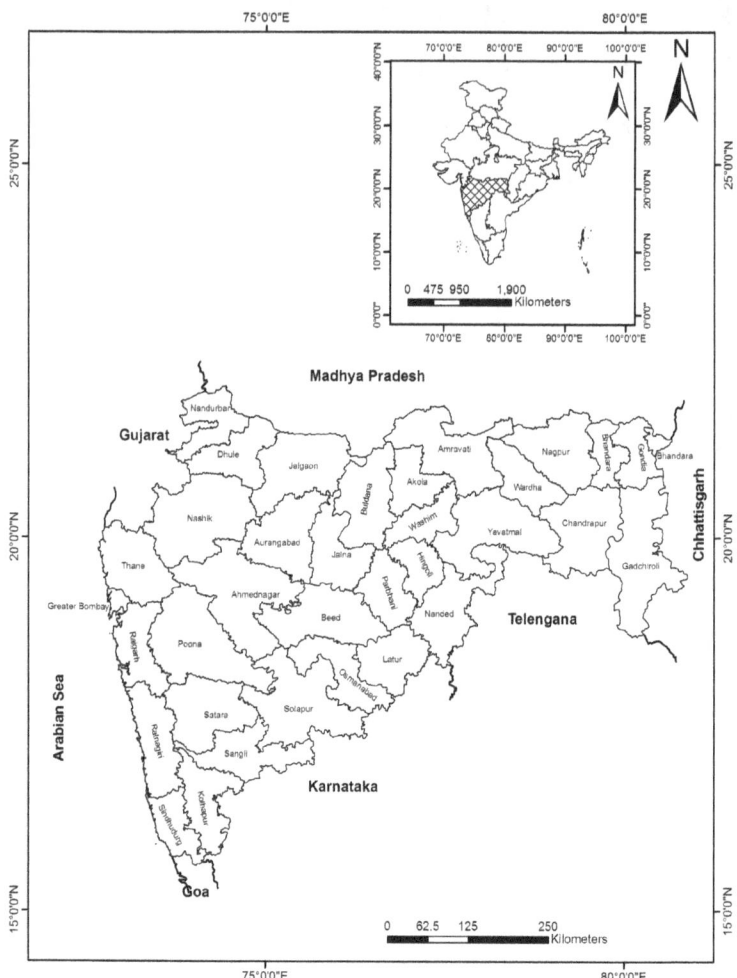

Map 1.1 Map of Maharashtra

Source: Drawn by the author.

Colonial beginnings

Though sporadic efforts were made during the pre-colonial period to bring changes in agriculture, the modernisation of the agrarian economy and society started with the advent of British colonialism. However, the effects of colonial policies were not uniformly felt across the regions of the state. While WM came under direct rule of the British in 1818, Vidarbha[1] fell under British rule in 1853; Marathwada was not directly under the control of the British, as it was part of the Nizam's territory.

The agrarian reform of the British Raj began with the introduction of a new land tenure system, which was the lynchpin of the colonial policy designed to undermine the collective institutions of rural society by creating a climate of individualism and competition among the cultivators and a demand for higher revenue, which ultimately laid the foundation for commercial agriculture. Keeping in view the state of crisis in the area under permanent settlement, the *ryotwari* system was introduced in 1820 in the districts under the Bombay Presidency, which provided formal property rights to the actual holders, and the *malguzari* system, resembling a village-based land revenue system, was adopted in the 1860s in the districts belonging to the Central Provinces.[2]

The next crucial phase in the process of agricultural transformation started with an increasing emphasis on the cultivation of cash crops. To transform agrarian economy into one oriented towards markets and exports, cultivators were motivated to grow commercial crops like sugarcane and cotton instead of a diversity of staple food crops, as they practised traditionally. Of all crops, the government took the greatest interest in the improvement of cotton cultivation. The East India Company began its first experiment on cotton cultivation as early as 1829 and continued for several years thereafter, which was eventually abandoned due to unsatisfactory results. However, demand from cotton manufacturers in England following commercial crisis in 1836 stimulated the company to develop India as source of cotton, which could compete with the United States. Consequently, the East India Company launched a cotton improvement programme in 1840 to encourage cotton cultivation and improve methods of growing, picking, and cleaning, as well as introducing new varieties of long-stapled cotton, which was suitable for Lancashire cotton mills (Harnetty1972: 82–83). In order to promote cotton cultivation, the Bombay government declared concessions in land tax to cotton growers in 1836 and even received the rents in cotton instead of in cash for a few years. There were instances where cultivators were offered large advances to grow cotton, even to the extent of exempting land revenue for cotton fields.[3] To conduct experiments to grow improved varieties of foreign cotton and to introduce American field implements and ginning presses, American planters were brought to the Bombay Presidency in 1840 (Divekar 1983: 338). A factory for agricultural implements was established in the Deccan (Choksey 1955: 75). However, the attempt made to "convert the whole of Bombay Presidency into one great cotton field" failed. The formation of the Cotton Supply Association in 1857 in Manchester revived again the hope of developing Maharashtra (a substantial part of the then Bombay Presidency) as a new source of raw cotton.

With the outbreak of the American Civil War in 1861, the association campaigned for making India into a major supplier of good quality cotton, and under the pressure of the Lancashire cotton lobby, the secretary of state sanctioned measures for improving the quality of Indian cotton (Harnetty 1972: 83). The development of a communication network started from the 1850s, with mounting pressures on the government from Lancashire and European trading interests for extending the means of communication among the cotton-growing districts of western India, especially in Berar and Central Provinces. As a result, the first railway line of Asia, from Bombay to Thane, was opened in 1853, the Great Indian Peninsular Railway in 1866 and the Daund-Manmad Railways in 1878, which ultimately connected a large number of cotton-growing districts of Maharashtra.[4]

Progress was also registered in sugarcane cultivation, resulting in the boom following the First World War as prices of *gur* (jaggery) remained at a high level. Although there was a break in its prices in the later part of the 1920s, it again picked up after 1935. However, the area under sugarcane was not very large. It was reported that between 1886 and 1940, sugarcane covered 40,000–70,000 acres annually in the Bombay Deccan (Guha 1985: 100). Water supply being requisite for its cultivation, sugarcane was grown only in the Bombay Deccan, particularly in Ahmednagar, Poona, Satara, and Nashik, where irrigation was perennial. The limited canal water, made available under the irrigation projects by the end of the nineteenth century, was not even fully exploited by the farmers on large-scale. Those who did use the water emerged as agrarian entrepreneurs in the regional sugar economy with the support of urban finance capital (Ludden 1999). To increase productivity, new varieties mostly grown in Europe, were introduced in several districts as early as the 1830s, and wooden sugar mill technology was replaced by iron mill technology, which was imported from the United States (Guha 1985; Mohanty 1999).

All these developments led to the expansion of the area under commercial crops. The area under cotton increased significantly in Khandesh region and the whole of Bombay Deccan (Choksey 1955; Divekar 1983; Guha 1985). Even districts like Ahmednagar, where cotton cultivation was negligible, started growing and exporting cotton on a large scale. This was only during the years of the American Civil War (1860–65) when supplies of American cotton stopped and the textile industry in the United Kingdom suffered, resulting in a short-lived rise in demand for Indian cotton (Kumar 1968). Besides the Bombay Deccan, the area under cotton increased dramatically in districts like Amravati, Yavatmal, and Nagpur[5] of Vidarbha region. Even though the cotton boom of the 1860s was short-lived, the area under cotton continued to increase due to the growing demand for Indian cotton in the domestic[6] and international markets (Choksey 1955; Tomlinson 1993). The establishment of agricultural experiment stations in Poona in 1841 was instrumental in promoting the cultivation of potato, sugarcane, and varieties of rice and wheat. From about the 1840s, potato cultivation spread in almost all districts, replacing the local variety of sweet potato. Oilseeds like groundnut, sesamum, and castor seed were grown in many parts of Maharashtra, which were largely exported to France, England, and the United States (Choksey 1955: 63–65).

The first technology to emerge for agricultural transformation was canal irrigation, which started in the 1870s. Initially, the government neglected irrigation and left it to various landed groups considering investment in irrigation was not its responsibility.[7] However, the building and repairing of wells was encouraged by the government through a *takavi*[8] advance, a loan the cultivator took from the government, which was repaid through the land revenue machinery. Following a bad famine in 1877–78, *takavi* loans worth Rs. 0.3 million were advanced to the cultivators, mostly for digging new wells and repairing old ones (Attwood 1987: 343). But when it was realised that the recurring droughts and famines caused by scanty and uncertain rainfall were affecting the prospects of high revenue and the very basis of commercial agriculture, the colonial administration undertook construction of a series of centralised large dams with large-scale, canal-based structures shifting away from the traditional irrigation systems. However, the British canal-building activity in India was mainly fuelled by the search for revenue.[9]

Begun in 1869, the Mutha canal came into operation in 1874 as the first storage dam, which was capable of irrigating 17,000 acres of land. The first large-scale irrigation work in Maharashtra was the Krishna canal, which was opened in 1870. A few years later, the first big modern irrigation project, the Nira Left Bank Canal, was undertaken in 1876, as a famine relief work; it went into operation in 1885 with a capacity of irrigating 0.113 million acres of land (Attwood 1987: 344; Ludden 1999: 204). This was followed by the opening of the Girna canal in 1910, the Godavari canals in 1911, the Pravara canals in 1920 and the Nira Right Bank canal in 1930. In Vidarbha, irrigation through the Ramtek, Ghorajheri, Asola-Mendha, and Naleshwar tanks were started early in the twentieth century (Maharashtra State Irrigation Commission 1962). Besides, many other major irrigation works were undertaken in the 1890s. At the end of the decade (1889–1900), while 91,000 acres of land were irrigated by government canals, 41,000 acres were getting irrigation from wells, tanks, and other sources in the Bombay Deccan (Choksey 1955: 81). Though many of these water infrastructures were mainly developed for protection against famines and droughts through extensive irrigation of subsistence crops, they were later diverted largely for sugarcane cultivation following a new set of irrigation policies laid down by the First Irrigation Commission in 1901. The construction of river-fed canals in the Deccan region in the first quarter of twentieth century, particularly the Godavari and Pravara canals, watering areas mostly in the northern part of Ahmednagar district, played a significant role in the development of sugar industry in the region (Chithelen 1985: 606). The government gave leverage to industrial sugarcane development with the establishment of the first sugarcane factory in the Krishna Valley in 1932.

Until the 1870s, institutional credit system was almost non-existent and the government was unwilling to extend credit facility to the farmers due to the risk of recovery associated with uncertain agricultural production in the dry land. The availability of irrigation, which reduced the chances of crop failure, motivated the government to extend credit facilities to the cultivators. Moreover, the Deccan Riots, which were caused by usurious moneylending practices (as described in Chapter 7) alarmed and compelled the government to reconsider its credit policy. The Deccan Agriculturists' Relief Act was passed in 1879 to relieve the agricultural classes from indebtedness and to improve relations between the agriculturists and moneylenders. After a few years, to support the cultivation of cotton and other cash crops, the government decided to advance loans to the cultivators; this led to the enactment of the Land Improvement Loans Act, 1883 and the Agriculturists' Loans Acts of 1884.[10] However, such credit facilities benefitted only a few privileged farmers (Brahme and Upadhyaya 1979; Mohanty 2001a). Even the cooperative credit societies, which were run as a virtual panacea in the years following the 1904 Act, did not meet the long-felt credit requirement of the farmers (Catanach 1966: 76). The Bombay government passed its own cooperative societies Act, the Bombay Cooperative Societies Act, in 1925, under which cooperatives expanded in the state.

Post-colonial developments

After independence, the state witnessed significant changes in the agrarian scene. The post-colonial transforming measures included land reforms, cooperatives, community development programmes, adoption of new farm technology, and other infrastructural developments intended to maximise agricultural production on one hand and to ensure social justice on the other. Following the national guidelines, land reforms were introduced in the erstwhile Bombay state, Vidarbha and Marathwada regions, and other parts of the country. It began with the abolition of revenue-collecting intermediaries[11] leading to the abolition of *khoti* in the coastal districts, *malguzari* and *izardar* in Vidarbha, *jagirdar* in Marathwada and *vatans* in WM. However, Maharashtra took longer time to complete the process of abolition of intermediaries compared to many other states. To bring the tenancy reforms, the Bombay Tenancy and Agricultural Lands Act were passed in 1948, which repealed the earlier Bombay Tenancy Act of 1939, though it incorporated most of its provisions. This act was amended several times and, finally, the Act conferred upon the tenant a right to ownership of land from a landlord with effect from 1 April 1957, called the 'tillers day'. To set right the weaknesses and to clear the loopholes, the Act was amended many times later on. Similarly, the Bombay Tenancy and Agricultural Lands (Vidarbha Region and Kutch Area) Act of 1958 and the Hyderabad Tenancy and Agricultural Lands Act of 1950 (amended in 1954 and 1956) provided for the transfer of ownership of the land in Vidarbha and Marathwada, respectively.

Besides, rapid and substantial progress was also made in the cooperative field with the support of the state, particularly during the tenure of Y. B. Chavan (1956–1960) as the chief minister of erstwhile bilingual Bombay state, in the spirit of Nehruvian vision embedded in Fabian socialism. The progress of Pravara Cooperative sugar factory established in 1948 encouraged the state government to launch a chain of cooperative sugar factories in the state. The government then decided, in 1954, to grant industrial licenses for the cooperative sugar industry alone. During the period between1955 and 1960, 13 new cooperative sugar factories were established in the canal-irrigated areas of Poona, Ahmednagar, Satara, and Nashik, and lift-irrigated areas of Kolhapur. Prior to the commencement of the Five-Year Plans in 1952, the state had only 36 irrigation projects, which were mostly in WM. Irrigation from eight reservoirs was started in Vidarbha, and the first significant irrigation project in Marathwada was completed in 1954. Apart from this, massive irrigation schemes in river valley regions were planned and carried out. Long-term projects like the Krishna Valley project were initiated, and the potential of the Shahyadri Mountains and eastward-flowing rivers for irrigation was explored (Kamat 1980: 1630). To generate participatory grassroots democracy, the Bombay Village Panchayats Act was introduced in 1958.

It was during this time that Maharashtra was formed as a separate state (in 1960). By this time, the First Five-Year Plan was over and the Second Plan was at the verge of completion. During this period, Bombay state, which was already industrialised, witnessed further growth with the establishment of larger industrial units like oil refineries and engineering/manufacturing sectors in the Thane-Pune-Nashik triangle. Industrialisation was diversified, from traditional textile and food processing industries to chemicals, metals, and alloys; machinery and machinery tools; electrical machinery and appliances; and transport equipment (with spare parts), which, ultimately, led to the emergence of a new class of business and industrial elite entrepreneurs (Kamat 1980: 1670). In view of concentration of industrialisation in Bombay, the first chief minister of the newly formed state, Y. B. Chavan, took initiatives to set up a few industries in backward pockets, mostly utilising private capital, and introduced the Maharashtra Industrial Development Corporation (MIDC) to promote industrialisation. Though Chavan opted for industrialisation to fulfil Nehru's dream and to accommodate the interests of the Bombay-based urban-industrial capitalist class, his interest centred more on agricultural and rural development, which he sought to improve through cooperatives and Panchayati Raj institutions. He strengthened the cooperative movement and implemented the new three-tier Panchayati Raj system in 1961, in pursuance of the recommendations of the Balwantrai Mehta Committee, as recommended by the Naik committee.

The Cooperative Societies Act, 1960, was passed to gear up the cooperative movement towards the promotion of industrial development. Many more cooperative sugar factories were added within and outside of canal-irrigated areas, where required sugarcane could be grown. Larger financial outlays were earmarked during the Third Plan as compared to the earlier two plans formulated by erstwhile Bombay state for accelerating and intensifying the programmes under irrigation and agrarian development. Put together, about 45 per cent of the total plan expenditure of the state was diverted to the development of agriculture and irrigation (Kalamkar 2011). However, in the state's subsequent Five-Year Plans, the share of agriculture declined in percentage terms, though irrigation continued to remain a prioritised sector.[12] In response to the growing concern on land concentration, the Maharashtra Agricultural Lands (Ceiling on Holdings) Act, 1961, was passed, prescribing the ceiling on landholdings for both irrigated and non-irrigated areas 1962 onwards. However, the ceiling limit was kept at a higher level apprehending opposition from the sugar industries, which was later lowered further in 1975–77 (during the national emergency). Subsequently, other measures related to tenancy, land ceiling, consolidation of holdings, and distribution of land to SCs and STs were taken. It is commonly believed that Maharashtra was one of the few states where land legislations were better implemented, at least with regard to the tenancy reforms (Kamat 1980). It is indicated that the collective impact of these measures enhanced agricultural productivity particularly in WM (Rajasekaran 1996; Deshpande 1998). However, as regards the impact on land redistribution, while earlier studies (Dandekar and Khudanpur 1957; Brahme and Upadhyaya 1979) highlighted the poor performance of the state, recent studies have reported its noticeable achievements compared to other states (Rajasekaran 1996; Mohanty 2001a, 2001b).

The policy shift at the national level prioritising agriculture during the mid-1960s boosted tremendously agricultural development in Maharashtra. In the meantime, Y. B. Chavan, who pioneered agricultural development through cooperatives and Panchayati

Raj, left the state in 1962 to join Nehru's cabinet at the centre. After him, it was V. P. Naik who played a key role in the transformation of rural economy and society during his 11½ years (1963–1975) of long tenure as chief minister. During this period, the state Congress party was tightly controlled by Y. B. Chavan, and Naik being his close follower, easily secured the support of all fractions within Congress in the state to pursue his agenda of development. In his first policy statement as chief minister, Naik declared that he would make every effort to make Maharashtra agriculturally prosperous.[13] Though Naik did not hold the agricultural portfolio, he literally functioned as a super minister for agriculture. Major policy decisions relating to agricultural development programmes were announced by him.[14] He revived the Panchayati Raj institutions, which later became a model for the whole country, and expanded the base of cooperatives. Sugar cooperatives were started in the less developed regions of Marathwada and Vidarbha, and spread to other fields like rural industries, agro-processing, agro-marketing, consumer stores, agro-services, etc. These cooperative societies emerged as the agencies of rural development and agricultural transformation as they were instrumental in the growth of mechanisation and improvement in methods of cultivation and rural infrastructure.

Following the government of India's adoption of HYVP programme and the ushering in of the green revolution in the mid-1960s, efforts were made by the government of Maharashtra to expand the area under this variety for increasing agricultural productivity. A programme for the commercial cultivation of hybrid *jowar*, *bajra*, maize, and HYVs of wheat and paddy seeds was undertaken in 1966–67. The chief minister himself, along with few other growers, being the major producers of hybrid cotton variety like H4 seed,[15] aggressively propagated the virtues of hybrid seeds and HYVP, and a handful of agriculturalists were commissioned to produce these seeds. He set up a high-powered cabinet subcommittee to promote cotton production, especially the production of H4 cotton. Cotton growers received generous government help in the form of easy credits and supply of fertilisers and hybrid seeds at concessional rates to grow these varieties. To go by the 11th report of the Estimates Committee of Maharashtra Legislature 1973–74, the state government had incurred an expenditure of 7 billion rupees between 1960 to 1961 and 1970 to 1971 on various agricultural schemes. In order to meet the growing credit need of the farmers, the network of primary agricultural credit societies, state cooperative agriculture and rural development banks, and commercial and regional rural banks were started; these provided short-term and long-term subsidised loans to cultivators. The nationalisation of over a dozen private banking institutions by Indira Gandhi in 1969, directing lines of credit away from the urban-industrial to rural sector with a substantial government loan programme, strengthened the rural credit that the green revolution areas needed badly.

In order to facilitate the production, storage, transportation, and distribution of seeds, the Maharashtra State Seed Corporation was established in 1976 under the National Seed Project. The State Seeds Subcommittee, constituted in 1969 under the Seeds Act of 1966, then released as many as 15 varieties of hybrids and HYV seeds between 1969 and 1974.[16] Several foreign companies in collaboration with Indian producers like MAHYCO started producing hybrid as well as genetically modified (GM) seeds. To encourage mechanisation as demanded by the new cropping pattern based on improved methods of cultivation, the government took initiatives to manufacture and distribute agricultural machineries in the state. The Maharashtra Agro Industrial Developmental Corporation Limited (MAIDC), known as *Krushi Udyog*, was set up in 1965. It started its Agro Engineering Division in 1969 to manufacture and sell tractors, electric pumpsets, combine harvesters, seed cum fertiliser drills, and power tillers at subsidised rates. As a consequence, there was a marked increase in mechanical appliances over the years.

To regulate the marketing of agricultural produce in the market, the Maharashtra Agricultural Produce Marketing (Regulation) Act was passed in the year 1963. Following this, all notified agricultural commodities, grown within the purview of a regulated market and sold wholesale, were marketed through regulated markets. Higher returns were ensured to farmers, thereby protecting them from exploitation by middlemen, eliminating superfluous charges, and minimising various costs of handling products. Fair market charges and fees and correct weight and measurements were ensured, and a system for grading agriculture products was also introduced. To promote cotton market, the Naik government passed the Raw Cotton (Procurement, Processing, and Marketing) Act 1971 to ensure fair and remunerative prices to cotton growers and to eliminate the role of middlemen in cotton trade. Under the provision of this act, the cotton monopoly scheme was implemented in 1972 through the Maharashtra State Cooperative Marketing Federation as a part of government's 15-point socio-economic programme, which was included in the Congress party's election manifesto for the 1972 assembly poll. Despite opposition from the central government, the Reserve Bank of India and later from the Cotton Federation of India, the state government pushed this scheme (Lele 1990: 196).[17]

The collective impact of this new package of modernising measures, however, did not enhance the quantum of agricultural productivity as desired. The output of food grains declined, particularly between 1969 and 1973, in many districts. Similarly, though the area and production under sugarcane increased just marginally, it virtually remained stagnant for crops like cotton and groundnut.[18] Prime Minister Indira Gandhi, who was unhappy with the Chavan-Naik dominance in Maharashtra's state politics, used this as an opportunity[19] to cut them to size. She replaced Sudhakarrao Naik with S. B. Chavan from Marathwada to challenge the dominance of Y. B. Chavan, who was well entrenched in the cooperative movement led by socially and economically powerful rich peasants. Soon after he became the chief minister, S. B. Chavan announced the decision to reduce the supply of canal water to sugarcane growers to only eight months in a year. To suggest that the government was committed to proceed against a class of 'VIP farmers', S. B. Chavan deliberately resorted to all sorts of methods and tricks to dodge the cooperative dues. He also announced the state cabinet's decision to auction the irrigated lands of the farmers in the event of their non-payment of the cooperative dues.[20] The Land Ceiling on Holdings Act 1961 was amended to further lower the ceiling limit despite strong opposition from the sugar lobby. To enhance agricultural productivity through effective and efficient use of water, the Maharashtra Irrigation Act of 1976 came into force in 1977, superseding the earlier acts[21] followed in various regions before the formation of the state. Moreover, with the declaration of national emergency in June 1975, many of the earlier developmental schemes, like the Employment Guarantee Scheme, introduced by V. P. Naik, were temporarily buried. Following the emergency, focus shifted to the implementation of the Prime Minister's 20-point programme.

Mrs. Gandhi's attempt to truncate the Maratha leadership led to intense factionalism and political instability in the state during the post-emergency period. Interestingly enough, the subsequent Congress-led government with Vasantdada Patil as chief minister also represented agrarian interests and, ironically, was part of a powerful lobby of sugar barons of WM. He institutionalised the Employment Guarantee Scheme by passing a bill in 1977 with the support of dominant *Maratha* elites. He also brought in 1984 the Maharashtra State Agricultural Marketing Board (MSAMB) under the provisions of the Maharashtra Agricultural Produce Marketing (Regulation) Act 1963; this was done to coordinate between the Agricultural Produce Market Committees for the development of markets and marketing area. Remarkably, throughout the post-independence period, the Congress party remained in power without interruption till 1991, baring a few years in between, and most of its council of ministers were the 'Cooperative Kings', for whom development of farm economy through cooperatives was key to consolidate their political base (Kamat 1980; Lele 1990). Even the non-Congress government led by Sharad Pawar's Progressive Democratic Front (1978–79) had a similar composition of council of ministers. It was remarked, "every politician worth his name has a cooperative in his empire".[22]

Despite reports of several committees (e.g. Gulabrao, Shivajirao, Godbole), which indicated the sickness of sugar cooperatives, the state government continued to protect them. Irrigation subsidy was increased further; it was reported that the growth rate of irrigation subsidy in Maharashtra was the highest in the country,[23] and a major part of it was appropriated by WM, which was growing sugarcane. It was reported that more than half the irrigation was going to sugarcane, which covered only about 3 per cent of the total cropped area of the state (Rath and Mitra 1989; Kale 2014). Moreover, the expansion of rural electrification enabled the farmers to pump groundwater and lift water from reservoirs using electrified pumps; this marked a significant shift in the irrigation pattern and agricultural production. To support the increasing electricity consumption of the farmers, Sharad Pawar declared a reduction in rates of electricity consumption in 1978 from 29 *paise* per unit to 20 *paise* per unit for farmers; similarly, a corresponding reduction in the rate for non-metered consumers from Rs. 180 to Rs. 125 (per hp) was announced (Kale 2014: 90–91). This subsidy continued through the 1980s.

In a nutshell, much of the transformation in the agricultural sector in Maharashtra was attributed to the growth of cooperative movements. As pointed out by Deshpande *et al.* (1992), the growth of cooperatives in Maharashtra went through three major phases. While in the first phase (1961–68) the growth of cooperatives was regionally concentrated, the second phase (1969–80) witnessed marginal reduction in the influence of sugar lobby, though cooperatives continued to grow. The third phase, which began in 1980, experienced greater prominence of sugar interest in government. It was observed that the government represented interests of sugar cooperatives more forcefully when in conflict with the central government. As regards the impact of cooperatives on rural society, while one group of scholars (Baviskar 1980; Attwood 1992) concluded that these cooperatives had enhanced the per acre yield of sugarcane and succeeded in improving the economic conditions of sugarcane growers and public in general, others argued that the *kulaks* had reaped the major benefits from this movement (Brahme and Upadhyaya 1979; Dhanagare 1994).

Nonetheless, Maharashtra witnessed a kind of agricultural revolution during the post-colonial period under the combined impact of cooperative movements, land reforms, and new farm technology, which is summed up in Table 1.1. Compared with the neoliberal economic programme for agriculture, as will be seen next, the agrarian sector in Maharashtra was on a better footing in the national developmentalist period, as there was an inclusive plan for the agricultural sector.

Table 1.1 Changes in agriculture during the post-colonial period

Sl. no.	Particulars	TE 1970–71	TE 1980–81	TE 1990–91
1.	Total geographical area in '000 hectares	92,296	92,275	92,275
2.	Per cent of grazing lands to total geographical area	4.8	5.1	4.5
3.	Per cent of culturable wasteland to total geographical area	3.2	3.2	3.3
4.	Per cent of non-agricultural use land to total geographical area	2.4	3.4	3.5
5.	Per cent of barren land to total geographical area	5.8	5.7	5.5
6.	Per cent of current fallow land to total geographical area	3.6	2.6	2.8
7.	Gross cropped area in '000 hectares	19,169.6	19,888.9	21,670.9

Sl. no.	Particulars	TE 1970–71	TE 1980–81	TE 1990–91
8.	Net sown area in '000 hectares	18,196.3	18,153.5	18,448.8
9.	Net irrigated area as per cent of net sown area	7.6	10.5	13.8
10.	Per cent of area under HYVs to gross cropped area	6.0	20.2	33.5
11.	Cropping intensity	105	110	117
12.	Per cent of area under food grains to gross cropped area	68.1	69.7	66.5
13.	Per cent of area under oilseeds to gross cropped area	9.1	9.1	12.8
14.	Per cent of area under sugarcane to gross cropped area	1.1	1.3	1.8
15.	Per cent of area under cotton to gross cropped area	14.4	12.7	12.3
16.	Total production of food grains in '000 tonnes	6,226.4	9,898.0	12,159.9
17.	Total production of oilseeds in '000 tonnes	678.3	801.3	1805.1
18.	Total production of sugarcane in '000 tonnes	1,585.5	2,294.0	3,255.6
19.	Total production of cotton (kgs./hectare)	182.0	237.6	309.4
20.	Average yield of food grains (kgs./hectare)	477	714	844
21.	Average yield of oilseeds (kgs./hectare)	387	445	652
22.	Yield rate of sugarcane (kgs./hectare)	7,591	8,803	8,575
23.	Yield rate of cotton (kgs./hectare)	66	94	116
24.	Per hectare fertiliser consumption in kg. per hectare	–	20.1 (1981–82)	48.0 (1991–92)
25.	Number of tractors per '000 hectares	–	–	3 (1991–92)

(Continued)

Table 1.1 Continued

Sl. no.	Particulars	TE 1970–71	TE 1980–81	TE 1990–91
26.	Number of pumpsets per '000 hectares	–	39 (1981–82)	74 (1991–92)
27.	Number of agricultural credit societies	20,420	18,577	19,565
28.	Number of members (000) in agricultural credit societies	3,794	5,416	7,942
29.	Working capital of agricultural credit societies (lakh rupees)	34,329	52,746	185,100
30.	Net state domestic product at factor cost (at 1999–2000) prices from agriculture and allied activities	1,365,690	1,866,159	2,680,716

Source: (1) *Season and Crop Report, Maharashtra State*, relevant issues. (2) *All India Report on Input Survey*, relevant issues. (3) *Statistical Abstract of Maharashtra State*, various issues.

Note: (1) Data only for particulars from sl. no. 1 to 24 for the years 1970–1971, 1980–1981, and 1990–1991 indicate triennium ending (TE). (2) Figures in parenthesis (sl. no. 24–26) denotes the reference year. (3) Data for particulars from sl. no. 27 to 30 relate to only specific reference years.

Changes during the neoliberal regime

Maharashtra is one of the few states that responded promptly to the demands of economic liberalisation and introduced a series of neoliberal measures for bringing changes in the agricultural sector. In fact, the liberalisation of the state's economy began three years before the national trend. When Sharad Pawar, known for his aggressive campaign for economic liberalisation,[24] became the chief minister in 1988, he pursued vigorously the economic policy of liberalisation and adopted a new approach to agriculture that emphasised export-oriented agriculture, horticulture, tourism, and new agro-based industries (Palshikar and Deshpande 2003). The subsequent Congress and non-Congress governments continued the same liberalisation policy. The initial steps taken by Sharad Pawar and policy continuity between Congress and the government led by Shiv Sena-BJP facilitated agrarian transformation in tune with the neoliberal order, with the support of farmer organisations like the *Shetkari Sanghatana* (discussed in Chapter 7). During the plan periods, the share of the state's expenditure on agriculture reduced substantially.[25] Within a short span of a few years, Maharashtra introduced new land reforms, reforms in sectors like power, irrigation, agricultural production, banking, and marketing and many others. Various committees and commissions were appointed from time to time to bring necessary changes in agriculture to meet the new demands of the changing global economic order.[26]

To give a boost to horticulture, an ordinance was passed in 1993 by the Sharad Pawar government – [the Maharashtra Agricultural Lands (Ceiling on Holdings) (Amendment) (Ordinance)] – to enable certain categories of individuals, firms, companies, etc., to hold land on lease beyond the prescribed ceiling limit fixed earlier by the Maharashtra Agricultural Lands (Ceiling and Holdings) Act 1961, amended in 1975. However, the ordinance was not approved by the central government. Additionally, certain types of private industrial units were permitted to acquire land beyond the existing ceiling limit under the provision of the revised Maharashtra Industrial Policy of 1993. With this policy in place, the MIDC acquired more than 30,000 hectares of land for purposes other than setting up its own government-operated industrial estates (Jenkins 1999: 105). This was followed by the Special Economic Zones (SEZs) Act 2005; the state acquired 6,658 hectares of land for private corporates to develop 55 SEZs in various parts of the state as of September 2016, giving them full freedom.[27]

Though the share of agriculture in the planned expenditure declined significantly in the post-reform period, the allocation for irrigation projects increased noticeably[28] as part of the strategy to generate electricity for the industrial entrepreneurs, as well as to appease the strong political groups, particularly from the sugarcane belt of the state. In order to accelerate the completion of irrigation projects, five irrigation development corporations were created for the different regions of Maharashtra[29] between 1996 and 1998, raising a major part of the funds from open market. It was reported that until March 2006, these corporations raised funds to the tune of Rs. 121 billion from the open market against the paid-up share capital of the state government of Rs. 110 billion.[30] According to the Central Water Commission (2015), a total number of 1,693 dams were completed as of 31 October 2014, which was the highest in the country, and 152 more dams were under construction. The irrigation water rates, which were fixed in 1975, were increased in 1991, and a government memorandum was issued revising the rates for supplying water to different regions (Dev and Mungekar 1996: A-41). Consequently, after the implementation of the State Water Policy in 2003, the Maharashtra Management of Irrigation Systems by Farmers Act and the Maharashtra Water Resources Regulatory Authority Act were passed in 2005 to improve the water-use efficiency and financial performance of irrigation projects, with operation and maintenance expenses being covered through water charges.

Keeping in view the high return from horticulture – i.e. fruits compared to the conventional food grain crops – the government implemented several programmes to promote horticultural development. Since 1991, the Employment Guarantee Scheme has been linked with horticultural development. In fact, following economic liberalisation, the government of Maharashtra launched an ambitious programme for converting the state into a horticultural state by the year 2000. In order to implement the schemes of National Horticulture Mission (2005–06) and National Medicinal Plant Board (NMPB), the Maharashtra State Horticulture and Medicinal Plants Board (MSHMPB) was established in 2005. Under the schemes of NMPB, 107 beneficiaries were covered for contract farming as of October 2008.[31] To promote sericulture, the Mulberry Silk Development Programme was implemented in 23 districts and the Tasar Silk Development Programme in four districts of the state.

In consonance with the economic reforms, the Maharashtra Agricultural Produce Marketing (Regulation) Act 1963 was amended in 2005 to facilitate direct marketing, private market, farmer consumers market, single licenses system, and contract farming. As a result, as many as 101 direct marketing licenses were issued, of which 54 were for purchasing cotton, and 47 for other agricultural commodities, including 24 private market licenses.[32] Besides, the state launched the Agri-business Infrastructure Development Programme with the assistance of the Asian Development Bank, and the land owned by Agricultural Produce Marketing Committees (APMCs)/Cooperatives was provided to private investors on lease to build and operate integrated value chains for agricultural produce on the basis of public-private partnership. Taking advantage of the new provisions, corporates like Reliance, Godrej, Deepak Fertilisers, and Petrochemicals Ltd., ITC, and Bhartiya Group entered agricultural markets for processing, marketing, and export of agricultural products. It is reported that 305 main markets and 603 sub-markets were there in the state, as of 31 March 2015.[33] The MSAMB recently established a network of computerised APMCs in the state called 'MARKNET' to provide information on price and daily arrivals to farmers through APMCs. About 294 main markets and 66 sub-markets were computerised and connected to MSAMB's website by March 2009.[34] A Memorandum of Understanding (MoU) was signed between Reuters and MSAMB in 2007 to disseminate information about market arrivals of produce, prices, weather forecast, and market guidelines through mobile phones.

To boost agricultural exports, Agri Export Zones (AEZs) were created in the state covering Alphonso mango, kesar mango, onion, oranges, banana, grapes, and pomegranate. Besides, MSAMB established export facilities centres in various pockets of the state through which mangoes, bitter gourd, pomegranate, banana, and other fresh vegetables and fruits were exported to the United States, Gulf nations, Japan, and Europe.[35] Additionally, contract farming emerged as a viable method of procurement. Many private companies like Mahindra Shubh Labh Services Ltd., Hindustan Unilever Ltd., Reliance Retail, Ion Exchange Enviro Farms Ltd., Marico Ltd., Nijjer, and PepsiCo started making contracts with farmers growing safflower, papaya, wheat, cotton, grapes, and many other crops.[36]

The distribution of hybrid, improved, certified-quality-assured seed varieties of various crops, which was earlier largely controlled by the Maharashtra Seeds Corporation and National Seeds Corporation, is now more or less dominated by the private sector. It is reported that the recent public and private sectors ratio in seed distribution went up to 42:58.[37] Nearly 333 private seed producers, such as *Krushi Vidnyan Mandal*, were involved in seed production and distribution.[38] Between 1991 and 2012, as many as 166 new hybrid and HYV seeds were introduced. The cultivation of Bt cotton started in 2003 and the Maharashtra Cotton Seeds Act came into force in 2009 to regulate its prices and distribution in the state. In the last two decades, total seed distribution increased fourfold. While the private sector was leading in the supply of hybrids, the public sector led the straight varieties.[39]

The rural and agricultural credit system also witnessed significant changes. In the 1990s, there was a slowdown in cooperative credit following the policy measures applied to credit cooperatives as a part of a financial liberalisation programme. However, the flow of agricultural credit increased in the 2000s due to a greater role played by commercial banks and regional rural banks. This was mainly due to the announcement of two relief packages by the central and state governments – namely, the prime minister's Vidarbha package and state package in 2006 in response to the reports of farmer suicides. In part, the revival of agricultural credit was inspired by the central government's announcement in 2004 to double the flow of agricultural credit between 2004 to 2005 and 2007 to 2008 (Chavan 2015). An analysis made by Chavan (2010) reveals three important dimensions to viewing the growth of agricultural credit critically: (a) increase in agricultural credit from commercial banks was accounted for by providing indirect finance to institutions that finance input dealers, irrigation equipment suppliers, and non-banking financial companies; (b) significant changes were made in the definition of agricultural credit to accommodate financing commercial, export-oriented and capital-intensive agriculture, and raising of the credit limit of existing agricultural financing; (c) much of the increase in the total advances to agriculture was due to the increase in the number of loans with a credit limit of Rs. 100 million and above, and especially Rs. 250 million and above. Also, between 1995 and 2005, the share of agricultural credit supplied by urban and metropolitan branches increased, accompanied by a decline in the share of rural and semi-urban branches. It was reported that in Maharashtra, the metropolitan branches of commercial banks provided nearly half the total agricultural credit in 2008, where Mumbai alone covers 43 per cent of it (Chavan 2010). Such an urban metro-centric supply of agricultural credit benefitted only city-based large corporations engaged in agricultural production. The major changes in agriculture in terms of key indicators during the post-reform period are illustrated in Table 1.2.

Table 1.2 Changes in agriculture during the post-reform period

Sl. no.	Particulars	TE 1994–95	TE 2000–01	TE 2010–11
1.	Total geographical area in '000 hectares	92,275	92,275	92,275
2.	Per cent of grazing lands to total geographical area	4.4	4.0	4.0
3.	Per cent of culturable wasteland to total geographical area	3.2	3.1	3.0

Sl. no.	Particulars	TE 1994–95	TE 2000–01	TE 2010–11
4.	Per cent of non-agricultural use land to total geographical area	3.9	4.3	4.7
5.	Per cent of barren land to total geographical area	5.6	5.2	5.6
6.	Per cent of current fallow land to total geographical area	3.1	3.8	4.5
7.	Gross cropped area in '000 hectares	21,283.6	21,742.1	21,747.0
8.	Net sown area in '000 hectares	18,003.2	17,713.1	17,409.7
9.	Net irrigated area as per cent to net sown area	13.9	18.5	18.7
10.	Per cent of area under HYVs to gross cropped area	29.8	29.4	32.5
11.	Cropping intensity	118	123	131
12.	Per cent of area under food grains to gross cropped area	65.0	61.6	53.6
13.	Per cent of area under oilseeds to gross cropped area	12.6	12.2	16.8
14.	Per cent of area under sugarcane to gross cropped area	2.0	2.6	3.6
15.	Per cent of area under cotton to gross cropped area	12.1	14.5	15.4
16.	Total production of food grains in '000 tonnes	13,010.9	11,887.4	13,061.6
17.	Total production of oilseeds in '000 tonnes	1,974.0	2,451.8	34,51.6
18.	Total production of sugarcane in '000 tonnes	3,433.5	5,015.6	7,016.6
19.	Total production of cotton (kgs./hectare)	385.0	420.6	982.4
20.	Yield rate of food grains (kgs./hectare)	941	888	1,072
21.	Average yield of oilseeds (kgs./hectare)	736	922	901
22.	Average yield of sugarcane (kgs./hectare)	8,140	8,758	8,458
23.	Average yield of cotton (kgs./hectare)	150	133	281
24.	Per hectare fertiliser consumption in kg. per hectare	–	101.1 (2001–02)	130.1 (2011–12)
25.	Number of tractors per '000 hectares	–	20 (2001–02)	107 (2011–12)
26.	Number of pumpsets per '000 hectares	–	176 (2001–02)	220 (2011–12)

(Continued)

Table 1.2 Continued

Sl. no.	Particulars	TE 1994–95	TE 2000–01	TE 2010–11
27.	Number of Agricultural credit societies	–	20,551	21,343
28.	Number of members (000) in agricultural credit societies	–	10,125	13,853
29.	Working capital of agricultural credit societies (lakh rupees)	–	698,824	1,390,596
30.	Net state domestic product at factor cost (at 1999–2000) prices from agriculture and allied activities	–	3,525,725	4,190,129 (2009–10)

Source: (1) *Season and Crop Report, Maharashtra State*, relevant issues. (2) *All India Report on Input Survey*, relevant issues. (3) *Statistical Abstract of Maharashtra State*, various issues.

Note: (1) Data only for particulars from sl. no. 1 to 24 for the years 1994–95, 2000–01, and 2010–11 indicate TE. (2) Figures in parenthesis (sl. no. 24–26) denotes the reference year. (3) Data for particulars from sl. no. 27 to 30 relate to only specific reference years

Conclusion

To recapitulate, agriculture in Maharashtra has experienced a series of significant changes from the colonial through the post-colonial period till the ongoing current phase of neoliberalism. While colonial Maharashtra had emphasised commercial agriculture with the least investment, the Indian state gave priority to introducing new technologies to achieve faster growth, and the neoliberal policies stressed market liberalisation, which quintessentially meant structural adjustments. Although the focus in all three phases was to augment agricultural production, enhance growth and productivity, and promote the link between agriculture and industry, the colonial and post-colonial approaches embraced state-led measures, whereas the neoliberal state blindly accepted market fundamentalism, which defined its programmes and targets. Thus, the neoliberal reforms since the 1990s have apparently transformed the forces and relations of production on a global scale, marking a shift from planned but slow growth processes to market-driven rapid growth accompanied by changes in government policies for agricultural development (Table 1.3).

Table 1.3 Changing paradigms of agricultural transformation

Particulars	Colonial	Post-colonial	Neoliberal
Land	New land tenure system for enhancement of land revenue, agricultural growth and individual competitiveness	Land reform for promoting growth as well as social justice	Slowing down of state-led land reforms and shifting towards market-led reforms for agricultural growth and efficient management of land
Crop	Expansion of area under commercial crops	Priority for cultivation of food grain crops	Emphasis on the promotion of high-value crops
Credit	Promotion of private moneylending and little emphasis on institutional credit	Expansion of formal credit network mainly through cooperative societies	Weakening of institutional credit system and emphasis on privatisation of credit
Irrigation	Limited emphasis on irrigation expecting huge revenue in return	Emphasis on irrigation to enhance agricultural growth	Declining public expenditure on irrigation projects and emergence of new water policies for efficient water management
Seed	Introduction of a few new varieties of foreign seeds for selected commercial crops	Introduction of high yielding variety of seeds for food grain crops, which gradually shifted to non-food crops; seed production and distribution are controlled by the private and public sectors	Increasing emphasis on genetically modified, pest-resistant varieties which required high doses of inputs, dominance of private sector in seed supply, liberalised import of seed
Fertiliser	Negligible application of chemical fertilisers	Increased use of fertiliser backed by subsidy	Application of high doses of chemical fertiliser and withdrawal of subsidy

(Continued)

Table 1.3 Continued

Particulars	Colonial	Post-colonial	Neoliberal
Mechanisation	Limited adoption of advanced agricultural implements	Gradual growth in use of tractors and other advanced machineries	Rapid mechanisation of agriculture
Subsidy	Absence of subsidy, no remission of water tax and land revenues, even at the time of crisis	Provision for subsidy for machinery, fertiliser, pesticides, electricity, etc.	Withdrawal of all kinds of subsidies in a phased manner
Marketing	Marketing through traders and middleman, opening of international markets for major commercial crops like cotton	State regulated marketing, protectionist marketing policies	Free-market; small farmers incorporated into the global economy, intense competition amidst expansion of global trade
Industry-agriculture linkage	Emergence of agriculture industry link for selected crops like cotton	Limited linkage between agriculture and industry	Link between agriculture and industry became the basis for agricultural production
Policy mechanisms	Export-oriented growth of selected agricultural commodities with least investment in public infrastructure and welfare programmes	Public investment in agriculture and rural development	Export-led growth, competitive specialisation, and trade liberalisation through reduction of tariffs and non-tariff barriers

Although the overall impact of this transformation contributed to the enhancement of agricultural growth and productivity, it is in no way spectacular compared with the nationalist period considering the application of inputs (Tables 1.2 and 1.3). The recent literature on agricultural growth in Maharashtra also highlights the same trend (Kalamakrar 2011; Sawant *et al.* 1999). As per the estimate of the World Bank, while the index of output prices rose by 50 per cent during 1993–94 to 1999–2000, the index of prices of all inputs taken together went up by 56 per cent during the same period.[40] On the other hand, the social consequences of this rapid transformation adversely affected the weaker sections and led to larger problems, such as regional disparity, farmer suicides, social inequality, and labour migration, which are illustrated in greater details in the following chapters.

Notes

1 Vidarbha includes (a) Nagpur sub-region (i.e. East Vidarbha) and (b) Berar sub-region (i.e. West Vidarbha). Nagpur sub-region became a part of the East India Company's administration after its annexation policy of 1853; this was supported and legitimised by the British parliament. However, it would be debatable whether it became a part of the British kingdom. The crown took over the reins of Indian colony from the East India Company's rule in the late 1857 after the Sepoy Mutiny was crushed. The Berar area was conquered by the Duke of Wellington in a war with the Marathas in 1803. At this war, the Nizam supported the British. Therefore, as a gift, the company gave Berar (i.e. four districts of West Vidarbha) to the Nizam. These were called Hyderabad assigned districts though administration for all practical purposes was looked after by the East India Company from 1853 onwards, but sovereignty over Berar remained with the Nizam till 1937. See Longford (1971).
2 However, there is a dispute among the economic historians regarding the land tenure system prevalent in the Central Provinces. While Banerjee and Iyer (2005) classified it as a predominantly landlord revenue system, Iversen *et al.* (2013) suggested that for the most part the *malguzari* settlement of the Central Provinces was very different from the Permanent (*zamindari*) settlement introduced in Eastern India by Cornwallis 70 years earlier (1793).
3 See *Ahmednagar Gazetteer*, Revised edition, 1976, p. 271.
4 The opening of the Great Indian Peninsula and Dhond and Manmad Railways connected Nashik, Nagpur, Yavatmal, Amravati, Wardha, Satara, Poona, Ahmednagar, and many other districts with the major Indian cotton-trading centres and ports.
5 As reported by the *Central Provinces District Gazetteer of Nagpur* (1908), the area under cotton in Nagpur steadily increased from

70,000 acres in 1863 to 140,000 at the settlement of 1892–94 and to 476,000 in 1905–06. Similarly, while the area under cotton in Amaravati increased from 38 per cent in 1891–92 to 57 per cent in 1925–26, in Yavatmal it rose from 30 per cent to 45 per cent during this period (Mohanty 2001a). In the whole of the Bombay Presidency, excluding Sind, the coverage under cotton cultivation, which was 2,702 square miles in 1881, went up to 5,581 square miles in 1904. See *Imperial Gazetteer of India Provincial Series: Bombay Presidency*, Vol. I, 1909, p. 135.

6 By 1861, 12 cotton mills had been established in the Bombay Presidency. Although during the American Civil War, due to increase in raw cotton exports, the expansion of cotton industry was affected in the later period, many more new industries came up in Bombay and Nagpur. For details, see Schmidt (2015: 110)

7 A draft code prepared in 1851 by Wingate, the then revenue survey commissioner, proposing governmental contribution in aid for privately undertaken irrigation work was sent to the Bombay government, but it was finally turned down by the government of India after a long correspondence. See Divekar (1983: 335).

8 It was an important agricultural institution since the Mughal times through which an advance was made to the peasants by their superior tenure holders during the period of distress.

9 Realising the enormous revenue potential of irrigation, Arthur Cotton, the architect of canal construction in British India noted that water in India is more valuable than gold of Australia. The provision was made to pay canal water tax regardless whether or not the use was made of the canal in a particular year or whether or not there was a reliable supply form the canal in canal commands compelled the cultivators to cultivate commercial crops. The Parliament of England on the recommendation of a Select Committee appointed in 1879 decided that the results of irrigation works in India should be tested by their financial returns.

10 See the *Imperial Gazetteer of India Provincial Series: Bombay Presidency*, 1909, pp. 55.

11 A number of enactments were passed for the elimination of intermediaries like the Bombay Khoti Act 1949, the Bombay Taluqdari Tenure Abolition Act 1949, the Bombay Pargana and Kulkarni Vatan Abolition Act 1950, the Bombay Personal Inams Abolition Act 1952, the Bombay Merged Territories (Janjira and Bhor) Khoti Tenure Abolition Act 1953, the Bombay Merged Territories Avas (Jagir Abolition) Act 1953, the Bombay Inams (Kutch Area) Abolition Act 1958, the Bombay Inferior Village Watan Abolition Act 1958, the Bombay Tenancy and Agricultural Land (Vidarbha Region) Act 1958, and the Maharashtra Lands Act 1961.

12 It is estimated that the plan-wise share of expenditure on agriculture and allied activities, which was 31 per cent in the Third Plan, declined gradually in the subsequent plans. The decline was substantial from the Fifth Plan. In the Fifth Plan, it came down to 13 per cent, and by the Seventh Plan, it was as low as 6 per cent. However, the percentage

share for irrigation, which was 15 per cent in the Third Plan, increased continuously till the Sixth Plan and declined marginally thereafter. See Kalamakrar (2011: 160).

13 Even he went to the extent of saying that if he fails to make Maharashtra self-sufficient in food grains, he shall hang himself. See "Rs. 700 crores down the drain", *Economic and Political Weekly*, 1974, Vol. 9, No.36, p. 1527.

14 See "Rs. 700 Crores Down the Drain", *op.cit.*, p. 1527.

15 Ibid., p. 1528.

16 See the GoM (2016).

17 Also see "Tussle Over Cotton Monopoly Procurement", *Economic and Political Weekly*, 1981, Vol.16, No. 28/29, p. 1181.

18 "Rs. 700 Crores Down the Drain", *op.cit.*, p. 1528.

19 In a meeting convened on August 3, 1974 in Poona, Indira Gandhi, the then prime minister, sought answers for fall in agriculture production in the last ten years, but none of the officials and ministers, including the chief minister, had a ready answer. She expressed her displeasure at Naik's handling of Maharashtra agriculture, and on her return to Delhi, she appointed a committee of central government officials to enquire into the cause of the state's dismal performance in agriculture. See Venkatesan (2002). Also see "Rs. 700 Crores Down the Drain", *op.cit.*

20 See "The 'VIP Farmer'", *Economic and Political Weekly*, 1976, vol. 11, no. 46, p. 1774.

21 Prior to the formation of Maharashtra, the state was governed by three separate irrigation acts: (i) the Bombay Irrigation Act 1879 amended in 1950, applied to western Maharashtra; (ii) the Hyderabad Irrigation Act of 1952 for Marathwada; and (iii) the Central Provinces Irrigation Act of 1931 for Vidarbha.

22 This remark was made by the former cabinet secretary Madhav Godbole, who submitted a report on sugar cooperatives in Maharashtra. See *India Today*, June 23, 2003.

23 The irrigation subsidy in Maharashtra increased from Rs. 339 million in 1974–75 to Rs. 2107 million in 1987–88, amounting to 6 per cent annual growth rate. See Dev and Mungekar (1996: A-41).

24 It was reported that the former Karnataka chief minister Veerappa Moily considered Sharad Pawar a 'guru' among the chief ministers in matters concerning liberalisation, as Pawar was the first to start liberalising Maharashtra's economy well before it began in India in 1991 (Jenkins 1999: 148).

25 The share of agriculture, which was 5.6 per cent in Seventh Plan, declined to only 3.5 per cent in the Tenth Plan. However, the allocation for irrigation increased considerably. See Kalamakrar (2011).

26 For example, a high-level committee (Swaminathan Committee) was constituted in 2001 to prepare an action plan for agriculture for the next 25 years. The Khole Committee was set up to look into the various aspects of the agricultural marketing.

27 See Special Economic Zones in India, the Ministry of Commerce and the Industry, Government of India. Accessed from www.sezindia.nic.in/writereaddata/pdf/List of Formal approvals.pdf on 17 September 2016.

28 For details of plan-wise (from First to Ninth Plan) expenditure and potential created in major, medium, and minor irrigation in Maharashtra, see Narayanamoorthy and Deshpande (2005: 93).

29 These corporations are Maharashtra Krishna Valley Development Corporation, Vidarbha Irrigation Development Corporation, Tapi Irrigation Development Corporation, Konkan Irrigation Development Corporation, and Godavari Marathwada Irrigation Development Corporation.

30 See *Economic Survey of Maharashtra 2006–07*. Government of Maharashtra, 2007. Mumbai: Directorate of Economics and Statistics, p. 207.

31 See *Annual Plan*. Planning Department, 2009. Mumbai: Government of Maharashtra, p. 95.

32 See *Economic Survey of Maharashtra 2013–14*. Government of Maharashtra. Mumbai: Directorate of Economics and Statistics, p. 94.

33 See Annual Report (2014–2015), MSAMB, p. 13.

34 *Economic Survey of Maharashtra 2010–11*. Government of Maharashtra, 2011. Mumbai: Directorate of Economics and Statistics, p. 86.

35 Ibid., p. 15.

36 Mahagrapes, a marketing company, which came into existence in 1991 with the support of several government agencies like the National Cooperative Development Corporation (NCDC), the National Horticultural Board (NHB), the Agricultural and Processed Food Products Export Development Authority (APEDA), the Department of Cooperation, the government of Maharashtra, and MSAMB, is projected as a model for contract farming.

37 *Economic Survey of Maharashtra, 2015–16*. Government of Maharashtra, 2016. Mumbai: Directorate of Economics and Statistics, p. 83.

38 *Economic Survey of Maharashtra 2012–13*. Government of Maharashtra, 2013. Mumbai: Directorate of Economics and Statistics, p. 89.

39 See "Maharashtra State Seed Scenario", Government of Maharashtra, 2016. Accessed from www.mahaagri.gov.in/level3PdfDisp.aspx?Id=2&subid=1&sub2id=1&FileName=Seed_G_info.pdf on 15 September 2016.

40 See *India Promoting Agricultural Growth in Maharashtra*, Vol. I and II, Report No. 25451-IN, Rural Development Unit, Asia Region, World Bank June 2003.

Chapter 2

Regional disparity in agricultural development[1]

The theoretical debate on unequal regional development has its origin in the history of development theory extending from the classics of eighteenth and nineteenth century political economy to the many different streams of development ideas. In the immediate post-war period, regional backwardness was largely posed by the economists in terms of vicious circles of poverty and backwardness that seemed to affect many parts of Africa, Asia, and Latin America. Later, in the 1950s and 1960s, cultural explanations influenced by 'the modernisation paradigm' became popular in development discourse. It tended to identify internal cultural 'defects' as the reasons for regional inequalities. However, with the failure of modernisation programme in Latin America, the neo-Marxist dependency school emerged as the alternative paradigm for understanding unequal economic development. It argued that the peripheral regions within a nation remain poor because of their dependence on core regions for expertise, technology, and capital investment. However, the dependency theory is criticised for its relatively static and core-centred analysis. It does not give due importance to the regional and local forces while analysing economic development (Attwood 1992). The pattern of regional or national economic incorporation into the world economy depends not only upon the investment decisions of the national or international capital but also upon local class relations and economic interests of local elites (Tomaskovic-Devey and Roscigno 1997). Moreover, the dependency perspective restricted its attention almost exclusively to the economic mechanisms of domination and control, and to a lesser extent on the socio-political mechanisms (Tucker 1999).

At the time of independence, Indian society was dominated by two powerful classes, which emerged under the social and economic impacts of British colonialism. While the literate, educated, and urban-drawn middle class employed in the government services and business elites mostly belonging to the higher castes like Brahmins constituted one class, the other was the commercial peasantry, the class of rich farmers who adopted new technologies and strategies for production. After independence, the first group became more powerful through its control of public- and private-sector industries and bureaucracy. The latter expanded its base further due to the greater importance given to agricultural commercialisation and associated modernising measures, and started dominating power politics through the newly created democratic institutions such as cooperatives and Panchayati Raj bodies using caste and kinship relations. These two classes, known as intermediate classes (Bardhan 1984; Nayar 1989), remained the most powerful forces in India in the post-independence period. The demands on the Indian state from a coalition of these classes led to the management functions of the public economy (Bardhan 1984). The state has become the grand arena of the accommodation of interests of these groups, and their bargaining determines the final allocation of state resources to the region (Migdal 2001: 92). As the internal cohesion and relative power of the intermediate classes vary from region to another (Attwood 1992: 15), the regions, which represent the interests of these classes in most articulated manner, are prioritised by the state for development.

Against this background, the present chapter attempts to analyse regional disparity in Maharashtra, with reference to its agricultural sector, which is always a case in point.

Politics of regional development

Historically, the state of Maharashtra is divided into four main regions (Map 2.1) – namely, Konkan, Marathwada, Vidarbha, and WM. However, Konkan is considered a part of WM because of its proximity to Mumbai. Prior to the arrival of the British, though the agricultural sector witnessed very minor changes in terms of land revenue, the mechanism of revenue collection and in the appointment of intermediaries, these regions were almost at the same level of development. Subsistence farming with the use of simple technologies was the common practice, as elsewhere in the country. The rulers, by and large, did not undertake any significant initiatives for agricultural development (Brahme and Upadhyaya 1979; Mohanty

1999). However, WM emerged as the centre of Maratha-Brahmin dominance during the rule of Shivaji (Lele 1990; Sirsikar 1995). He established a Maratha kingdom, imbuing it with a Maratha identity, and also promoted Brahmins by offering them official positions and titles.

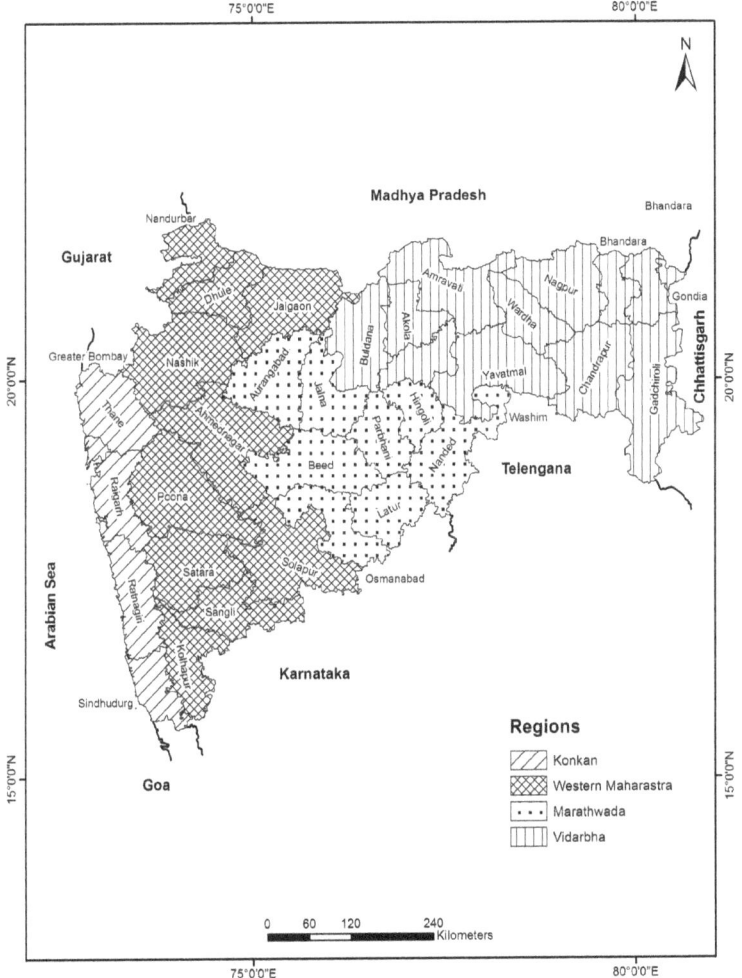

Map 2.1 Regions of Maharashtra

Source: Drawn by the author.

With the establishment of British colonial rule the *Marathi*-speaking people were geographically divided into three political regions as a part of the 'divide and rule' policy. Konkan and Deccan remained within the Bombay Presidency, Vidarbha-Nagpur became a part of the Central Provinces and Berar, and Marathwada came under the rule of the Nizam state of Hyderabad. Different kinds of land tenures were introduced. WM and Marathwada were mostly under *ryotwari* land tenure system. There were two different systems of land tenure in Vidarbha. The districts which were under the Central Provinces had a *zamindari* system of land tenure, and in Berar districts, *ryotwari* was prevalent, similar to the WM. In parts of Konkan, a form of *zamindari* settlement was made with the *Khots*.

Western Maharashtra

During the British rule, agriculture in WM witnessed profound changes. The new land tenures increased the fondness for land investment among the relatively well-off farming communities, such as the *Marathas* and the *Kunbis*. The construction of major irrigation works in the famine-prone districts of Poona, Satara, and Ahmednagar opened up opportunities for irrigated agriculture (Brahme and Upadhyaya 1979; Mohanty 1999). Sugarcane cultivation became widespread in the boom following the First World War, and many sugar factories were established in the canal-irrigated area. The *Marathas* and *Kunbis* largely reaped the benefits of these expanded forces of production due to their association with cultivation and landownership. The Brahmins were the only 'obstacle' to their complete dominance. The Deccan Riots of 1875 in Poona, Ahmednagar, and Satara, led by *Marathas* in a sense, were an attempt to challenge the dominance of the Brahmins, who were mostly the moneylenders (Lele 1981: 51). In the later phase of British rule, the differences between the Brahmins and the *Marathas* became pronounced with the rise of the nationalist movement due to the subtle efforts of the British.[2] Sahu Maharaja of Kolhapur and other activists began to promote non-Brahmins. The *Marathas* joined the *Satyasodhak Samaj* and other non-Brahmin organisations to spearhead anti-Brahmin agitations. Though the *Marathas* and *Kunbis* were claiming to be separate castes despite their common social background, the effects of colonialism brought them together at least in the political arena. It is argued that the *Satyasodhak* movement was dominated by the *Maratha-Kunbis* asserting themselves as a community against the Brahmins (Omvedt 1976; Rodrigues 1998). The widespread riots following the assassination of Mahatma Gandhi in 1948 compelled Brahmins to withdraw from the political arena. Added to this, moneylending legislations and land reforms (in particular), the abolition of intermediaries, and transfer of land to the tillers (a move introduced following independence) weakened their dominance in rural areas. They mostly migrated to urban and industrial centres of Poona and Bombay (Rodrigues 1998: 157).

With the growth of Bombay as a centre of capitalist development, a class of big bourgeoisie emerged, who were mostly non-Maharashtrians. The educated and urban-based Brahmins of WM dominated politics in the state in collaboration with these industrial and business elites. Until 1930, the Congress party was under the control of these educated, urban-based Brahmins and business elites of Bombay (Attwood 1992). Gradually, they tried to absorb *Maratha* elites, considering their rising dominance in the countryside. To accommodate the interests of *Maratha* elites, rural populism became a strong theme in the campaign rhetoric. Brahmin intellectuals, along with the businessmen from Bombay, helped to establish agricultural cooperatives, educational institutions, sugar factories, and other organisations in the countryside (Attwood 1992). The increasing modernisation of agriculture in WM was also in the interest of the industrial bourgeoisie of Bombay, as it created a market close by for modern agricultural appliances and inputs, and opportunities for the establishment of agro-processing industries. The modernisation of agricultural practices pushed a large chunk of the labour force from rural areas to Bombay to meet the requirements of the industrial sector. The growth of industrial establishments in Bombay and Poona also benefitted the rural elites. Both the rural rich and urban-based elites helped each other in protecting their interests (Sirsikar 1995: 99). Gradually, the *Maratha-Kunbis*, who were united through their kinship ties, brought the ruling Congress party under their control by accommodating the elites from other competing caste groups, such as the *Malis, Dhangers, Vanjaris, Mahars*, and institutionalised their own ideology of agrarian development (Jadhav 2006).

Vidarbha

Vidarbha also witnessed prosperity during the colonial period due to expanded cotton cultivation, particularly following the American

Civil War and the Lancashire Cotton Famine (Guha 1972; Mohanty 2005). However, by and large, the region remained agriculturally backward owing to the absence of any irrigation and due to erratic monsoons.[3] Moreover, cotton cultivation was limited only to the districts in Berar. With the increasing cultivation of cotton, famine was also a regular feature in this region (Mohanty 2001a). Nevertheless, cultivators belonging particularly to *Maratha-Kunbis, Rajputs,* and *Telis* who had surplus food grains prospered very well. Unlike WM, the *Maratha-Kunbis* were not in a position to dominate the society and politics of Vidarbha. *Marwaris* and *Komtis* were the archetypal moneylender-cum-traders with near complete stronghold on the regional economy (Phansalkar 2005: 612). To put it somewhat differently, the region was dominated by the Hindi-speaking areas of Madhya Pradesh, and there was a kind of animosity between the Hindi and *Marathi*-speaking people in the region. Beneath this, the *Kunbis, Marathis,* and *deshmukhs* of this region were rivals to each other (Lele 1981; Sirsikar 1995). The conflict between the interests of rich cotton traders and cash-crop farmers, and the smaller grain producers, peasants, and labourers also contributed to *Maratha-Kunbi* disunity. In addition, tension between the lower castes (SCs and nav-Buddhists) and the higher castes following Ambedkar's socio-political movement and the consequent conversion to Buddhism was a challenge to the *Maratha-Kunbis'* hegemony (Mohanty 2001a). In a sense, the society in Vidarbha was a fragmented one in terms of adherence to caste, ethnic, and class loyalties. Though a tiny business and urban-oriented industrial class emerged in Nagpur following the establishment of textile and other industries, this class did not show interest in agricultural development, as it was drawn from non-*Marathi* communities that had almost no linkages with rural Vidarbha.

Marathwada

Marathwada did not witness substantial agricultural development in contrast to WM during the colonial period. It was a neglected part of the Hyderabad state ruled by the Nizam of the Asafiaya dynasty. Though this region was characterised by *ryotwari* land tenures which were conducive to the growth of enterprise and generated incentives for work, the social structure perpetuated by the oppressive rule of the Nizam was inimical to agricultural development. Moreover, the *ryotwari* settlement did not characterise the Marathwada districts until Salar Jung became the chief minister of Nizam in 1853. Though the *Maratha-Kunbis* improved their position through cotton and sugarcane cultivation due to their large holdings of fertile lands, the region remained backward because of the negligence of the Nizam. The *Marathi* elites were continuously stifled by the policies of an autocratic and theocratic Muslim regime (Lele 1990: 164). The domination of the Muslims, as well as other non-*Marathi* (Telugu and Kannada-speaking) communities, was an obstacle for the *Maratha* dominance. Unlike the rest of Maharashtra, Marathwada did not enjoy the contingencies of democracy and capitalist modernisation along with British colonial rule; rather, it had to continue within the framework of a community based on caste and religion (Tambe 2004: 686). The industrial and business elites of this region were mostly based in Hyderabad, having roots in Telugu-speaking areas.

Konkan

In Konkan, the exploitation of rich mineral, water, and forest resources during the colonial period impoverished people and led to continuous migration to Bombay city (Brahme and Upadhyaya 1979: 8). It experienced the disadvantages rather than the advantages of the twin forces of commercialisation and expanded communication. Only the *Khots* (chiefly the chitpavan Brahmins and *Marathas*) emerged as rich farmers (Rodrigues 1998: 23). People mostly depended upon employment opportunities available in Bombay.[4] Politics here either followed the elite *Maratha* pattern or was influenced by politics in Bombay city (Lele 1981: xviii).

The continuous rise of WM

After 1960, the introduction of community development programmes and the Panchayati Raj system, along with cooperatives provided opportunity structures and encouraged the *Maratha-Kunbi* elites of WM to increasingly participate in state politics. The cooperatives emerged as a source of state power and served as a training ground for the emerging political leadership (Carter 1975; Baviskar 1980). The *Maratha* elites monopolised cooperative institutions, especially the sugar factories, and used the policy processes of pluralist democracy with great sophistication to their maximum advantage (Dandekar 1973; Rosenthal 1977). True, the 'cooperative sugar factory', together with the networks of 'cooperative credit societies', also provided a basis for the prosperity of other rich landowning communities, such as the *Malis*, but being small communities, they could not rise to leading positions like those of the *Marathas*. The symbiotic relationship between cooperatives and politicians enabled the *Marathas* to consolidate their position in the economic and political fields, and contributed significantly to the higher rate of overall development in WM (Baviskar 1980; Attwood 1992). Though the sugar cooperatives came up gradually all over the state, including in Vidarbha and Marathwada, the preponderance of these ventures was confined to WM. The landowning classes of other regions had a sense of exclusion from the benefits of development as enjoyed in the western region (Tambe 2004). Even the Employment Guarantee Scheme (EGS), which was implemented in the state from the 1970s, served the landed interests of WM over the years (Jadhav 2006).

Thus, the *Maratha-Kunbis* of WM emerged as the dominant political class in Maharashtra. They were relatively united without any distinct sub-class within themselves (Carter 1975: 39). The internal cleavages in Marathwada and Vidarbha helped them to gain support from their *Maratha-Kunbi* counterparts in these two regions. The *Maratha-Kunbis* of these two regions were desperately looking for an alignment with their fellow members of their community in WM to challenge the dominance of other communities. In this context, the *Samyukta* Maharashtra movement,[5] which became intense following the appointment of the State Reorganisation Commission in 1953 and the subsequent formation of the unilingual state of Maharashtra gave the *Marathas* of WM a near-monopoly over state politics.

Regional movements

The Maha-Vidarbha movement, which started in the early part of the twentieth century demanding a separate statehood, finally culminated in the Nagpur Agreement.[6] Though the Nagpur Agreement was accepted as the basis for addressing regional inequality after the formation of Maharashtra State in 1960, excepting some symbolic changes, such as the shifting of the legislative session to Nagpur and setting up of a high court bench, nothing substantial took place on the developmental front till the elapse of five successive Five-Year Plans. Rather, attempts were made to distract from the issue of regional disparity. It is strongly felt that leaders, who claimed that they struggled for justice for Vidarbha, ended up compromising the cause for their personal gain (Phansalkar 2005). The statement made by the chief minister V. P. Naik, a *Vanjari* landlord from Vidarbha and the only established alliance leader to speak for the region after almost a decade of the formation of the new state, in the state assembly on 20 August 1969 is worth quoting:

> We should now reject the view that a certain district or a certain region is underdeveloped and hence should be given additional assistance. Instead we should direct our efforts to secure a balanced development of all the regions of the state, the whole of which is more or less underdeveloped.

On the other hand, the Marathwada region was not conceived as an arena of contestations and negotiation for radical transformation. The Marathwada Janata Vikas Andolan, which put forward the issue of economic backwardness and remained active from 1970 to 1974 demanding more budgetary allocations for the region, the then ruling Congress party, strategically diluted this sub-regional protest by co-opting the dominant landowning *Maratha* elites who were mostly associated with the Congress party at that time (Tambe 2004: 685). Shankarrao Chavan was made the chief minister at that time on the basis of his being from Marathwada. The leaders, as well as people, seldom took an adversarial position *vis-à-vis* Maharashtra. Leaders like S. B. Chavan, Patil-Nilangekar, and Shivajirao Patil worked through the same kinship network without overt tantrums or confabulations (Phansalkar 2005). Moreover, caste-based divisions, alignments, and consequent tension were also the characteristic feature of this region.[7]

The resurgence of interest in regional imbalances had come at the beginning of the Sixth Five-Year Plan, and until then, most elites from the backward regions had silently accepted the switch in focus from region to district. However, the Sixth Plan itself contained no specific, time bound provisions for fulfilling all or any of those glibly announced promises. In response to some kind of agitation against regional disparity, the government appointed a Fact Finding Committee (FFC) in 1983 (known as the Dandekar Committee), which submitted its comprehensive report in 1984, indicating a huge developmental backlog in Vidarbha, Marathwada, and Konkan. Though the committee suggested removing disparity through appropriate allocation of resources in the three regions within a period of five to seven years, the government did not formally accept the recommendations till 2001–02, using some or other pretexts.[8] However, it started allocating 'special' outlays for the removal of backlogs from 1985 to 1986 onwards. As a result, low budgetary outlays for backlog removal and inequitable allocation of funds for non-backlog schemes continued. A high-level committee on balanced regional development constituted by the government of Maharashtra in 2011 (known as the Kelkar Committee) indicated in its report (submitted in 2013) that across the regions, a sense of relative deprivation and 'lagging behind' has hardly disappeared and suggested elaborate strategic interventions to reduce disparity.[9]

Stated precisely, in the absence of a well-articulated structure of factions and alliances, politicians from Marathwada and Vidarbha were unable to compete effectively for a larger share of the state's resources. Though these regions have produced many powerful leaders, there have been no leaders of a significant stature and widespread influence, like Y. B. Chavan, Vasantdada Patil, and Sharad Pawar, all of whom hail from WM. Even leaders like V. P. Naik from Vidarbha and S. B. Chavan from Marathwada, who were the chief ministers of the state, found it difficult to push up the interest of their own backward regions. Instead, their attempts to challenge the dominance of WM invited their own downfall.[10] The *Shiv Sena*, which came up in late 1960s, propagating *Marathi* nationalism by the industrial bourgeois and urban-oriented higher caste elites based in Bombay and Thane, also tried to integrate dominant rural interests of the western region. In addition, farmer organisations, such as the *Shetkari Sangathan*, which tried to protect the interest of rich farmers by demanding higher prices for crops like onion and sugarcane, (Dhanagare 1994) were also rooted in WM. As Sirsikar (1995: 26) rightly observed,

> The politics of Maharashtra is in one way an expression of the dominant interests of the sugar barons and the big bourgeoisie of Bombay. The hierarchy has strict rules of allegiance and obedience. Any person who tries to build support structure independent of the linkage network is never able to remain in power.

Trend of disparity in agricultural development

After independence, the government of Maharashtra made systematic attempts to expand the area under irrigation in the state as whole. During the first three plans, about 40 per cent of total outlay under the state plan was spent on irrigation and power, which increased subsequently in successive plan periods.[11] Even though the net result of these planned efforts was an increased irrigated area in all regions, the most significant progress in this respect was made in WM (Table 2.1). The proportion of irrigated area, which was 10 per cent in the TE 1970–71 in this region, gradually rose to 24 per cent in TE 2000–01.

Irrigation continued to be negligible in Konkan, as it was before. Though the trends in Vidarbha and Marathwada indicated a continuous rise, the gap between them and the western region remained almost at the same level. Over 54 per cent of the total irrigated area of the state was in WM. However, the progress in Marathwada in this respect is relatively better. It is noteworthy that, covering over 10 per cent of the total irrigated area in the state, sugarcane (which is mostly grown in western region) gets 50–60 per cent of the total irrigation of water (Rath and Mitra 1989: 49). The question raised by Attwood (1992: 286) is worth mentioning here:

> Should a semi-arid region invest so much irrigation in sugarcane, when the same volume of water could be spread over wider areas to protect seasonal crops from drought?

The FFC (1984) noted the disparity in the irrigation sector and recommended a budgetary allocation amounting to 138.60 billion rupees to wipe it out. The physical backlog in irrigation as per the FFC report as of 2001 was 0.3 million hectares in the three regions (Vidarbha, Marathwada, and the rest of Maharashtra) in which Vidarbha and Marathwada account for 1,15,000 and 1,06,000 hectares, respectively. This figure had escalated further to 9,45,000 hectares for three regions (5,71,000 and 2,85,000 hectares for Vidarbha and Marathwada, respectively) as of April 2000 according to the estimate of the Indicators and Backlog Committee (IBC). A recent estimate reported that though the state average for irrigation potential had increased and the

Table 2.1 Information on irrigation

Regions	Irrigation intensity (% of irrigated area to net sown area)				% to total irrigated area of the state			
	TE1970–71	TE1980–81	TE1990–91	TE2000–01	TE1970–71	TE1980–81	TE1990–91	TE2000–01
Konkan	2.7	3.56	4.33	6.71	1.64	1.49	1.52	1.96
WM	10.44	13.42	17.87	23.63	56.23	51.8	51.64	54.18
Marathwada	5.14	9.21	14.11	15.47	18.04	23.48	27.86	25.84
Vidarbha	6.77	8.66	9.28	12.71	24.56	23.18	18.85	18.02
All	7.62	10.52	13.8	17.06	100.00	100.00	100.00	100.00

Source: Season and Crop Report of Maharashtra, various issues.

residual backlog of FFC had been taken into account in the IBC report, the presence of 57 per cent of the outstanding backlog in Vidarbha and 31 per cent in Marathwada indicates a rising trend of regional disparity in this sector.[12] Both regions also experienced a substantial shortfall in actual expenditure in the recent (three) annual plans (2002–03 to 2004–05) for the irrigation sector as opposed to excess expenditure in the rest of Maharashtra.[13] The Kelkar Committee also emphasised the removal of the irrigation backlog as a top priority while allocating the funds and noted that the lack of irrigation deprived the farmers of Vidarbha and Marathwada in availing the consequential benefits like subsidies on fertilisers, pump sets, etc.

Due to a greater proportion of area under irrigation, the gross cropped area in the western region remained at a higher level (Table 2.2). Nearly 40 per cent of the gross cropped area in the state belonged to this region as opposed to 28–29 per cent in Vidarbha and Marathwada, respectively. However, WM witnessed a marginal decline over the years. The share of the Konkan region was negligible.

The disparity is also evident in terms of consumption of electricity for agricultural purposes (Table 2.3). The use of electricity in agriculture showed a rising trend across regions. WM was much ahead of other regions, both in terms of its share in total agricultural consumption of the state as well as per hectare (cropped area) consumption. It consumed 65 per cent of total agricultural consumption, with about 40 per cent of the total cropped area of the state in TE 2006–07 as opposed to 24 per cent for Marathwada and 10 per cent for Vidarbha (each accounting nearly 30 per cent share in the total cropped area). As electricity is provided at a subsidised cost through the provision of a flat-rate tariff for agricultural use,[14] farmers of WM appropriated a lion's share of this benefit due to their greater consumption.

In order to meet the demands of the farmers for more working capital, the credit network was expanded throughout, largely through the establishment of agricultural cooperative credit societies. The information on outstanding loans and overdue payments in these credit cooperative societies also indicated the state's preference for agricultural development in WM. Over 64 per cent of total outstanding loans in the state from these cooperative societies belonged to the western region in TE 2009–10, and its share

Table 2.2 Gross cropped area and cropping intensity (%)

Regions	Share in gross cropped area					Cropping intensity				
	TE 1970–71	TE 1980–81	TE 1990–91	TE 2000–01	TE 2010–11	TE 1970–71	TE 1980–81	TE 1990–91	TE 2000–01	TE 2010–11
Konkan	4.52	4.31	4.11	4.80	3.87	103.36	106.95	104.47	107.66	110.28
WM	40.84	40.78	40.01	40.52	39.32	105.91	109.84	115.88	126.17	120.72
Marathwada	26.81	26.98	28.3	29.19	29.28	105.92	110.23	124.16	147.15	126.51
Vidarbha	27.43	27.85	27.43	25.49	27.53	104.93	108.13	115.47	125.82	124.63
All	100.00	100.00	100.00	100.00	100.00	105.58	109.34	117.47	130.66	122.79

Source: Same as in Table 2.1.

Table 2.3 Electricity consumption

Regions	Share in agricultural consumption		
	TE1990–91	TE 1996–97	TE 2006–07
Konkan	2.07	2.5	0.72
WM	53.46	57.89	65.43
Marathwada	24.13	23.22	23.75
Vidarbha	20.33	16.39	10.10
All	100.00	100.00	100.00

Source: Statistical Abstract of Maharashtra, relevant issues.

went on increasing, accompanied by a continuous decrease in that of other regions. The same could be said about the overdue payments of members of these societies (Table 2.4). The western region accounted for as much as 51 per cent of the total overdue payments of the state as opposed to 26 per cent in Marathwada and 19 per cent in Vidarbha.[15] According to an estimate made in the mid-1970s, major sugarcane growing *talukas*, which accounted for 5 per cent of the gross cultivated area in the state, secured as much as 25 per cent of the loan amount.[16]

As a result, the farmers of the western region apply greater doses of agricultural inputs and also utilise the low-cost credit capital for acquiring high-valued agricultural assets. The data on ownership of high-valued agricultural machinery, such as tractors and oil engines, indicate rising regional inequality (Table 2.5).

The ownership of tractors in WM was as high as 7.8 per 1,000 hectares of gross cropped area in 2003, which was many times more than that of the other regions. While the farmers in Vidarbha owned three tractors per 1,000 hectares, in Marathwada, it was only two tractors. Though tractor ownership position improved during the course of time in these backward regions, the pace of growth in WM was much higher. The ownership of oil engines exhibits almost a similar pattern. The substantial decline in the number of oil engines in WM is mainly due to the increased use of electricity-operated pumpsets. As regards the position of agriculture pumpsets, this region is much ahead of others in terms of both total numbers as well as per 1,000 hectares of cropped area (Table 2.6).

Table 2.4 Loans from primary agricultural credit cooperative societies

Regions	Share of outstanding loans					Share of total overdue payments				
	TE 1970–71	TE 1980–81	TE 1990–91	TE 1996–97	TE 2009–10	TE 1970–71	TE 1980–81	TE 1990–91	TE 1996–97	TE 2009–10
Konkan	2.25	2.72	1.89	1.34	1.98	4.52	4.32	2.81	2.43	3.89
WM	51.91	52.93	60.73	62.55	63.89	45.63	45.57	52.79	52.55	51.20
Marathwada	25.33	17.21	23.29	22.78	20.13	25.16	19.35	23.95	26.02	26.27
Vidarbha	20.5	27.15	14.09	13.33	14.00	24.68	30.76	20.45	19.48	18.64
All	100.00	100.00	100.00	100.00	100.00	100.00	100.00	100.00	100.00	100.00

Source: Statistical Abstract of Maharashtra, various issues.

Table 2.5 Agricultural machinery (per '000 hectare of gross cropped area)

Regions	Tractors				Oil engines			
	1972	1982	1992	2003	1972	1982	1992	2003
Konkan	0.13	0.17	0.02	0.98	2.45	1.38	0.01	9.96
WM	0.64	2.03	1.86	7.94	18.44	12.44	2.45	6.36
Marathwada	0.11	0.46	0.72	2.44	8.75	4.13	1.02	2.34
Vidarbha	0.16	0.57	1.08	2.83	2.45	3.60	3.20	6.58
Maharashtra	0.33	1.12	1.24	4.69	10.23	7.24	2.15	5.42

Source: Relevant issues of (a) *Statistical Abstract of Maharashtra*, (b) *Season and Crop Report of Maharashtra*.

Table 2.6 The estimate on region-wise position of agriculture pumpsets as of 31 March 2005

Division	Agriculture pumps	No of pumps per 1,000 hectare cropped area
Konkan	2.35	68.93
WM	54.42	191.25
Marathwada	23.32	126.13
Vidarbha	19.91	100.32
Maharashtra	100.00	142.44

Source: Planning Commission, government of India, 2006.

Crop patterns

Regional inequality in agricultural development may be analysed by looking into the regional shares in area under major crops. Sugarcane, which is a highly valued cash crop and enjoys maximum state support, was mostly grown in WM. Nearly 75 per cent of the total area under this crop in the state was in WM (Table 2.7).

Though a major part of the cropped area in Vidarbha was under cash crops, the area under sugarcane was almost non-existent. Cotton continued to be the predominant cash crop in this region. As the area under irrigation was negligible, cotton cultivation was mostly left to the vagaries of monsoon. Due to variations in rainfall, crop failure was a frequent visitor to this region (Mohanty 2001a). The recent spate of suicides of farmers in this region reflects the plight of

Table 2.7 Region-wise shares in area under major crops

Regions	TE 1970–71	TE 1980–81	TE 1990–91	TE 2000–01	TE 2010–11
Food grains					
Konkan	6.04	4.31	4.37	4.35	4.46
WM	59.5	43.4	43.28	44.72	43.10
Marathwada	35.89	28.01	28.21	31.06	29.50
Vidarbha	32.64	24.27	24.11	19.87	22.94
All	100.00	100.00	100.00	100.00	100.00
Sugarcane					
Konkan	0.21	0.11	0.06	0.01	0.00
WM	83.97	80.01	71.9	70.94	74.90
Marathwada	13.99	18.23	24.48	25.89	23.80
Vidarbha	1.87	1.65	3.56	3.15	1.30
All	100.00	100.00	100.00	100.00	100.00
Cotton					
Konkan	*0.00*	*0.00*	*0.00*	*0.00*	*0.00*
WM	13.91	12.33	9.3	16.21	21.78
Marathwada	27.74	25.6	27.14	33.36	40.58
Vidarbha	58.36	62.07	63.55	50.43	37.63
All	100.00	100.00	100.00	100.00	100.00

Source: Same as in Table 2.1.

cotton growers (Mohanty and Shroff 2004; Mohanty 2005; Mishra 2006b). Cotton also covered a significant part of the cropped area in Marathwada. It is also evident that the share of WM to gross value of all crops was nearly 50 per cent (Table 2.8).

While Marathwada and Vidarbha contributed only 25 per cent each, Konkan's share was only about 5 per cent. However, WM's share decreased by less than 2 per cent between TE 1980–81 and TE 1990–91.

Backward regions like Vidarbha witnessed more disadvantages in recent years owing to the adverse impact of economic liberalisation on cotton growers. The domestic subsidies given to farmers in the United States caused significant price suppression of cotton in the international market between 1999 and 2002.[17] Between 1994 to 1995 and 2001

Table 2.8 Region-wise shares in gross value of all crops

Regions	TE 1970–71	TE 1980–81	TE 1990–91
Konkan	5.78	5.02	4.42
WM	49.34	49.71	47.76
Marathwada	20.44	22.27	23.43
Vidarbha	24.42	23.01	24.38
All	100.00	100.00	100.00

Source: Sawant et al. (1999: 113).

to 2002, the Maharashtra Cotton Federation incurred losses to the tune of 39.85 billion rupees (Shroff 2003). In view of this, the scheme of monopoly procurement of cotton which was at work since 1971 to ensure fair and remunerative price to cotton growers has finally given up its monopoly character and has allowed private agencies to procure cotton. On the other hand, sugarcane growers remained in a relatively better position due to state intervention. As of 2001–02, of the total 160 sugar factories in the state, 147 are in the cooperative sector. Many of these cooperative sugar factories pay sugarcane price to farmers, which is much beyond their financial capacity (Shroff 2003). Though three-fourths of these mills were incurring huge losses and accumulating heavy debts, adding to the fiscal burden, the government continues to support them due to the strength of the sugar lobby in state politics. The Godbole Committee (1999) appointed to look into the sickness of cooperative sugar factories noted the unlimited state support enjoyed by these cooperatives. In addition, the agricultural activities in WM have been well adjusted with the global market by integrating its agricultural activities with floriculture, horticulture, viticulture, and food processing with state support. Being in tune with the economic change, agriculture, and allied activities of this region is effectively linked with service sectors such as information technology, banking, insurance, and leisure industry.

Conclusions

The process of agricultural development in Maharashtra over the last four decades indicates regional inequality in which WM remained much ahead of other regions in terms of major developmental indicators. However, compared to Vidarbha, the Marathwada region experienced better improvement in some respects. The rapid development in WM is attributed to the rise of the *Maratha-Kunbi* peasants as a unified political class that dominated state politics through caste and kinship networks during the colonial, as well as post-colonial periods and thereafter. The increasing modernisation of agriculture in this region was also in the interest of the industrial bourgeoisie of Bombay, as it created a nearby market for modern agricultural appliances and opportunities for the establishment of agro-processing industries. The Marathwada and Vidarbha regions were unable to compete effectively for a larger share of state's resources due to the absence of a well-articulated structure of factions and alliances. As a result, the influential elites of WM remained in an advantageous position to divert the developmental resources of the state to their region. The relatively better performance of the Marathwada region was mainly due to the socio-cultural proximity of its local elites with those of WM.

Notes

1 This is the revised version of the article published earlier (with the same title) in *Economic and Political Weekly*, Volume 44, No. 6, February 7, 2009, pp. 63–69.

2 The British used the elite *Marathas*, the numerically strongest community known for their military record during Shivaji's time, as a counterweight to the Brahmins, who were questioning their rule. See Sirsikar (1995).

3 The variation in rainfall was frequent, and it had its corresponding effect on the output of crops. Though there were some occasional good harvesting years, crop failure took place regularly in this region. See Mohanty (2001a).

4 The people in this region live mostly on remittances by the workers in Mumbai. It is popularly described as 'money order economy'.

5 The *Samyukta Maharashtra Movement*, which began in 1946 in Pune by projecting *Maratha* nationalism, had support in WM. In the 1957 general election to the Legislative Assembly, it won 102 seats from 135 seats in WM (excluding Bombay), amounting to 76 per cent of total seats. In Vidarbha and Marathwada, it won only 17 per cent of their total seats. See Sirsikar (1995: 40).

6 The eminent political and social workers of three regions signed an agreement in September 1953 which constituted the basis for uniting the three *Marathi*-speaking areas in one single state. This is known as the Nagpur Agreement.

7 Recent polarisation following the renaming of the Marathwada University into two warring camps (those who support renaming the university after Babasaheb Ambedkar and those who oppose it) provides firm evidence on deep-rooted, caste-based opposition and tension in the region.

8 The FFC submitted its report in April 1984. An empowered committee was appointed to study the recommendation of the FFC, which submitted its report in April 1987. After the establishment of the development boards in 1994 to work out the revised backlog of the different regions of the state, IBC was set up in November 1995. The committee submitted its report in July 1997. Considering the development that took place till March 1994–95, the committee worked out a fresh backlog. But the physical backlog calculated by the committee was not acceptable. The IBC was reconstituted, and the committee submitted its report in September 2000. Finally, the government accepted the report and decided to provide funds for the removal of backlog from the years 2001 to 2002.

9 See *Report of the High Level Committee on Balanced Regional Development Issues in Maharashtra*, 2013, Planning Department, Mumbai: Government of Maharashtra.

10 V. P. Naik was replaced by S. B. Chavan in 1974. He could not return to power politics apart from becoming a member of parliament in 1977.

11 See *Maharashtra State Development Report*, Planning Commission, Government of India 2005.

12 See *Performance Evaluation of Statutory Development Boards (SDBs) in Maharashtra 2003*, Programme Evaluation Organisation, Planning Commission, Government of India 2003, p. 27.

13 See *Report of the Fact-Finding Team on Vidarbha: Regional Disparities and Rural Distress in Maharashtra*, with particular reference to Vidarbha, Planning Commission, Government of India 2006.

14 Though agriculture accounted for 26 per cent of electricity consumption, it contributed only 10 per cent to revenue receipts of Maharashtra State Electricity Board in 2002–03. While the current average cost of power supply per unit is Rs. 3.26 average realisation per unit in agriculture comes to only 1.10. It is observed that ever since the unmetered billing was introduced for agriculture, it became impossible to segregate legitimate consumption based on flat-rate tariff. For details, see *Economic Survey of Maharashtra 2003–04*, p. 209.

15 Keeping in mind the low interest rate of the cooperatives, the farmers prefer to postpone the repayment and make other profitable investments, as result of which overdue payments increase. This practice is common, particularly among the rich farmers in many parts of rural India.

16 *Final Report of the Study Team on Cooperative Agricultural Credit Institutions in Maharashtra*, Reserve Bank of India, as cited in Brahme and Upadhyaya (1979: 105).

17 See *Agricultural Subsidies and Negotiations Strategies and Options*, Centre for Trade and Development, Hong Kong Series-2, 2005, pp. 49–50.

Chapter 3

Agricultural modernisation and social inequality[1]

Agriculture being the mainstay of rural Indian economy around which socio-economic privileges and deprivations revolve, any change in the structure of its production is likely to have a corresponding impact on the existing pattern of social inequality. Indian agriculture has experienced a series of substantial changes since the British Raj days. The Raj attempted to transform the agrarian economy with the aim of increasing the agricultural production that could be used (as raw material) for Britain's industrialisation on one hand and on the other for polarising Indian rural society into classes for pursuing the 'divide and rule' policy. The changes in the post-independence period were significantly different from those in the pre-independence period in the nature and extent and approach to them. The government of independent India set upon a path for higher growth along with social justice, and it was in operation in all the Five-Year Plans, though with varying degrees of emphasis, with the state playing the central role. The 1990s, however, saw fundamental changes with the adoption of neoliberal policies. The emphasis shifted to high economic growth through efficient allocation of resources, free market, and free trade, with minimal state intervention. However, the emerging implications of these wide-ranging changes for the professed goal of social equality are not very clear.

Agrarian studies in India on the effects of agricultural modernisation vary immensely in style and temper, but they generally subscribe to three important views in this regard: (a) one group holds that agricultural modernisation measures, instead of removing the earlier inequalities, have created further inequalities (Parthasarathy 1970; Frankel 1971; Epstein 1973; Griffin 1974; Mencher 1978; Dasgupta 1980; Dhanagare 1987); (b) another group argues that by improving the economic conditions of poor peasants and landless agricultural labourers, the new measures have reduced the earlier inequalities (Mellor 1976; Ahluwalia 1978; Blyn 1983; Hazell *et al.* 1991); and (c) a third group views the modernising measures as a package having mixed effects (Bhalla and Chadha 1983; Agarwal 1983; Harriss 1991).

This debate, however, tapered off inconclusively in the 1990s due to the decline of agrarian studies in the neoliberal period and the shifting focus towards agrarian crisis and distress. Moreover, a review of the earlier studies revealed that while some of the studies had been undertaken during the early phase of agricultural modernisation, many of them confined their analysis to the macro level, concentrating on the individual components of modernisation. The most important feature common to these studies is that they are largely based on the ground realities in Punjab, Haryana, Tamil Nadu, and western Uttar Pradesh. It appears, as observed by Bradnock (1984: 138), that the irrigated wheat lands of the north-west and the rice-growing deltas of the south-east have been seen as laboratories for experimenting with agricultural changes in preparation for the transfer of successful models across the country. Given the diversity of natural and social environments in South Asia, Farmer (1986) remarks, "It is prima facie not to be expected that the 'new technology' would operate in the same way or have the same social and economic effects all over south Asia, or even all over any one of its countries". In such a context, the findings of studies conducted in other parts of the country may not fit neatly into the situation in Maharashtra where the agro-climatic conditions, socio-cultural features, and historical specificities are different. So far as studies on Maharashtra are concerned,[2] they have not devoted attention to the effects of agricultural modernisation measures on social inequality.

In light of this, the present study attempts to renew and extend the earlier debates on social inequality in the context of agricultural modernisation in rural Maharashtra with reference to Satara, the most highly modernised[3] in the sphere of agriculture of all districts in the state. The main objectives of the study are (a) to examine the nature and extent of agricultural modernisation[4] in the historical perspective, (b) to analyse the impact of modernisation measures on the various categories of rural population, and (c) to find out the pattern and trend of social inequality.[5]

The chapter is organised into five sections. While the following section provides the broad perspective, the next section makes an inquiry into the process of agricultural modernisation in Satara district during colonial, post-colonial and post reform period, the third section analyses the impact of this modernisation on social inequality at the district level. The fourth section presents a comprehensive account of the structure of social inequality as it exists at the village level. The last section presents the conclusion.

The perspective

Social inequality, structurally or functionally, is a universal feature of all societies regardless of its complexities. However, the nature, extent, and pattern of inequality vary from society to society and are conditioned by the structure of economic production and distribution. Traditionally, the Indian rural society was hierarchical and iniquitous structured in terms of caste relations. Agricultural production, which was mostly at a subsistence level, was organised around caste structure. Though the attempt of British colonialism to 'modernise' the economy in the capitalist line generated a class structure in the countryside, instead of replacing the earlier caste-based inequalities, it reinforced them. Caste-inbuilt mechanism remained within the class structure. While the large landowners invariably belonged to the upper castes, the cultivators belonged to the middle castes and agricultural labourers largely to the lower castes.

The new agricultural development strategy began after independence within this colonial class structure, keeping growth with equity as its major thrust. In the early stages of planning, there was no explicit emphasis on the distribution of gains, though land reform measures were introduced with the professed aim of removing inequality. Subsequently, when doubts were raised about the 'trickling down' of the benefits of growth, emphasis was placed on distributive justice. The paradigm of development was in a sense based on the idea of 'welfare capitalism' (Chakrabarti et al. 2016). A new institutional mechanism (like cooperatives and panchayats) was created to allocate state-funded subsidised resources. Given their dominance in the rural society grounded in caste and kinship identities, the upper-caste large landowners remained in an advantageous position to control these newly created institutions for siphoning off the benefits of the new measures. As a result, in the era of this 'welfare' and 'subsidised capitalism', a new breed of entrepreneurial capitalist farmers emerged in the countryside that consolidated their economic position with the state's support.

The introduction of neoliberal reforms in the 1990s created conditions for reversing the flow of income from upper to lower classes, which had occurred earlier during the era of state-led planned development. As the conditions of neoliberalism favour producers having more resources, information, and capacity to cope with stringent market demands (Young and Hoppe 2003; Joekes 1999), these large farmers with their land, capital, and technical resources and superior links with national and international markets get new opportunities to emerge as dominant groups in the countryside. As a result, the accentuation of social inequality, which already exists strongly in the rural society, is in the making (Patnaik 1996: 2448). This can be substantiated by any empirical study.

Agricultural modernisation in Satara

British impact

Satara is one of the western-most districts of Maharashtra. The district found its place on the map of the Raj in 1848 when it lapsed to the British upon the death of Shahaji Raja.[6] Prior to its British occupancy, it was under the control of the *Marathas*, the Moghals, and other Muslim rulers. Before the Muslim invasion of the Deccan, Satara was under the control of the Yadavs of Devagiri, who inherited the region from the earlier Hindu rulers.[7] Our knowledge of the agrarian economy of the district before the British period suffers for want of systematic and elaborate information. However, it is evident from the available sources that the early rulers did not take much interest in agricultural development. There were successive phases of wars and conquests, and the rulers confined their attention to revenue collection, which often amounted to a lion's share of the output. Agricultural production was carried on with primitive technology, and the area under cultivation was adjusted to the increases and declines in population. The rulers spent whatever surplus could be extracted largely to meet military and administrative needs. The system of revenue management then prevailing could never lead to improvement in agriculture.[8]

The agrarian structure of Satara witnessed profound changes during the colonial period owing to the introduction of new land tenures, commercialisation of agriculture, and growth of Bombay as a centre of capitalist development. Satara came under the direct control of the British in 1848, though the relevant regulation was not introduced until 1863. At the outset, the British regime, keeping in view the high revenue assessment of the *Marathas*, only regularised and systematised the earlier system. Although the assessment was made high by the *Marathas*, they could not always collect it in full.[9] Nevertheless, this served as a model for the British.

The first revenue survey and settlement of different *talukas* in the district was completed between 1852 and 1863, and revision settlements were undertaken in 1881–82 and 1926–27. Land revenue rates were revised in each settlement. The Survey Act of 1865 abolished all the previous tenures and introduced the *ryotwari* tenure of heritable and transferable occupancy rights, which had been extended to the Bombay Presidency in 1830. The new tenure differed from the old *mirasi* tenure in four important respects:[10] (a) the occupant was given the right to sell or otherwise alienate his land without government permission; (b) the occupancy right was liable to forfeiture if the holder failed to pay the assessment; (c) the assessment was liable to revision once every 30 years; (d) a guarantee was given that no additional taxation would be imposed on account of improvements made by the occupant. The provisions of the new land tenure enhanced the propensity to invest in land. The privileged and affluent sections that were earlier reluctant to increase their holdings started acquiring more land, as a result of which landholding disparities assumed significant proportions.[11] It is found that in 1882–83, more than 57 per cent of the holdings were of less than 10 acres each, and the rest of the holdings ranged from 10 plus to more than 400 acres each.

The agricultural policy of the Raj was designed to feed Britain's industry and Britain-based industrial units in Bombay and elsewhere. Whatever agricultural modernisation was initiated kept colonial interests in view. Up to the end of the *Maratha* rule, there was no major source of irrigation in Satara, except a limited number of tanks and wells (most of which remained dry during the years of low rainfall). To support and extend the commercial trend of agriculture, the British undertook the construction of a series of irrigation works. Within 30 years (1848–76), around 7 major canals were constructed and these could irrigate more than 6,000 acres of land.[12] A number of wells, temporary dams, water lifts, ponds, reservoirs, etc., were also constructed.

As a result, a significant percentage of the net sown area of the district was irrigated (Table 3.1), which facilitated the cultivation of sugarcane and other cash crops. The total cultivable area and net sown area were also increased, and the cropping intensity went up (Table 3.1).[13] The traditional cropping pattern was changed, and emphasis was placed on the cultivation of cash crops, especially cotton. During 1850–51, native cotton was grown only on 11,155 *bighas*. Its production increased gradually, but the quantity varied with the state of the foreign market. Though the greater part of the produce was used within the borders of today's Maharashtra, a significant portion was exported to Manchester.[14] Attempts were made to introduce new varieties of cotton, such as Orleans and Broach cotton, to increase productivity.[15] In some parts of the district, the peasants were forced to adopt alien seed varieties.[16] The area under cotton production increased more rapidly during the period of the American Civil War (1861–65)[17] when the supply of the fibre from the United States (to Britain) was interrupted.[18] Later, however, the growing of cotton fell (Table 3.1) in tandem with local demand, following an increase in imports of European and Bombay piece goods,[19] and perhaps also because of the emergence of Khandesh and Nashik as major exporters[20] of the crop.

From the beginning of British rule over Satara, sugarcane cultivation was also encouraged.[21] New varieties such as Mauritius and Chinese sugarcane, which were mostly grown in Europe and which had very short intervals between sowing and reaping, were introduced in the district. Though the area under sugarcane increased gradually during the initial phase of the British rule, the falling of sugar prices in the later phase discouraged the peasants from increasing the tillage area for this crop. With a rise in grain prices and improvements in agricultural infrastructure, the area under food grains increased during the early period of British rule. It started declining later owing to the expansion of the area under oilseeds (Table 3.1).

Such commercialisation of agriculture along with higher land revenue created a need for more working funds, as a result of which peasants increasingly fell into debt. The indebtedness also grew between 1862 and 1865, the years of the American Civil War, with money becoming extraordinarily cheap and higher prices being fetched by field produce.[22] The economic condition of the peasants was further aggravated by frequent famines. Widespread and recurring famines became a feature of the British period owing to increasing cultivation of cash crops in preference to food crops and growing population.

But those who had land and other resources became wealthier by taking advantage of the demand for cash crops. The landholders

Table 3.1 Land utilisation, irrigation intensity, cropping intensity, and cropping pattern (1891–92 to 1945–46)

Year	Total cultivable area (in acres)	Net sown area (per cent of total)	Ratio of irrigated land to net sown area (per cent)	Cropping intensity (per cent of gross cropped area)	Area under food grains	Area under oilseeds	Area under sugarcane	Area under cotton	Total area under fibre
1891–92	2,864,381	60.1	5.87	104.26	92.71	5.5	0.77	0.44	1.02
1895–96	2,859,362	55.5	7.17	107.12	90.47	7.02	0.76	0.36	1.75
1901–02	2,866,675	51.94	7.12	106.35	94.11	4.08	0.63	0.59	1.18
1905–06	2,880,299	49.67	5.54	107.04	92.52	5.75	0.66	0.64	1.07
1911–12	3,089,481	51.48	6.37	105.54	88.44	9.3	0.75	0.9	1.52
1915–16	3,089,378	62.88	5.51	102.7	91.35	6.82	0.75	0.46	1.08
1920–21	3,142,842	61.59	5.28	101.96	92.95	5.42	0.58	0.55	1.05
1925–26	NA	NA	NA	NA	NA	NA	NA	NA	NA
1931–32	3,133,616	67.02	5.5	100.54	86.51	11.59	0.61	0.94	1.3
1935–36	3,133,701	67.97	6.14	101.38	85.82	12.47	0.75	0.61	0.96
1941–42	3,134,340	65.87	5.57	101.16	84.8	13.94	0.48	0.58	0.79
1945–46	3,134,950	63.4	5.29	100.5	86.13	12.98	0.61	0.06	0.28

Sources: 1. Agricultural Statistics of British India, various issues
2. Agricultural Statistics of India, various issues

Note: NA = Not Available

and traders belonging to the Brahmin, *Gujarati vani*, *Marwarvani*, and *Maratha* communities emerged as professional moneylenders who accumulated large quantities of land by trapping peasants in debt. Prior to the British rule in Satara, land was not liable to sale for meeting debt, and the lenders were also fewer in number.[23] During the British period, the provision of the Survey Act 1 of 1869, which made land disposable either by sale or mortgage, induced the moneylenders to encourage the peasants to take loans secured by land. The usurers had an eye on the debtors' lands. When the debt amount reached a breaking point, they insisted that the debtors dispose of their lands. Due to this exploitative moneylending system, agrarian riots broke out in some parts of the district in 1874 and 1875.[24] As recommended by the Deccan Riots Commission, the Deccan Agriculturists' Relief Act (XVII of 1829) was made applicable to Satara district. However, though the Act restricted the creditor's power of recovery, this restriction was easily evaded through fictitious sales with oral agreements.[25] It was revealed by a survey[26] that the total land owned by moneylenders increased from 5.3 per cent in 1876–77 to 6.4 per cent in 1888–89.

Land was mortgaged either with or without possession being given to the moneylender. In the case of a mortgage without possession, the debtor had to hand over a part of the produce to the moneylender by way of interest till the mortgage was redeemed.[27] But in cases where land was mortgaged with possession, the moneylenders generally allowed the borrowers to cultivate the land on the condition that three-fifths of the produce was surrendered to them, and the lenders paid the revenue.[28] Provisions of the Bombay Land Revenue Code of 1879 left the decision over the amount of rent to be charged for land entirely to the landlords, allowed them to increase the rent at any time and recover it with the support of government machinery, and gave them the right to evict the tenants for non-payment. These provisions helped the moneylenders to take the upper hand. The process of landownership passing from the cultivators to the moneylenders through the operation of a debt mechanism and of the cultivators becoming tenants in their own land gathered momentum rapidly. Besides, moneylenders gave landless labourers loans against the mortgage of labour services. The labourers took loans and were thus bound to toil on the moneylenders' accumulated lands till the debts were cleared.[29] In case a labourer failed to clear the debt in his lifetime, it was imposed on his son or other members of his family.

Under the British rule Satara was well provided with communication facilities, which systematically linked the agriculture of the district with the wider economy. During the early years, cart traffic was negligible, pack bullocks accounting for the bulk of the traffic. The opening between 1857 and 1876 of the Varandha, Kumbharli, and Fitzgerald passes, which formed the chief outlets to the coast, facilitated wheeled traffic across the *Sahyadris*. The construction of an extensive railway network between 1860 and 1890, especially the construction of Poona-Londa Line (which passed south and south-east along almost the whole centre of the district) connected the region with the rest of India, and the formerly isolated segments of the district came under the net of British free trade. Unlike Western Europe and America where railways catalysed industrial revolutions, in India, they catalysed the completion of colonisation (Habib 1995: 328).

The exports from the district increased,[30] and the cultivation of cash crops became widespread. Merchants-cum-financiers from Bombay extended their tentacles to the villages and devised numerous means to siphon away the agricultural produce of the district. Millet, wheat, chilli, turmeric, and tobacco were sent to Bombay from Satara, Karad, and Valva by local and *Gujarati vanis* who collected the produce from the cultivators either in payment of earlier debts or in cash. Cotton was sent from Valva and Tasgaon by *Bhatias* and *Gujarati vanis*. On the other hand, silk goods, wool manufactures, and specially the products of British heavy industry, such as metal manufactures, appliances, and tools, began to be imported in large quantities.[31] This in turn adversely affected the incomes of the artisan castes.

Such vast changes in agriculture had a polarising effect on the rural population of the district. While the large landholders and moneylenders with superior irrigated land and cheap labour at their disposal prospered; the condition of the poor peasantry and the rural proletariat became more critical. The exploitative land, labour, credit, and market relations made the poor peasants vulnerable to land sales, and in fact, a regular land market took shape. Each year, a substantial quantity of land (both revenue-paying and revenue-free) came into the market (Table 3.2). However, as data on land transfers before 1901–02 and after 1918–19 are not available, a detailed picture on land transfers during the British rule cannot be provided. Systematic and continuous land transactions accelerated peasant marginalisation. Many of the impoverished peasants

Table 3.2 Number of transfers of land and area transferred 1901–02 to 1918–19

Year	Revenue-paying land		Revenue-free land (partially or wholly)	
	No of transfers	Area (in acres)	No of transfers	Area (in acres)
1901–02	12,503	32,695	1,444	5,640
1902–03	13,045	35,759	1,902	6,068
1903–04	13,727	42,374	1,412	6,324
1904–05	14,109	42,758	1,752	5,749
1905–06	14,078	53,800	1,294	4,595
1906–07	15,784	56,592	2,204	7,221
1907–08	13,233	39,974	1,672	5,953
1908–09	14,066	41,495	1,790	5,022
1909–10	12,665	35,770	1,556	8,235
1910–11	9,935	31,598	1,119	3,926
1911–12	9,209	27,048	1,136	3,345
1912–13	8,379	27,017	1,117	4,399
1913–14	9,680	28,883	1,222	3,816
1914–15	11,369	33,587	1,569	5,898
1915–16	10,311	29,169	1,398	5,611
1916–17	10,147	30,086	1,543	6,395
1917–18	11,400	33,946	1,774	5,769
1918–19	15,594	39,253	2,095	7,427

Source: *Agricultural Statistics of British India*, various issues.

and labourers migrated to Bombay and other places in pursuit of survival. The 1881 census reveals that 0.1 million people born in the district were living outside the region, nearly half of them in Bombay. It was reported that in 1911, emigrants from the district accounted for about 20 per cent of the district's total population.

On the whole, the cumulative impact of the British contribution to land revenue administration, land tenures, cropping pattern, and infrastructure and institutional arrangements necessitated by them widened the social disparities in Satara. As pointed out by Scott (1976), the imposition of capitalism and the development of the modern state under colonialism had a profound effect on the agriculture of the district. The first disrupted the agrarian order by transforming land and labour into commodities for sale; the second did so by enforcing the imposition of a market economy and by creating a new environment for the generation of peasant income. The privileged and the affluent remained in an advantageous position to consolidate their socio-economic position, and the lower rungs of the peasantry led miserable lives.

Post-colonial situation

Following the merger of the princes' territories in 1947, Satara district was enlarged and divided into North Satara and South Satara. In 1960, North Satara was renamed Satara, South Satara being designated as Sangli district. Satara district was included in the Bombay state with the reorganisation of states in 1956, and it has been part of Maharashtra since 1960. In the post-independence period, the agrarian structure of Satara entered a new phase under the impact of land reforms, cooperatives, large-scale irrigation, mechanisation, adoption of high-yielding varieties (HYV), and associated infrastructural development. Agricultural modernisation started more elaborately, in a planned way.

In pursuance of policies laid down in the country's Five-Year Plans, a package of land reform measures were launched by the erstwhile Bombay state and Maharashtra. Most of them were applicable to Satara. The major objectives of the land reforms cover abolition of intermediaries, tenancy reforms, fixation of size ceilings on holdings distribution of ceiling surplus land, and consolidation of holdings. During the early 1950s, steps were taken to eliminate all intermediaries between the tiller and the state. The *zamindari* system, along with the system of the rent-collecting intermediaries prevalent in the district, was abolished. However, the abolition of the *zamindari* system had little impact, and the *ryotwari* tenures were widely prevalent in the district. All 'inams' except *Deosthan inams* (held by religious institutions or for religious services) were abolished.

To protect the tenants and ameliorate their condition, the Bombay Tenancy and Agricultural Lands Act 1948 was enacted. An earlier move on tenancy reform had been made through the Bombay Tenancy Act 1939. The act had been applied to a few selected areas as an experimental measure, and its extension to other areas was delayed due to the intervention of war. It came into force in Satara from 11 April 1946, and the act was amended later that year. The 1948 act repealed the earlier one but retained many of the provisions regarding protecting the tenants. The major provisions of the 1948 act were concerned with, among other things, the protection of tenants against eviction by landlords and the creation of conditions for promoting the transfer of land to the tillers.

The study of Dandekar and Khudanpur (1957), which evaluated the implementation of the Act in the different districts of Maharashtra, including Satara, reported that owing to inherent weaknesses, the Act became largely ineffective in practice and failed to achieve the desired objectives. A 1956 amendment of the Act provided for a more radical measure. It imposed on the landlords the obligation to sell land to the tenants, effective from 1 April 1957 (known as tiller's day), based on the payment of the price in 12 instalments. However, micro-level and macro-level studies undertaken in Satara as well as in other parts of Maharashtra have found that the Act of 1956 met with a very limited success, mainly on account of the ignorance of the tenants, improper maintenance of land records, socio-political and economic pressures on the tenants, etc. (Shah and Sawant 1973; Dandekar 1978; Brahme and Upadhyaya 1979).

To bring about a more equitable distribution of land and to make the tenancy reforms more effective, the Maharashtra Agricultural Land (Ceiling on Holding) Act 1961 was enacted. It came into force on 26 January 1962. So far as the achievements of this Act were concerned, it was reported from a source[32] that 9,352.63 hectares of land were declared a ceiling in Satara, the surplus of which 5,291.18 hectares were taken under possession by December 1995. But a few studies (Dandekar 1978; Rajasekaran 1996) note that many large holders distributed the lands among their family members for land-record purposes while cultivating jointly. Moreover, factors, such as the exclusion of leased-out land from total holdings, problems of appropriate categorisation, and prevalence of 'benami' transfer, let the large landholders off the hook (Deshpande 1998). To prevent further fragmentation of holdings and to consolidate the fragmented holdings through 'mutual' exchange of small and scattered plots, the Bombay Prevention of Fragmentation and Consolidation of Holdings Act 1947 was applied to Satara district in 1949, and operations under the Act started in one *taluka* as an experiment in 1950. The act was later extended to other *talukas*. It is reported that though consolidation of holdings were achieved to some extent, the beneficiaries were the village elites, who could transfer and attach fertile fragments to their holdings in collusion with the implementing officials (Dandekar 1978; Deshpande 1998).

Besides adopting these land reform measures, the government made continuous efforts during the planning era for the expansion of irrigation facilities and the improvement of land utilisation and cropping intensity. As a result, the net sown area and the gross cropped area of Satara district increased consistently over the years, along with a rise in cropping intensity (Table 3.3). Between 1970 to 1971 and 1990 to 1991, the net sown area and the gross cropped area increased by 35.3 per cent and 40 per cent, respectively. The cropping intensity went up to 113.18 in 1990–91 from 107.85 in 1970–71. Similarly, the percentage of irrigated area to net sown area increased significantly (Table 3.3) from 13.47 per cent in 1970–71 to 22.52 per cent in 1990–91. The percentage of the irrigated area to net sown area was more in the case of the small and marginal farmers' holdings, which might be one of the reasons for their higher cropping intensity (Table 3.3).

Following the expansion of irrigation facilities and the growth of commercial trends in agriculture, the cropping pattern of the district underwent noticeable changes (Table 3.3). Though the area under food grain cultivation continued to be higher in the district as a whole, there was little variation across the size classes. While among the smaller landowners the area was relatively constant over time, there was a declining trend among the larger landowners. The area under sugarcane increased over the years. The rising trend was more prominent in the case of the small and marginal holders. It is primarily due to the establishment of major sugar factories in Phaltan and Karad *taluka* (both in the Satara district), which started production in 1957 and 1961, respectively. The subsequent growth and expansion of these factories and the establishment of other factories[33] necessitated augmentation of the area under sugarcane. The factories offered the farmers various incentives to increase the acreage under sugarcane. In addition, the government provides crop-specific loans and other facilities for expanding cane cultivation. Though the area under oilseeds declined at the district level, it showed an upward trend among the larger landowners.

Table 3.3 Land utilisation, irrigation, and cropping pattern by size class

Year	Size classes	Net sown area	Cropping intensity	Per cent of irrigated area to net sown area	Area under food grains	Area under sugarcane	Area under cotton	Area under oilseeds
1970–71	0–1	41,117	112.18	15.53	73.67	2.09	1.42	14.93
	1–2	66,321	109.59	14.70	73.64	2.47	1.34	15.81
	2–4	121,824	108.67	13.71	74.15	2.22	1.40	14.79
	4–10	183,882	106.79	12.64	77.74	1.93	1.16	11.78
	>10	103,661	105.91	13.05	80.18	2.76	1.07	7.74
	All	516,805	107.85	13.47	76.50	2.26	1.24	12.48
1980–81	0–1	72,052	117.93	18.07	73.00	5.44	1.05	10.07
	1–2	116,318	112.80	16.24	71.45	4.94	0.95	10.34
	2–4	173,638	112.63	13.98	71.51	3.42	0.96	9.30
	4–10	186,671	110.97	11.36	69.74	1.75	0.86	5.99
	>10	81,507	109.08	9.31	56.97	1.87	0.58	3.69
	All	630,186	112.25	13.49	69.33	3.26	0.89	7.91
1990–91	0–1	130,554	120.19	27.09	73.28	8.82	0.16	7.69
	1–2	179,994	116.80	27.73	72.58	7.93	0.13	6.95
	2–4	188,576	111.58	21.48	71.98	4.56	0.11	6.23
	4–10	145,295	108.22	16.01	67.30	2.26	0.12	5.92
	>10	54,925	103.25	15.32	55.45	2.55	0.04	8.06
	All	699,344	113.18	22.52	70.28	5.70	0.12	6.78
2000–01	0–1	210,987	111.74	27.98	78.20	8.32	0.08	10.63
	1–2	186,547	108.58	27.51	77.58	8.20	0.14	10.08
	2–4	173,247	106.83	20.45	83.58	3.91	0.05	7.04
	4–10	81,154	105.63	17.14	84.15	1.94	0.03	5.80
	>10	19,188	102.22	9.41	59.15	3.02	0.00	7.75
	All	671,123	108.58	24.06	79.58	6.27	0.08	8.92
2010–11	0–1	233,808	119.99	32.72	59.27	11.06	0.18	18.96
	1–2	163,152	115.08	25.36	66.00	10.18	0.27	16.26
	2–4	107,803	113.04	22.12	71.32	7.05	0.29	15.50
	4–10	60,644	113.50	26.77	70.37	5.15	0.24	16.55
	>10	28,027	128.75	50.27	64.20	10.95	0.12	8.74
	All	59,3434	117.13	28.99	64.56	9.53	0.23	16.85

Source: Government of Maharashtra, Report on Agricultural Census, for the respective years.

Along with the extension of cash crop cultivation, extensive coverage of HYVs was encouraged. It was reported[34] that more than 90 per cent of the area under food grains was covered by such varieties. Owing to the increasing cultivation of cash crops and the widespread adoption of HYVs, chemical fertiliser consumption per hectare went up tremendously in the district. It increased from 14.1 kg in 1971–72 to 32.6 kg in 1981–82 and 84.7 kg in 1991–92.[35] Consequently, upon the introduction of a new cropping pattern based on improved cultivation methods and the adoption of new technology and implements, a growing trend of mechanisation became widely visible in the district. The number of mechanical appliances used in the district increased several fold over time (Table 3.4).

The number of tractors, which was only 37 in 1961, increased to 2,956 in 1992. The number of sprayers, dusters, oil engines, pumpsets, etc., also increased. To meet the demands of the peasants for more working funds in view of the changing conditions, the credit network was expanded, largely through the establishment of agricultural cooperative societies. The growth in the number of such cooperative credit societies in the district was phenomenal, as was the rise in membership and the outstanding loan amounts.[36]

Taken together, these wide-ranging changes considerably enhanced the overall agricultural production of the district (Table 3.5).

The production and average yields of food grains and oilseeds showed a constant and significant uptrend. However, in the case of sugarcane, which needed relatively higher investment, average yield per hectare declined, though the total production kept increasing. The increase in production was mainly on account of the extension of cropped area. There was no marked trend in the production of cotton. Such uptrends in the production of the principal crops must have had differential impacts on the different categories of landholders. While the declining trend of sugarcane productivity must have affected the small and marginal holders, who continued to increase the tillage area, the larger landowners must have prospered owing to the increasing productivity of food grains and oilseeds, as they were the major cultivators of these crops. In addition, analysis of price of trends[37] in the district showed that while the harvest prices of food grains and oilseeds increased fast year by year, the sugarcane price showed a fluctuating trend owing to unpredictable market situations and the operations of the sugar lobby in state politics.

Table 3.4 Major agricultural machinery and implements (per 100 hectares of gross cropped area)

Year	Plough		Seed-drillers	Threshers		Sprayers and dust-ers	Sugarcane crushers		Oil engines with pumps for irrigation	Electric pumps for irrigation	Trac-tors	Persian wheels	Two wheeled walking tractors or power tillers	Gha-nies	Carts
	Wooden	Iron		Bullock operated	Power oper-ated		Power oper-ated	Bullock operated							
1961	4.727	2.052	–	–	–	–	0.05	0.059	0.358	0.005	0.005	0.001	–	0.047	4.656
1966	4.958	4.068	0.28	0.34	–	0.342	0.06	0.056	0.928	0.279	0.028	0.002	0.015	0.034	5.227
1972	7.226	4.775	6.324	0.209	0.052	0.346	0.046	0.008	1.533	0.889	0.06	0	0.002	0.003	8.204
1978	6.181	4.627	9.342	0.348	0.004	0.44	0.064	0.013	1.617	2.663	0.116	0.003	0.001	0.005	6.526
1982	6.371	5.477	9.987	0.462	0.031	0.501	0.065	0.005	1.407	2.689	0.145	0.002	0.002	0.005	5.239
1987	6.347	4.385	–	0.462	0.185	0.815	0.074	0.011	1.898	2.947	0.339	0.093	0.004	0.002	4.525
1992*	6.178	6.802	8.644	–	–	1.159	0.053	0.01	1.394	3.641	0.378	0.007	0.008	0.002	–
1997	2.890	2.567	2.855	0.295	0.356	1.182	0.051	0.014	1.068	5.577	0.769	0.009	0.033	0.026	4.019
2003	4.363	4.580	5.758	–	0.348	0.662	0.027	0.076	1.443	4.881	0.982	–	0.045	0.045	4.521

Sources: 1. Statistical Abstracts of Maharashtra, various issues
2. Livestock and Farm Equipment Census: Maharashtra State, various issues

Note:* As the data on gross cropped area of 1992 is not available, the per hectare agricultural machineries and implements has been calculated on the basis of the gross cropped area of 1991.

Table 3.5 Total production and average yield per hectare of major crops (production in '00 tonnes, yield kg)

Year/crops	1960–61		1970–71		1980–81		1990–91		2000–01		2010–11	
	Total	Per hectare	Total	Per hectare	Total	Total	Per hectare	Total	Per hectare	Total	Per hectare	Total	Per hectare
Rice	230	887	301	1,024	631	1,850	718	1,722	513	1,169	822	1,619
Wheat	118	864	102	649	291	1,207	351	1,263	516	1,568	865	1,979
Jowar (kharif)	508	849	682	981	1,140	1,145	1,423	1,530	1,171	1,757	761	1,706
Jowar (rabi)	998	685	598	362	1,080	687	1,295	838	732	510	1,052	786
Bajra	353	233	317	314	203	239	429	369	526	585	592	783
Total cereals	2310	554	2072	523	3,547	852	4,475	983	3,694	922	4,585	1,214
Total pulses	262	377	156	325	297	449	272	457	331	446	529	622
Total food grains	2572	529	2,228	502	3,844	797	4747	922	4,025	848	5,113	1,106
Groundnut (kharif)	753	1,250	577	912	529	862	551	1,155	729	1,238	816	1,628
Total oilseeds	NA	NA	NA	NA	551	814	975	1,061	1,198	1,360	1,574	1,565
Sugarcane (dressed cane)	NA	NA	8,671	93,236	16,033	91,109	31,903	87,167	35,040	91,488	69,634	94,100
Cotton	38	251	111	262	142	359	13	300	106	332	13	142

Source: District-Wise Agricultural Statistical Information of Maharashtra, various issues.

Post-reform changes

With the introduction of neoliberal reforms, the agrarian economy experienced significant changes which integrated the district with the global order. The cropping intensity increased significantly across size classes, and cropping patterns shifted more towards high-value commercial crops like sugarcane and oilseeds, accompanied by a corresponding decline in the area under food grains (Table 3.3). Among the size classes, the change was noticed prominently among the large holders, which was mainly due to the substantial increase in the area under irrigation. To meet the global demand, farmers moved increasingly towards horticulture, floriculture, and sericulture. Satara became a major fruit-producing district in the state due to the recent promotional activities undertaken by the National Horticulture Mission (2005–06). The average gross cropped area under horticultural crops accounted for 9 per cent between 2009 and 2013, of which fruit crops only covered 63 per cent. Mango, pomegranate, *sapota*, and strawberry were the largest grown crops by acreage. It was reported that while the average gross cropped area (2009–13) under mango covered 15,939 hectares, the area under pomegranate came to 4,422 hectares.[38] The area under strawberry increased from 870 in 2009–10 to 1,012 hectares in 2012–13. It was estimated that Satara contributed towards 83 per cent of the total production and market supply of strawberry in India.[39] Floriculture emerged as an important initiative, with around 900 poly houses (40 per cent of the total in Maharashtra) dedicated to flower production across the district. Out of 4,368 green houses in Maharashtra, 1,520 green houses were in the Satara district, producing flowers like Gerbera and Carnation. Sericulture activity also picked up, with 751 farmers producing on 830 acres of land in 157 villages.

Taken all crops together, the area under HYVs covered more than 80 per cent of gross cropped area in 2011–12.[40] The application of purchased inputs like fertilisers and pesticides went up. The per hectare consumption of fertilisers increased from 85 kg in 1991–92 to 173 kg in 2010–11. There was a phenomenal growth in the number of tractors and other advanced mechanical appliances. The number of tractors (per 100 hectares of gross cropped area), which was only 0.339 in 1987 increased to 0.982 in 2003 (Table 3.4). Similarly, the number of electric and oil engine pumpsets increased noticeably. As a result of these changes, the total production per hectare yield of major crops went up considerably, as compared to the pre-reform period, particularly for commercial crops like sugarcane (Table 3.5). The per hectare yield of sugarcane increased from 87,000 to 94,000 kg between 1990 to 1991 and 2010 to 2011. The district witnessed the practice of contract farming in a big way. Contract farming for procuring potato was done in Khatav, sweet corn in Karad and ladyfingers (okra) in Phaltan taluka. It was reported that in the past few years multinational agencies like PepsiCo and Indian conglomerate ITC moved in to the rural areas of Satara to establish potato contract schemes for particular varieties of chipping potatoes (Vicol 2015: 11). Similarly, strawberry procurement in Mahabaleshwar was carried out by Mapro. Mapro was also into procurement of mangoes for their mango-based beverages. Private license holders and contract farming agents were important market participants, though they seemed to be more active in horticulture and spices. Many well-known agro-based companies, having their manufacturing facilities in Pune, operated in the Satara markets. Some of the prominent names include Gits Foods, Cargill Foods, Pravin Chordia Foods, Dohler India, Weikfield, Frito Lays, and others. The APMC Act was amended, allowing private markets and direct transaction of food commodities between buyers and sellers.

Impact on social inequality

One way of assessing the impact of these changing conditions on the various categories of landowning groups is to analyse the trend of operational holdings in the district. Data on operational holding patterns in the district from 1970 to 1971 through 2010 to 2011 are presented in Table 3.6.

It is seen that the proportion of small and marginal holdings is increasing gradually in terms of number and total area, and that the opposite process is taking place in the case of large holdings. This gives the impression of a decreasing trend of land concentration. Actually, the trend is just the reverse, for the average size of the larger holdings is seen to be increasing consistently, with a corresponding decrease in that of the small and marginal holdings.

The increase in the proportion of small and marginal holdings, as it was reported by some studies conducted in Maharashtra (Dandekar 1978; Rajasekaran 1996) as well as in other parts of the country (Rajasekar 1988; Mohanty 2000a), was possibly due

Table 3.6 Breakup of operational holdings by size class

Size class	1970–71			1980–81			1990–91			2000–01			2010–11		
	Per Cent of total holdings	Per cent of total area	Average holding size (ha.)	Per cent of total holdings	Per cent of total area	Average holding size (ha.)	Per cent of total holdings	Per cent of total area	Average holding size (ha.)	Percent of total holdings	Per cent of total area	Average holding size (ha.)	Per cent of total holdings	Per cent of total area	Average holding size (ha.)
0–1	39.89	6.37	0.43	47.15	10.33	0.44	57.49	17.24	0.41	67.89	29.01	0.42	78.20	37.39	0.36
1–2	20.04	10.78	1.45	22.31	16.97	1.52	23.38	24.39	1.44	19.66	26.53	1.34	14.79	27.17	1.38
2–4	19.96	21.08	2.85	18.47	26.27	2.82	13.25	26.46	2.75	9.68	25.95	2.66	5.30	18.86	2.67
4–10	15.79	35.62	6.08	10.28	30.77	5.96	5.2	21.71	5.76	2.55	13.68	5.32	1.49	11.30	5.68
>10	4.33	26.14	16.27	1.8	15.66	17.36	0.68	10.2	20.61	0.21	4.83	23.26	0.21	5.27	18.89
Total	100.00	100.00	2.69	100.00	100.00	1.99	100	100	1.38	100.00	100.00	0.99	100.00	100.00	0.75

Source: Government of Maharashtra: Report on Agricultural Census, various issues.

to partitioning of the large holdings within families for avoiding the ceiling laws and getting the benefits of the developmental measures meant for small and marginal peasants. The rapid growth of rural population in Satara and the declining land-man ratio of the district[41] could be another cause of the increase in the number and total area of small and marginal holdings. Besides, the displacement caused by land acquisition in the post-reform period marginalised a number of peasants.[42] Permission was accorded to many industrial units to acquire land over the ceiling limit as per the provisions of Maharashtra Industrial Policy, which was revised in 1993. Following the SEZs Act, 2005, recently, the MIDC acquired 101.25 hectares of cultivable land for setting up a sector-specific SEZ for engineering in the villages of Suvadi and Nandal of Phaltan *taluka*, which affected 51 landowners.[43] More recently, the government of Maharashtra acquired 348.6 hectares of cultivable land to set up SEZs in Satara, from which the Kesurdi SEZ was allotted 111 hectares to make the district a defence hub.[44]

So far as the operational holdings of the SCs in the district are concerned,[45] their percentage of the total in terms of number and area decreased to 6.34 and 3.22, respectively, in 1990–91 from 6.84 and 3.41 in 1985–86. However, in 2010–11, the number and area of their holdings increased to 7.23 and 6.64 per cent, respectively. On the contrary, the size of the average holdings of the SC decreased from 0.92 hectare in 1985–86 to 0.84 hectare in 1990–91 and 0.69 in 2010–11. As for the operational holdings of the STs,[46] while the percentage of the total increased from 0.17 in 1985–86 to 0.50 in 1990–91 in terms of number, it increased from 0.08 to 0.28 in terms of area within this period; the average size of the tribal holdings also increased from 1.14 hectares in 1985–86 to 1.34 hectares in 1990–91. However, in the post-reform period, the total number of operational holders declined to 0.30 per cent in 2010–11, accompanied by a decrease in average size of holdings to 0.81 hectares, though the area increased marginally (0.32 per cent). The gender disparity in landholdings in the district is striking.[47] The women's share of the total holdings in terms of number as well as area is only about 10 per cent, and it shows a declining trend more particularly among the lower landowning groups, between 2000 to 2001 and 2010 to 2011.

The district-level impact of new agricultural practices on the agricultural population can also be assessed by analysing the relative changes in the populations of agricultural labourers and cultivators. The proportion of agricultural labourers and cultivators

Table 3.7 Ratio of cultivators and agricultural labourers to main workers *(per cent)*

Year	Cultivators			Agricultural labourers		
	Male	*Female*	*Total*	*Male*	*Female*	*Total*
1961	61.31	79.59	70.08	7.53	11.99	9.67
1971	55.87	56.90	56.15	12.82	33.24	18.32
1981	51.10	56.73	52.90	12.56	30.85	18.40
1991	47.17	53.31	49.28	14.50	35.09	21.58
2001	44.56	58.06	49.30	12.99	26.47	17.73
2011	43.58	44.98	44.05	14.80	31.74	20.57

Source: *District Census Handbook of Satara*, various issues.

among the main workers of the district from 1961 to 2011 is shown in Table 3.7.

We found that the proportion of cultivators decreased rapidly over the years and that the proportion of agricultural labourers rose steadily. The proportion of cultivators, which was more than 70 per cent in 1961, came down gradually to 44 per cent in 2011. On the other hand, the proportion of agricultural labourers increased to 21 per cent in 2011 from 10 per cent in 1961. It seems that the modernising measures have had the cumulative effect of augmenting the agricultural labour force in the district. It is interesting to note that the proportion of female agricultural labourers among the district's main workers increased almost threefold (from 11.99 per cent in 1961 to 31.74 per cent in 2011), and the growth rate was much faster than in the case of their male counterparts (from 7.53 per cent in 1961 to 14.80 per cent in 2011). The proportion of female cultivators came down from 79.59 per cent in 1961 to 44.98 per cent in 2011. It appears that the growing impoverishment of the poor peasants and landless agricultural labourers compelled the women to take to agricultural labour to supplement their family income. The fact that the rates of wages[48] payable to field labourers in the district were low up to 1990–91 might have contributed to this impoverishment.

It is evident from the preceding analysis that the new conditions and forces of modernisation created, instead of removing the earlier inequalities, further inequalities. However, the effects of the new measures could be substantiated in more detail by looking at the situation at the village level.

Village-level impact

For examining empirically the impact of agricultural modernisation on the pattern of social inequality at micro level, two villages were selected, one representing the district's agriculturally advanced *talukas* and the other the less advanced *talukas*.[49] These villages are, as it were, diametrically opposed to each other from the point of view of agricultural practices and resource ownership (Table 3.8). The study considers Atke village, representing the advanced *talukas* of the modernised village. The village had to its credit extensive and assured irrigation facilities, a relatively high degree of mechanisation, extensive adoption of HYV, and other infrastructural arrangements. On the other hand, Khatval village, representing the less developed *talukas*, was treated as a traditional village. Its cultivation of technology was largely traditional, based on broadcasting of seeds and the use of predominantly human and bullock labour.

For detailed analysis of the agricultural practices prevalent in the two villages and the respective impacts on the socio-economic condition of the peasants, a sample of 100 households (50 from each village) was drawn from various categories of households (landless agricultural labourers, marginal farmers, small farmers, medium farmers, and large farmers), classified on the basis of landownership.[50]

It was found at the village level that the inroads of agricultural modernisation had drastically changed the cropping pattern. While in the traditional village, *bajra* and *jowar* were the principal crops, the agriculture of the modernised village was characterised by the predominance of cash crops, such as sugarcane, soybean. The cropping patterns of all categories of farmers in the modernised village were heavily dependent on HYV seeds. The essential condition for HYVs being the availability of assured irrigation in terms of both quantity and timing, the farmers of the village, well served by irrigation facilities, adopted these varieties extensively. Higher productivity and larger margins of profit were the results. On the other hand, the increasing cultivation of cash crops, along with the coverage of a large area by HYVs, made the application of high doses of fertiliser, pesticides, and other inputs an integral part of the cultivation process, as a result of which the cost of cultivation of all categories of farmers was very high vis-à-vis traditional agriculture (Tables 3.9a and 3.9b).

In view of the higher cost of cultivation, the formal credit agencies provided more liberal credit facilities to the farmers of Atke. The per hectare entitlement of loans from formal credit agencies for

Table 3.8 General information on selected villages

Particulars	Atke (modernised)	Khatval (traditional)
No of households (2011)	930	282
Total population (2011)	4,497	1,385
Percentage of male population (2011)	50.95	49.89
Percentage of female population (2011)	49.05	50.11
Percentage of SC population (2011)	8.87	18.12
Percentage of scheduled tribe population (2011)	0.29	–
Percentage of general literacy (2011)	78.50	72.20
Percentage of male literacy (2011)	82.19	77.86
Percentage of female literacy (2011)	74.66	67.57
Percentage of cultivators to main workers (2011)	51.73	69.26
Percentage of agricultural labourers to main workers (2011)	27.36	22.24
Total area (in hectares) (2011)	946.09	1149
Percentage of cultivable area (2001)	87.74	82.07
Percentage of irrigated area to cultivable area (2001)	98.78	12.41
Percentage of gross cropped area under cash crops (1996–97)	89.23	0.08
Percentage of area under HYV* (1997)	95	20
No of tractors (1997)	350	7
No of threshers (1997)	60	–
No of oil engines and pumpsets (1997)	666	11
Livestock population (1997)	5730	684
No of primary agricultural cooperative societies (1997)	2	1
No of milk cooperative societies (1997)	2	1

Sources: 1. *District Census Handbook, Satara,* 2001 and 2011
2. Gram Panchayat Office of the respective village
3. Talathi Office of the respective village
*Based on the estimate of Talathi and Agricultural Development Officer of the respective village.

Notes: Figures in parentheses indicate the reference years.

Table 3.9a Size-class-wise distribution of resources among selected households of modernised village (Atke)

Sl. no.	Particulars	Size classes					
		Landless	Marginal	Small	Medium	Large	All
		N = 6	N = 16	N = 13	N = 9	N = 6	N = 50
1.	Average family size	5.50	5.06	5.54	5.89	6.67	5.58
2.	Average landholding in hectares	—	0.25	0.68	1.27	4.18	1.12
3.	Percentage of irrigated land to total holding	—	93.60	96.60	98.95	98.80	98.01
4.	Cropping intensity*	—	109.00	125.00	111.21	105.58	110.63
5.	Percentage of area under cash crops to gross cropped area	—	82.35	82.36	82.20	84.92	83.56
6.	Percentage of area covered by HYV	—	85.88	96.36	96.85	100.00	97.43
7.	Per hectare ownership of tractors	—	—	0.18	0.47	0.34	0.31
8.	Per hectare ownership of oil engines	—	—	0.18	0.31	0.15	0.18
9.	Per hectare ownership of electric pumpsets	—	0.71	0.36	0.39	0.26	0.35
10.	Per hectare consumption of chemical fertiliser (in kgs)	—	1,948.24	2,054.55	2,005.51	2,043.40	2,029.38
11.	Per hectare use of labour days	—	659.76	369.91	421.97	360.00	399.85
12.	Percentage of family labour days to per hectare labour days	—	75.61	61.05	72.10	28.30	51.30
13.	Per hectare cost of cultivation (in Rs.)	—	25,516.94	2,4231.36	26,550.39	25,671.70	25,757.25
14.	Percentage of total produce sold	—	99.02	99.50	99.80	99.80	99.70

No.	Indicator						
15.	Per hectare net income (Rs.)	–	42,824.24	43,586.82	47,642.52	47,252.07	46,073.42
16.	Average household income from all sources in (Rs.)	20,215.00	25,388.00	55,424.31	96,550.00	245,011.67	71,740.68
17.	Percentage of agricultural income to total income	–	44.81	66.54	69.63	85.18	70.22
18.	Average labour days hired in	–	42.75	121.92	166.11	1,140.00	212.08
19.	Average labour days of family members in own land	–	132.50	191.08	428.89	450.00	223.36
20.	Average labour days hired out	466.33	113.19	20.31	23.33	–	101.66
21.	Per hectare loan from formal credit agencies (Rs.)	–	2,1082.35	4,227.18	11,401.57	28,490.57	20,437.10
22.	Average amount of loan from formal credit agencies (Rs.)	–	5,600.00	9,492.31	16,088.89	125,833.33	22,256.00
23.	Average transaction cost as per cent of average loan	–	4.12	1.97	1.12	0.65	1.14
24.	Percentage of literacy	51.52	79.01	81.94	84.91	92.50	79.57
25.	Percentage of literate persons having post matric qualification	0.00	3.13	27.12	17.78	29.73	16.67
26.	Percentage of households owning pucca houses	0.00	18.75	38.46	55.56	83.33	36.00
27.	Percentage of households owning television sets	–	50.00	76.92	77.78	100.00	64.00
28.	Average ownership of trucks/cars/jeeps	–	–	–	0.11	0.83	0.12
29.	Average ownership of mopeds/scooters/motorcycles	–	0.13	0.62	1.11	2.00	0.64

Source: Field survey.

Notes: Per hectare calculation is based on gross cropped area. Though sugarcane is a one-year crop, the area under this crop is taken as a single cropped area.* Cropping intensity of the modernised village looks lower than that of the traditional village due to the extensive cultivation of sugarcane.

Table 3.9b Size-class-wise distribution of resources among selected households of traditional village (Khatval)

Sl. no.	Particulars	Size classes					
		Landless	Marginal	Small	Medium	Large	All
		N = 6	N = 15	N = 14	N = 11	N = 4	N = 50
1.	Average family size	6.00	6.07	5.64	5.82	4.75	5.78
2.	Average landholding in hectares	—	0.64	1.50	2.60	5.63	1.85
3.	Percentage of irrigated land to total holding	—	—	—	14.07	14.88	9.02
4.	Cropping intensity*	—	156.90	141.91	131.36	104.18	129.94
5.	Percentage of area under cash crops to gross cropped area	—	—	—	0.27	5.41	1.25
6.	Percentage of area covered by HYV	—	33.20	4.71	19.74	15.31	16.46
7.	Per hectare ownership of tractors	—	—	—	—	0.05	0.01
8.	Per hectare ownership of oil engines	—	0.13	0.10	—	0.05	0.06
9.	Per hectare ownership of electric pumpsets	—	0.20	0.10	0.13	0.27	0.16
10.	Per hectare consumption of chemical fertiliser (in kg.)	—	63.67	60.88	46.63	110.36	66.68
11.	Per hectare use of labour days	—	78.13	54.25	33.81	56.49	50.81
12.	Percentage of family labour days to per hectare labour days	—	89.76	55.67	70.06	64.11	68.63
13.	Per hectare cost of cultivation (in Rs.)	—	1,742.00	1,664.98	1,130.03	3,219.82	1,814.28
14.	Percentage of total produce sold	—	40.75	57.21	38.49	89.25	72.06
15.	Per hectare net income (Rs.)	—	3066.00	2366.30	2564.35	4232.43	2932.75

#		1	2	3	4	5	6
16.	Average household income from all sources in (Rs.)	15,225.00	15,433.27	26,240.00	17,208.20	54,001.50	21,910.10
17.	Percentage of agricultural income to total income	1.81	19.87	19.15	50.84	43.50	28.13
18.	Average labour days hired in	—	8.00	57.07	34.55	112.50	33.30
19.	Average labour days of family members in own land	—	70.13	64.14	80.82	134.00	72.86
20.	Average labour days hired out		267.00	121.86	104.55	85.00	196.82
21.	Per hectare loan from formal credit agencies (Rs.)	—	2,340.00	945.17	644.82	2855.86	1443.61
22.	Average amount of loan from formal credit agencies (Rs.)	—	2340.00	2,007.14	2,200.00	15,850.00	3,016.00
23.	Average transaction cost as per cent of average loan	—	1.64	0.43	0.41	0.16	0.60
24.	Percentage of literacy	50.00	60.44	73.42	70.31	68.42	64.71
25.	Percentage of literate persons having post matric qualification	61.11	52.73	53.57	73.33	69.23	59.90
26.	Percentage of households owning pucca houses	—	—	21.43	36.36	75.00	20.00
27.	Percentage of households owning television sets	—	20.00	21.43	18.18	50.00	20.00
28.	Average ownership of trucks/cars/jeeps	—	—	—	—	—	—
29.	Average ownership of mopeds/scooters/motorcycles	—	—	—	—	0.50	0.04

Source: Same as in Table 3.9a.

Notes: Same as in Table 3.9a.

all categories of farmers was many times more than for the farmers of Khatval (Table 3.9a and 3.9b). However, in the modernised village, the major beneficiaries of the credit facilities were the large farmers, whose per hectare loan entitlements were much higher than their costs of cultivation. On the other hand, the per hectare loan entitlements of the marginal farmers in the traditional village were higher than their costs of cultivation. This is mainly because their costs were very low, and they used a significant portion of the loans to meet dire needs not directly related to agriculture.

In both the villages, the marginal and large farmers had a higher rate of per hectare indebtedness to formal agencies. While the marginal farmers' indebtedness was caused by perpetual scarcity of funds, the large farmers' indebtedness was attributed to their propensity to utilise the low-cost credit for other profit-generating endeavours. Moreover, the hectarage-based lending policies of the credit agencies helped the large farmers to appropriate the lion's share of the credit, and this share was even larger in a modernised setting. The average indebtedness of the large farmers of Atke was remarkably higher than that of other categories of farmers (Table 3.9a).

The credit transaction costs borne by the farmers of the two villages provided some other interesting insights. The ratio of transaction cost to the amount of a loan was more in the modernised village than in the traditional one. As the cultivation process in Atke was very expensive, the farmers of the village cultivated the goodwill of the officials of the credit agencies by entertaining them in various ways, which raised the transaction costs. In both the villages, the marginal farmers bore a relatively heavy transaction cost; the burden became lighter with an increase in holding size. The larger loan entitlements of the large farmers, along with their wider network of political contacts, reduced the ratio of transaction costs to loan amounts.

In addition to this, the availability of large amounts of low-cost credit helped the large farmers of the modernised village to acquire high-value, income-generating agricultural machinery, such as tractors. As most modern agricultural operations were tractorised, the large farmers earned a lot of additional income by hiring their tractors out to small and marginal farmers. They also hired out the tractors during lean periods for non-agricultural use.

The diversity of agricultural practices between the two villages is matched by radically different labour relations. The advent of modern agriculture altered the pattern of labour relations. The demand for hired labour was much greater in the modernised village. The cropping pattern and the water management system associated with modern agriculture, which envisage quick completion of all operations within a short period, make the hiring of labour inevitable. The small and marginal farmers who earlier managed mostly with family labour were now compelled to depend on hired labour to a considerable extent. The heavy demand for hired labour in the modernised village attracted migrant labour from other parts of the district and the state on a large-scale. Now there were two sets of labourers – migrants and natives. The migrant labourers generally worked as contract labourers. Under the contract labour system, a group of labourers was required to complete a given agricultural operation within a stipulated period, and the wage amount depended on the nature of the operation, the time limit, the location of the land, and the relative bargaining positions of the group leader and the employer. This type of labour was widely preferred among the landowners because it helped meet deadlines for operations. The large-scale cultivation of sugarcane increased the demand for such labour services. Around 80 per cent of labour requirements of the modernised village were met by contract labour.

The farmers preferred the migrant labourers to the native ones. The migrants accounted for 89.82 per cent of the hired labour days in the modernised village. As a result, the native labourers who earned their livelihood mainly through agricultural labour became poorer and were forced to toil in building and road construction works. More women belonging to the landless labour category took to wage labour because of the reduced employment opportunities. In traditional agriculture, on the other hand, the use of family labour predominates across size classes (Table 3.9b). The marginal, small, and medium farmers of the village exchanged labour services during peak periods and the large farmers used, besides family labour, casual and attached labour.

Collectively, the new practices made overall agricultural production much higher in the modernised village than in the traditional one. The per hectare gross return and the net incomes of all categories of farmers were much higher in the modernised village. Every additional hectare added to the income stream of a farmer, which in turn increased household income disparities among the various categories of the rural population on account of unequal distribution of land. The household incomes of the large farmers of the modernised village were many times more than those of the other categories of farmers. In the traditional village, on the other hand, household income disparity was minimal. An analysis of the distribution of resources among sample households of the two villages (Table 3.9a and 3.9b) on the basis of the size class of holdings reveals that the modernised village exhibited higher incidence of inequality than the traditional one in terms of landownership, land-based resources, and high-value household assets. Though the concentration of land was associated with concentration of resources in both villages, the magnitude of concentration was higher in the modernised village. An analysis of caste group-wise[51] distribution of resources reveals a similar trend (Table 3.10a and 3.10b).

Table 3.10a Caste-group-wise distribution of resources among the selected households of modernised village (Atke)

Sl. no.	Particulars	Caste groups			
		Upper	Medium	Lower	All
		N = 38	N = 5	N = 7	N = 50
1.	Average family size	5.55	5.80	5.57	5.58
2.	Average landholding in hectares	1.25	0.25	0.09	1.12
3.	Percentage of irrigated land to total holding	98.99	63.49	93.75	98.01
4.	Cropping intensity	110.89	105.45	100.00	110.63
5.	Percentage of area under cash crops to gross cropped area	83.65	74.14	93.75	83.56
6.	Percentage of area covered by HYV	97.72	86.21	93.75	97.43
7.	Per hectare ownership of tractors	0.32	–	–	0.31
8.	Per hectare ownership of oil engines	0.13	2.59	–	0.18
9.	Per hectare ownership of electric pumpsets	0.34	0.86	–	0.35
10.	Per hectare consumption of chemical fertiliser (kg)	2,035.19	1,562.50	2,398.41	2,029.38

Sl. no.	Particulars	Caste groups			
		Upper	Medium	Lower	All
		N = 38	N = 5	N = 7	N = 50
11.	Per hectare use of labour days	403.29	274.14	351.56	399.85
12.	Percentage of family labour days to per hectare labour days	50.23	100.00	82.22	51.30
13.	Per hectare cost of cultivation (Rs.)	25,735.65	32,758.62	14,843.75	25,757.25
14.	Percentage of total produce sold	99.80	95.83	95.98	99.70
15.	Per hectare net income (Rs.)	46,851.81	36,120.69	15,703.12	46,073.42
16.	Average household income from all sources (Rs.)	87,411.42	25,672.00	19,577.14	71,740.68
17.	Percentage of agricultural income to total income	73.96	32.64	7.33	70.22
18.	Average labour days hired	278.00	–	5.71	212.08
19.	Average labour days of family members in own land	280.66	63.60	26.43	223.36
20.	Average labour days hired out	40.79	56.00	464.71	101.66
21.	Per hectare loan from formal credit agencies (Rs.)	20,575.50	14,439.66	19,921.88	20,437.10
22.	Average amount of loan (Rs.)	28,507.89	3,350.00	1,821.43	22,256.00
23.	Percentage of literacy	74.88	93.10	94.87	79.57
24.	Percentage of literate persons having post matric qualification	21.52	7.41	2.70	16.67
25.	Percentage of households owning pucca houses	47.37	–	–	36.00
26.	Percentage of households owning television sets	68.42	100.00	14.28	64.00
27.	Average ownership of trucks/cars/jeeps	0.15	–	–	0.12
28.	Average ownership of moped/scooters/ motorcycles	0.79	–	0.28	0.64

Source: Same as in Table 3.9a.

Notes: Same as in Table 3.9a.

Table 3.10b Caste-group-wise distribution of resources among the selected households of traditional village (Khatval)

Sl. no.	Particulars	Caste groups			
		Upper	Medium	Lower	All
		N = 30	N = 2	N = 18	N = 50
1.	Average family size	5.23	3.50	6.94	5.78
2.	Average landholding in hectares	2.02	–	1.17	1.85
3.	Percentage of irrigated land to total holding	11.58	–	1.70	9.02
4.	Cropping intensity	127.63	–	136.41	129.94
5.	Percentage of area under cash crops to gross cropped area	1.72	–	–	1.25
6.	Percentage of area covered by HYV	19.83	–	7.64	16.46
7.	Per hectare ownership of tractors	0.01	–	–	0.01
8.	Per hectare ownership of oil engines	0.08	–	–	0.06
9.	Per hectare ownership of electric pumpsets	0.19	–	0.10	0.16
10.	Per hectare consumption of chemical fertiliser (kg)	72.24	–	52.08	66.68
11.	Per hectare use of labour days	55.08	–	39.60	50.81
12.	Percentage of family labour days to per hectare labour days	64.60	–	83.35	68.63
13.	Per hectare cost of cultivation (Rs.)	1987.71	–	1358.89	1814.28
14.	Percentage of total produce sold	83.35	–	20.43	72.06
15.	Per hectare net income (Rs.)	3,454.52	–	1,536.63	2,932.75
16.	Average household income from all sources (Rs.)	26,238.17	10,375.00	16,070.06	21,910.10
17.	Percentage of agricultural income to total income	33.20	7.95	15.30	28.13
18.	Average labour days hired	49.17	–	10.56	33.30
19.	Average labour days of family members in own land	89.73	–	52.83	72.86

Sl. no.	Particulars	Caste groups			
		Upper	Medium	Lower	All
		N = 30	N = 2	N = 18	N = 50
20.	Average labour days hired out	124.60	250.00	311.28	196.82
21.	Per hectare loan from formal credit agencies (Rs.)	1,853.27	–	367.93	1,443.61
22.	Average amount of loan (Rs.)	46,73.33	–	588.89	3,016.00
23.	Percentage of literacy	74.52	71.43	52.00	64.71
24.	Percentage of literate persons having post matric qualification	58.97	60.00	61.54	59.90
25.	Percentage of households owning pucca houses	30.00	–	5.56	20.00
26.	Percentage of households owning television sets	23.33	–	16.67	20.00
27.	Average ownership of trucks/cars/jeeps	–	–	–	–
28.	Average ownership of moped/scooters/ motorcycles	0.07	–	–	0.04

Source: Same as in Table 3.9a.

Notes: Same as in Table 3.9a.

Though the upper castes (*Marathas* and Brahmins) dominated the resource ownership structure in both villages, the extent of dominance was more in the modernised village. The resource distribution pattern clearly reveals an utter lack of resources with the people belonging to the lower castes.

The breakup by caste groups shows a familiar pattern in which the distribution of resources more or less reflected the caste hierarchy. In terms of landownership, investment in agriculture, income, ownership of agricultural machinery, and other assets, the upper castes were further ahead of the medium and lower castes. As the large farmers belonged to the higher castes, and the marginal farmers and landless labourers were generally from lower caste groups, agricultural modernisation aggravated caste disparities.

A comparative study of the major socio-political institutions of the two villages provides a more detailed picture of the patterns of social inequality in these villages. The gram panchayat, which is the apex socio-political body at the village level and which functions as a channel for most of the state-sponsored developmental measures, was completely dominated by the upper-caste large farmers in the modernised village (Table 3.11).

Most of the offices, including the key offices of *sarpanch* and deputy *sarpanch*, were held by them. Similarly, the cooperative societies, through which the state-supported credit facilities for agriculture development and allied activities were supplied to the village, were largely under their control.

It appears that in the modernised village, the large upper-caste farmers had taken control of the management of all village-level institutions and that the landless labourers and the poor farmers had almost no say in these institutions' affairs. Such domination helped the large farmers to corner the benefits of all developmental measures. Though it is claimed that the basis of power in these institutions is independent of caste and class, as it is based on numerical support, the upper-caste large farmers used their economic hegemony and caste ideology to mobilise the mass of poor farmers and landless labourers in their favour.

Though the dominance of the upper-caste farmers over these institutions persisted in the traditional village, the landless labourers and the lower landowning farmers were relatively well represented there (Table 3.12).

In the gram panchayat of the village, marginal farmers and landless labourers had occupied some important positions, including those of the *sarpanch* and deputy *sarpanch*. In the primary agricultural cooperative society and the milk cooperative society too, the small and marginal farmers and the landless labourers had better access than in the other village. The conditions of agricultural modernisation had induced the large farmers to take active part in the management of major village level institutions for the protection and promotion of their economic interests. As modern agriculture was heavily dependent on state-sponsored inputs and facilities, which flowed largely through these institutions, the large farmers showed a growing interest in local politics.

Table 3.11 Distribution of posts in socio-political institutions of Atke by size classes, caste groups (per cent)

Name of institutions	No of positions	Size classes					Caste groups		
		Landless	Marginal	Small	Medium	Large	Upper	Medium	Lower
Gram panchayat	16	6.25	12.50	12.50	6.25	62.50	87.50	6.25	6.25
Primary agricultural cooperative society	9	–	11.11	–	22.22	66.67	88.89	–	11.11
Bhairavnath agricultural cooperative society	9	11.11	–	11.11	11.11	66.67	88.89	–	11.00
Milk cooperative society	9	–	22.22	–	22.22	55.56	88.89	–	11.11
Hatkeshwar milk cooperative society	9	–	11.11	11.11	11.11	66.67	88.89	–	11.11
Total	52	3.85	11.54	7.69	13.46	63.46	88.46	1.92	9.62

Source: Comparative study.

Table 3.12 Distribution of posts in socio-political institutions of Khatval by size classes, caste groups (per cent)

Name of institutions	No of positions	Size classes						Caste groups		
		Landless	Marginal	Small	Medium	Large		Upper	Medium	Lower
Gram panchayat	9	22.22	11.11	22.22	11.11	33.33		77.78	11.11	11.11
Primary agricultural cooperative society	9	11.11	–	33.33	44.44	11.11		77.78	–	22.22
Milk cooperative society	9	44.44	11.11	22.22	–	22.22		88.89	–	11.11
Total	27	25.93	7.41	25.93	18.52	22.22		81.48	3.70	14.80

Source: Comparative study.

Conclusion

It may be seen from the foregoing analysis and discussion that the process of agricultural modernisation was initiated in Satara during the British period. As it was based on exploitative land, labour, and credit relations, and was guided by colonial interests, the process served to impoverish many and enrich a few. As a result, more social inequality was created over time. Agricultural modernisation was pursued more systematically and elaborately during the post-independence and post-reform periods. As modernisation measures were introduced without the earlier inequalities being removed, their coming into vogue generated further inequalities. It is true that the new measures brought prosperity for peasants of all categories. But as the degree of prosperity is related to ownership of resources, the large farmers emerged as the richer class dominating the economic and socio-political spheres of rural life, and the small and marginal farmers and the landless labourers remained disadvantaged. In fact, inequality breeds further inequality unless it is checked at the initial stage of growth and development.

Barraclough (1974) has rightly said,

> If we are really serious about wanting to encourage development policies that benefit the low-income rural classes, a great deal more attention must be paid to analysis of social structure and political process. Development is not just economics or sociology or technology but history. . . . Rural development for the low income majorities requires fundamental and often revolutionary reforms in social institutions.[52]

To conclude, the existing modernisation measures are desirable for agricultural growth but inadequate for the removal of social inequality.

Notes

1 This is the revised version of the article "Agricultural Modernization and Social Inequality-Case Study of Satara District" published earlier in *Economic and Political Weekly*, Volume 34, No. 26, June 26, 1999, pp. A50–A61.
2 So far as the agrarian structure of Maharashtra and its change in the post-independence period is concerned, there are a good number of studies. Notable among them are Brahme and Upadhyaya (1979),

Attwood (1979), Baviskar (1980), Walker and Ryan (1990), Rath and Mitra (1989), Dev and Mungekar (1996), and Deshpande (1998). While some of these studies have attempted to examine the specific aspects of the rural economy, others have confined their analysis only to the macro level. Studies describing the collective and cumulative effect of agricultural modernisation measures on social inequality are almost non-existent.

3 Satara has been ranked the most advanced in agricultural modernisation of Maharashtra's districts on the basis of four major parameters: area under irrigation, mechanisation, chemical fertiliser consumption, and area under HYV. Each district is given a rank on the basis of its position against each indicator separately, and the overall rank is obtained by adding all the ranks. Data on the area under HYV are available only for some crops: paddy, jowar, *bajra*, wheat, maize, cotton, etc. But there are districts with large areas under sugarcane, groundnuts, fruits, etc. The absence of information on the adoption of HYV for these crops might tend to distort the ranking. However, as the adoption of HYV is not the only indicator of modernisation, it would not have vitiated the process significantly.

4 Though the concept of agricultural modernisation is broad and complex, it refers to the extensive and intensive use of improved production technology and inputs for the maximisation of production.

5 Here social inequality means the social evaluation of differences existing among various categories of population (landholding size groups and caste groups) relating to the distribution of resources (land, land-based assets, household assets, etc.) and access to the institutions and agencies which provide or withhold these resources.

6 See *Gazetteer of Bombay Presidency: Satara*, Vol XIX, 1885, p. 328.

7 See *Maharashtra State Gazetteers: Satara District*, 1963, pp. 54–66.

8 This is based on observations made by Ogilvoy, the then commissioner of Deccan, as cited in Bombay Government Revenue Records, 1852, 22(27–28).

9 See *Maharashtra State Gazetteers: Satara District*, 1963, p. 602.

10 For details see Mann (1917).

11 The landholding pattern of Satara district in 1882–83 was as follows:

Land sizes (acres)	Percentage of the Holdings	Land sizes (acres)	Percentage of the holdings
Below 5	37.03	100–200	1.03
5–10	20.48	200–300	0.18
10–20	18.07	300–400	0.05
20–50	16.08	above 400	0.06
50–100	7.02	–	–

For more details see *Gazetteer of Bombay Presidency: Satara, op.cit.*, p. 150.

12 The major irrigation projects which were constructed during the British period are as follows:

Name of canal/ lake	Year of beginning of construction	Year of comple- tion	Total area irrigated (in acres)
Revari canal	1781	1849	519
Chikhli canal	1866–67	1870	217
Yerla canal	1867	1868	749
Gondoli canal	1867	1872	Not known
Mayni lake	1868	1875–76	742
Krishna canal	1863	1868	3023

Source: *Gazeteer of Bombay Presidency: Satara*, Vol. XIX, 1885, pp. 151–157.

The increase and decrease of the ratio of irrigated area to net sown area in certain years (Table 3.1) is largely due to variations in net sown area and rainfall (on which most of the irrigation sources depended).

13 While the decrease of net sown area during 1895–96 to 1911–12 was possibly due to the effect of bad seasons and famine years at the turn of the century, the decrease of cropping intensity after 1915–16 was mainly because of the changes in the criteria adopted for classifying areas under land utilisation. From 1915 to 1916, the occupied land under the grass and babul trees, previously categorised as fallow land, were classified as land under fodder crops. New area began to be included in the net sown area due to changed classification. For details see *Season and Crop Report*, 1915–16, p. 2.

14 See *Gazetteer of Bombay Presidency: Satara*, op.cit., p. 346.

15 From 1848, new varieties of cotton seeds were provided to the peasants. In 1850–51, about 60,000 pounds of New Orleans cotton seeds were given to the cultivators and about 3,200 acres were planted with these seeds. For details see *Gazetteer of Bombay Presidency: Satara*, op.cit., p. 166.

16 The information on forcible cultivation of certain alien varieties of cotton in some parts of Deccan has been given in Guha (1985: 107).

17 It was estimated that nearly 40,000 bighas, or about 36,727 acres, might grow cotton. For details see *Gazetteer of Bombay Presidency: Satara*, op.cit., p. 346.

18 See Benjamin (1973) and Borpujari (1973).

19 See *Gazetteer of Bombay Presidency: Satara*, op.cit., pp. 219–20.

20 During the latter part of the nineteenth-century, Khandesh and Nashik districts exported large quantity of cotton. About 85 per cent of the produce was exported and the local consumption was negligible. For details, see Guha (1985: 105–114).

21 As per the 1850–51 statistics, nearly 7,136 acres of native and 4,151 acres of Mauritius sugarcane were cultivated.

22 See *Gazetteer of Bombay Presidency: Satara*, op.cit., p. 186.

23 Ibid.

24 See Deccan Riots Commission Report 1878, Appendix A, pp. 40–41, Appendix C, pp. 10–12.
25 See Guha (1987: 132).
26 For details see Deccan Agriculturists' Relief Act Papers, vol II, p. 78.
27 See *Gazetteer of Bombay Presidency: Satara*, *op.cit.*, p. 189.
28 Ibid.
29 As per the then prevalent rate, a labourer had to serve five years to work off a loan of $10 (Rs. 100). See *Gazetteer of Bombay Presidency: Satara*, *op.cit.*, p. 189.
30 Ibid., p. 29.
31 Ibid.
32 See Rajasekaran (1996: 37).
33 According to the *Socio-Economic Review and District Statistical Abstract of Satara District*, 1991–92, there are seven sugar factories in the district.
34 Data on the adoption of HYV from 1980 to 1981 through 1993 to 1994 are available for a few crops like paddy, wheat, jowar, and *bajra*. It is seen that the area under HYV for these crops is increasing tremendously year by year, and in 1993–94, more than 90 per cent of cropped area was under HYV coverage. See *District-Wise Agricultural Statistical Information of Maharashtra* for the relevant years.
35 See *District-Wise Agricultural Statistical Information of Maharashtra* for the relevant years.
36 The progress of agricultural cooperative societies in Satara district is shown next:

Year	No. of societies	Membership (in 000)	Outstanding (in Rs.) in lakhs
1970–71	774	185	1444
1980–81	752	248	1236
1990–91	810	373	5421
2009–10	946	578	49949

Source: *Statistical Abstract of Maharashtra State*, various issues.

37 The analysis of the farm (harvest) prices of major crops from 1980 to 1981 through 1991 to 1992 reveals that the prices of paddy, wheat, jowar, *bajra*, and groundnut (per quintal) have been increasing sharply over the years. But the price of sugarcane was very low from 1981 to 1982 through 1989 to 1990 in comparison to its price in 1980–81. See *Season and Crop Reports* for the relevant years.
38 See *Maharashtra Agricultural Competitiveness Project (MACP) Marketing Strategy Supplement (MSS) District – Satara*, 2017, Government of Maharashtra, Pune: Department of Agriculture. Accessed from http://macp.gov.in/sites/default/files/user_doc/Satara%20MSS.pdf on 12 September 2017, p. 13.
39 The cultivation of strawberry and raspberry is a specialty of Mahabaleshwar in Satara District. Strawberries produced from this part have a unique brand recognition, having demand in national as well as international markets. Ibid, p. 13.

40 See *District-Wise Agricultural Statistical Information of Maharashtra* 1993–94 and *Input Survey Report* 2011–12
41 The changes in land-man ratio in Satara district, as reported in the Agricultural Census are given next:

Year	Rural population (in thousand)	Cultivated area (in 000 hectares)	Pressure on land
1961	1,272	761	0.60
1971	1,500	539	0.36
1981	1,773	590	0.33
1991	2,136	632	0.30
2001	2,411	718	0.30
2011	2,433	628	026

Source: (a) *Primary Census Abstract* (Satara District), various issues. (b) Report on Agricultural Census, various issues.

42 Protesting against the acquisition of 352.31 hectares of land by the MIDC, peasants of Satara under the banner of the *Bhumiputra Shetkari Sevabhavi Sanstha* filed a petition in the Bombay High Court. A film was also made on this issue to portray the discontentment of the displaced farmers. For more details, see *The Hindu*, 10 January 2010. Accessed from www.thehindu.com/todays-paper/tp-national/tp-otherstates/Farmers-oppose-acquisition-plan-in-Maharashtra-village/article15959470.ece on 14 September 2017.
43 See the *Gazetteer of India*, part ii, sec 3, sub-section (ii), 4 August 2010.
44 See *The Economic Times*, 31 August 2017. Accessed from http://economictimes.indiatimes.com/articleshow/46171672.cms?utm_source=contentofinterest&utm_medium=text&utm_campaign=cppst on 14 September 2017.
45 See *Agricultural Situation in India*, September 1994, p. 45.
46 Ibid.
47 The size class-wise percentage share of operational holdings among women in Satara District is presented next:

Size class	No		Area	
	2000–01	2010–11	2000–01	2010–11
Marginal	12.47	10.42	11.03	9.90
Small	9.89	9.08	10.13	9.05
Semi–medium	8.40	7.89	8.43	7.82
Medium	7.62	7.49	7.43	7.42
Large	4.42	5.77	2.95	5.09
All classes	11.43	10.03	9.23	8.74

Source: *Agricultural Census of India* (District Tables), relevant issues.

48 The analysis of the time series data on real wage rates of field labours (male, female, and child) in terms of constant prices on the basis of

consumer price index number for agricultural labourers in Maharashtra (Base: July 1960, June 1961 = 100) indicates that between 1961 to 1962 and 1980 to 1981, wage rates for all types of labourers were lower. The 1980–81 rates were lower than those of 1970–71. Though wage rates increased considerably after 1990–91, the increase was not proportionate to the growth in agricultural productivity and rising living standards. For details, see *Season and Crop Reports Maharashtra State*, various issues.

49 Looking at the agricultural situation of Satara district as described by various sources, we found that tremendous differences persist among its *talukas* as regards the pace of agricultural development. The irrigated belt is more advanced than the unirrigated belt. Given the situation, it would be proper to select villages representing both the advanced and the less advanced *talukas* of the district for comparative analysis. On the basis of talukwise general agricultural information available from the socio-economic survey, statistical abstracts, and information provided by the chief executive officer and other agriculture-related officials of the district, two *talukas* – Karad (relatively more advanced) and Katav (less advanced) – were selected, and one village from each *taluka* was chosen.

50 The households of the village were classified into landless agricultural labourers and marginal, small, medium, and large farmers on the basis of the capacity of holdings for the reproduction of an average household. The classification is as follows:

Categories	Holding size for irrigated village (ha)	Holding size for unirrigated village (ha)
Landless agricultural labourers		
Marginal farmers	Up to 0.5	Up to 1
Small farmers	0.5 to 1	1 to 2
Medium farmers	1 to 2	2 to 4
Large farmers	2 and above	4 and above

It was revealed from group interviews with some senior cultivators of the villages that on an average, a household of five to seven members (three to four adults and two to three children) belonging to the first category of holding size (marginal farmers) cannot possibly survive without adequate assistance from other sources of income. It would be difficult to maintain a similar family under the landowning group of the second category (small farmers) without additional income, but in exceptionally good harvesting years, they can manage with some wage incomes. A similar household belonging to the third category (medium farmers) can manage provided other factors of production and consumption remain unchanged. In contrast, a household of the same type but falling under the holding size of the last category (large farmers) can normally generate surplus and go in for expanded reproduction. It is also known from the same source that as regards productivity, one hectare of irrigated land is almost equal to two hectares of unirrigated land provided the rainfall and other climatic conditions are normal throughout the year. The value of irrigated land is estimated at double that of unirrigated land.

51 The caste groups and sub-groups of the studied villages are classified into three major groups on the basis of their position in the local hierarchy: upper castes, medium castes, and lower castes. While the Brahmins and the *Marathas*, who have historically been the most dominant and continue to be prominent in the upper echelons of both urban and rural society, belong to the upper-caste category, the scheduled castes form the lower castes, and the intermediary groups (other backward classes) come under the medium castes group.

52 Also quoted in Mencher (1978).

Chapter 4

Land and agriculture among SCs and STs[1]

Since agriculture continues to be the basic source of livelihood for a majority of population in India, land remains the pivotal property in terms of both income and employment. It is one of the most fundamental assets that serves as a gateway through which people gain access to many other assets and opportunities. This is more so for the SCs and STs that mostly reside in the countryside with multiple and intimate connections with land and agriculture. The importance of land has acquired greater significance for these communities in the context of recent global capitalist development. The entry of SCs and STs to the government sector, which was earlier facilitated through affirmative action, has lost its importance, as the ongoing process of neoliberal reforms downsized the employment opportunities available in this sector. On the other hand, the growing private sector also did not provide alternative sources of employment to these communities, as they lack the required level of skill and education set by the competitive market. The far-reaching livelihood diversification process that occurred in the countryside offers new opportunities only to those who own the necessary financial and social capital. In this context, therefore, effective control over land is more crucial now than before for the well-being of members of SCs and STs. It not only provides them a secured and stable livelihood but also enhances their bargaining power for higher wages and access to many other opportunities. In its 2006 *World Development Report*, the World Bank also underscored the importance of land access while focusing on the question of equity.

Ironically, SCs and STs are the most disadvantaged section with respect to land across the major states (Mohanty 2001b). In view of their poor landownership position, which is recognised as the principal reason for their perpetual poverty and vulnerability to social injustice and exploitation, land reforms and other supplementary measures were undertaken by the union and state governments after independence to protect and promote their landholding position, as well as to modernise agricultural practices. Although these measures suffered from several weaknesses, and the pace of their implementation was undeniably slow, they nonetheless constituted one of the key policy programmes till 1980s. However, since 1990s, a paradigm shift occurred with the adoption of the neoliberal economic policies. With their insistence on a prominent role for the market and a minimal role for the state, neoliberal policies restricted the scope of land reforms in redistributive justice and equality. The earlier reforms have been stalled in many states (Bhagat-Ganguly 2016: 3) and a discourse on new land reforms that can stimulate land-market and profit-oriented commercial agriculture has emerged. As noted by Saxena (2011: 184),

> The striking feature of new land reforms discourse is that it shifts focus from land redistribution to land management . . . what is advocated is the most efficient and productive use of land for accelerating agricultural growth irrespective of who the user is rather than its equitable distribution . . . the new land reform policy is riveted around parameters of production, efficiency, and manageability.

Many new acts like Special Economic Zones Act 2005; Scheduled Tribes and Other Traditional Forest Dwellers' (Recognition of Forest Rights) Act 2006; Right to Fair Compensation and Transparency in Land Acquisition, Rehabilitation, and Resettlement Act 2013, etc., have been enacted. Recently, a National Draft Policy on Land Reform (NDPLR) has been made. It states,

> The way land use has shaped up in recent years raises several issues and land reform has become much more relevant today than ever before. The increasing demand of land acquisition for housing, industrialization and several other purposes; diversion of agricultural land for non-agricultural purposes; the stagnating agricultural yields and the empirical evidence on the utility of small farms, all point towards the revisiting the issue seriously.[2]

In fact, land distribution and land reform have formed part of the (post) Washington Consensus over the past two-and-half decades. In countries across Latin America, Asia, and Africa, the international monetary and lending agencies have entered into a debate and attempted to influence the land reform policies. Agencies like the World Bank have advocated Market-Led Agrarian Reform (MLAR) through a market mechanism referred to as 'willing buyer-willing seller'. A new debate has emerged among most of the developing countries revolving around two dominant approaches to land reform: neoliberal and populist. While the neoliberal land reforms otherwise known as MLAR attempt to create or restore private rights to property for the purpose of improving the smooth functioning of rural markets (usually markets in land, credit, and agricultural inputs) and increasing efficiency and production through the security of title (Deininger and Feder 1998; Deininger 1999), the populist reforms try to restore the connection between the peasant communities and land, improving social justice by distributing resources to the poor (Borras 2003; Edelman 2003; Wright and Wolford 2003). India seems to follow a dual approach where it formulates policies that can meet the demands of the global capitalism on the one hand and shows concern on the other hand for the well-being of weaker sections like SCs and STs, which constitute about a quarter of its population.[3] While provisions are made for the protection of land rights of these weaker sections, resources are provided for government officials and private companies to legitimate land grabs, indicating a kind of policy ambivalence. To quote Levien (2015: 146), "The Indian state is caught between the land requirements of capitalist growth and the political compulsions of electoral democracy; whether, how and for whom this contradiction gets resolved constitutes India's 'new' land question".

Several studies indicated the adverse impact of neoliberal policies on SCs and STs (Teltumbde 1996; Narula and Macwan 2001; Saxena 2011; Still 2014; Vertika and Rodrigues 2015; Chakravarti 2016; Levien 2015). It is argued that without the forms of capital (financial, educational, social, and political) to succeed in a free-market economy, the population belonging to these categories would be marginalised all over again. However, all India statistics as reported by Agricultural Census reveal a marginal improvement in their landholding position. While the area operated by SCs increased from 7.90 per cent to 8.60 per cent, in the case of STs, it rose from 10.80 per cent to 11.42 per cent between 1990 to 1991 and 2010 to 2011. However, the estimates made by NSSO indicated contradictory trends. Its estimate on the share of land owned by SCs and STs in rural areas reveals that while the share for STs went up to 13 per cent in 2013 from 11 per cent in 2003, it remained almost constant for the SCs at 9 per cent during this period. On the other hand, the average area of land owned per household decreased from 0.767 hectare to 0.650 for STs and from 0.304 hectare to 0.272 for SCs between 2003 and 2013.[4] However, the NSSO estimate on distribution of cultivated land reveals that the proportion of all rural households without cultivated land increased from 30 per cent to 34 per cent among STs and from 53 per cent to 57 per cent in 2004–05 among SCs between 1993 to 1994 and 2004 to 2005.[5]

In view of the contradictory trends, an analysis of changes in landholding structure of SCs and STs and their agricultural practices is imperative to explore the extent to which the effects of earlier state-led measures have been continued, reversed, or modified under the impact of new market-led economic order and their impact on the agriculture of these disadvantaged communities. Though landholdings of SCs and STs, particularly the latter, was an important area of research and policy debate in the first two decades of planned development (for example, Fuchs 1972; Kulkarni 1974; Patel 1974; Murdia 1975; Rao 1987), it disappeared in the 1980s as the pace of land reforms slowed down. By the 1990s, one hardly finds a study on this area.[6] Recently, with the rising importance of land as a valuable resource not only for agriculture but also for industry, mining, and housing, and growing protest against land acquisition in the rural and tribal areas, landholdings of SCs and STs have received attention from some scholars (Mohanty 2001b; Saxena 2011; Chakravarti 2016). However, both the earlier as well as the recent studies were confined only to landownership and overlooked the agricultural practices of SCs and STs. Suffice it to say, though a lot has been written and said implicitly or explicitly about landholding structure of these communities, little is known about their agriculture either at the national or state level.

In this context, the present study attempts to look into the aspects of land and agriculture among SCs and STs with reference to the state of Maharashtra,[7] which has experienced rapid economic development following neoliberal reforms with a background of relatively better performance in land reforms[8] and organised movements by these communities.

The chapter is organised in three sections. While the following section provides an account of changes in landholding and agricultural practices of scheduled groups at the state level, the second section analyses these issues at the micro level. The last section presents the conclusion.

Land and agriculture of SCs and STs: a macro view

SCs and STs in Maharashtra constitute nearly 27 per cent of the total population and are mostly engaged in agricultural activities as labourers (Table 4.1).

Over 56 per cent of the population of SCs and 85 per cent of STs live in rural areas with agriculture as the principal source of their livelihood (as per the 2011 census). According to the recent NSSO

Table 4.1 Profile of SC and ST population in rural Maharashtra

Sl. no.	Particulars	SC	ST
1.	Total population (2011)	7,494,819	9,006,077
2.	Per cent to state population (2011)	12.18	14.63
3.	Sex ratio	959	984
4.	Number of households (2011)	1,707,321	1,861,647
5.	Per cent to total households (2011)	12.33	13.45
6.	Per cent of rural population (2011)	56.45	85.69
7.	Per cent of literates (population above 6 years)	75.79	63.20
8.	Per cent of main worker to total population (2011)	(43.30)	(43.12)
9.	Per cent of cultivators (2011)	16.17	29.53
10.	Per cent of agricultural labourers (2011)	66.52	60.16
11.	Per cent of agricultural work force (2011)	82.69	89.69
12.	Per cent of marginal workers to total population (2011)	6.78	9.66
13.	Per cent of non-workers to total population (2011)	49.92	47.22
14.	Per cent of households self-employed in agriculture (2011–12)	21.2	31.5
15.	Per cent of households self-employed in non-agriculture (2011–12)	8.9	3.3
16.	Per cent of households engaged as casual labour in agriculture (2011–12)	44.9	40.2
17	Per cent of landless households (2004–05)	3.0	5.7
18	Poverty ratio (2011–12)	23.80	61.60

Source: (1) *Primary Census Abstract* (Maharashtra State) 2011. (2) Rural Development Statistics 2011–12. (3) Employment and Unemployment Situation among Social Groups in India, NSS 68th Round (2011–12).

estimate, almost all the agricultural households across size classes belonging to these communities were engaged in cultivation and derived a major part of their income from agriculture.[9] According to the *Maharashtra State Development Report* (2007),[10] SC and ST households constituted nearly 40 per cent of the landless households.

Following the national guidelines, the erstwhile Bombay state and present Maharashtra launched a package of land reforms for promoting and protecting the land rights of scheduled communities. Keeping in view the widespread agrarian discontentment revolving around exploitative land and labour relations, some measures were also undertaken earlier by the colonial government to protect the land rights, particularly of the tribals. The important legislations were the Bombay Land Revenue Code 1879, Central Provinces and Berar Act 1939, Berar Alienated Villages Tenancy Law 1921, and Berar Forest Law 1886. However, as these measures were formally undertaken only to suppress the tribal agitations, the British did not show interest in their proper implementation because the provisions of these acts were affecting the interests of landlords, moneylenders, and other privileged groups that were the main agents of British administration.

The reform measures launched in the post-independence period were more elaborate and comprehensive. They were the Bombay Tenancy and Agricultural Lands Act 1948, Bombay Tenancy and Agricultural Lands Vidarbha Region and Kutch Area Act 1958, and Maharashtra Agricultural Land (Ceiling of Holding) Act 1961. The major objectives of these legislations were a) abolition of intermediaries, b) tenancy reforms, c) fixation of ceiling and distribution of surplus land, and d) consolidation of holdings. Despite inherent loopholes, as reported by many studies (Dandekar and Khudanpur 1957; Nanekar 1968; Shah and Sawant 1973; Rajasekaran 1998), many of these measures, particularly the imposition of ceiling laws and the distribution of ceiling surplus land, benefited the scheduled communities. It is reported that in the state as a whole, 729,600 acres of land were declared as ceiling surplus as of March 1999, and 76 per cent of it was distributed to the landless families. Of the distributed land, 29 per cent was given to the SC and 17 per cent to ST beneficiaries.[11] According to a more recent estimate, of the 634,000 acres of ceiling surplus land allotted to individual beneficiaries as of September 2012, 25 per cent and 16 per cent were distributed to the SCs and STs, respectively.[12] Though there has been relaxation of tenancy and ceiling laws in the present context of liberalisation, detailed information is not available on such changes.[13]

Besides the distribution of ceiling surplus land, a sizeable amount of land was obtained and distributed among SCs and STs through the Bhoodan Movement initiated by Vinoba Bhave. As of 31 March 1981, 88,500 acres of land were available for distribution through this scheme of which 57.17 per cent were allotted to SCs and 15.80 per cent to STs.[14] In addition, a package of specific measures was formulated to allot land to scheduled communities and to prevent it from being transferred to non-scheduled communities. The Maharashtra Land Revenue Code and Tenancy Laws (Amendment) Act 1974 provided for the restoration of agricultural land of a ST person if it was illegally transferred to a non-ST person before 6 July 1974. The other legislative measures like the Maharashtra Restoration of Land to the Scheduled Tribes Act 1974, Maharashtra Land Revenue Code 1966 (Section36), Bombay Tenancy and Agricultural Lands Act 1948, Bombay Tenancy and Assignment Lands (Vidarbha Region) Act 1958, and Hyderabad Tenancy and Agricultural Lands Act 1950 contained provisions for the restoration of agricultural land of STs lawfully transferred to non-STs by way of sale, mortgage, gift, etc. The government issued executive orders on 4 January 1961 for the disposal of land among SCs and STs as the first and second priorities. However, in areas other than the scheduled areas, they were accorded only the seventh priority. It was reported[15] that as of November 1999, 45,634 land alienation cases were filed in the court. Though a majority of them were rejected, and some of them were pending in the court 43.70 per cent of them were decided in favour of the tribals.

Changes in landholding of SCs and STs

The collective impact of these developments on the landholding position of SCs and STs can be assessed by analysing changes in their operational holdings. However, as data on the operational holdings of scheduled communities are available only from 1980 to 1981, it is difficult to assess the impact of land reform measures alone, because many of them were implemented in the late '60s and early '70s. Nevertheless, an analysis of the trend of operational holdings will be useful to interpret land distribution meaningfully.

The size class-wise information (Table 4.2) reveals that the operational holdings of both categories increased considerably in terms

Table 4.2 Operational holdings of scheduled groups by size class (per cent to total holdings of the state)

Size class	No.				Area			
	1980– 81	1990– 91	2000– 01	2010– 11	1980– 81	1990– 91	2000– 01	2010– 11
Scheduled castes								
Marginal	10.04	10.15	9.07	8.03	9.35	9.80	8.75	8.12
Small	7.48	8.28	7.85	7.68	7.26	8.14	7.72	7.59
Semi-medium	5.73	6.57	6.14	6.44	5.55	6.42	6.00	6.33
Medium	3.87	4.78	4.34	5.19	3.71	4.64	4.23	5.10
Large	2.50	3.45	3.12	4.01	2.40	3.26	2.99	3.72
All	6.81	8.02	7.78	7.51	4.48	6.03	6.17	6.59
Scheduled tribes								
Marginal	5.12	5.29	4.83	4.71	5.42	5.84	5.39	5.33
Small	6.48	7.35	7.16	7.20	6.36	7.38	7.10	7.25
Semi-medium	6.38	7.41	7.88	8.31	6.22	7.35	7.84	8.29
Medium	6.37	7.68	8.77	9.80	6.29	7.66	8.87	9.88
Large	5.81	7.59	9.52	10.02	5.56	7.18	8.67	9.20
All	6.02	6.69	6.41	6.30	6.08	7.32	7.63	7.88

Source: Agricultural Census Maharashtra State, relevant issues.

of area, as well as number, between 1980 to 1981 and 1990 to 1991.

The holdings of SCs, which were 6.81 per cent of the total holdings in the state in 1980–81, increased to 8.02 per cent in 1990–91, and the area also rose from 4.48 per cent to 6.03 per cent. Similarly, in the case of STs, the number and area of operational holdings went up from 6.02 per cent and 6.08 per cent in 1980–81 to 6.69 and 7.32 per cent, respectively, in 1990–91. Compared with the STs, the growth rate of operational holdings was relatively higher among SCs. Both categories witnessed an increase in the percentage of the large and medium operational holdings compared with that of the total large and medium holdings. Improvement of the operated area slowed down, and the number of holdings declined for both SCs and STs in the state as a whole after 1990–91. However, both medium and large holders experienced noticeable increases in operational holdings in terms of number and area.

Table 4.3 Operational holdings of scheduled groups by size class (per cent to their respective total holdings)

Size class	No.				Area			
	1980–81	1990–91	2000–01	2010–11	1980–81	1990–91	2000–01	2010–11
Scheduled castes								
Marginal	41.41	43.75	50.95	52.37	9.52	12.56	18.69	19.84
Small	24.68	29.73	29.99	30.26	17.73	25.68	31.90	33.44
Semi-medium	20.67	18.38	14.79	13.51	27.98	29.89	29.56	28.00
Medium	11.54	7.37	3.98	3.58	32.77	25.18	16.63	15.63
Large	1.70	0.78	0.29	0.26	12.00	6.69	3.22	3.10
All	100.00	100.00	100.00	100.00	100.00	100.00	100.00	100.00
Scheduled tribes								
Marginal	23.85	27.31	32.94	36.58	4.06	6.17	9.30	10.89
Small	24.17	31.62	33.19	33.80	11.42	19.19	23.72	26.70
Semi-medium	26.04	24.85	23.05	20.77	23.07	28.22	31.22	30.68
Medium	21.46	14.18	9.75	8.06	40.95	34.30	28.20	25.33
Large	4.48	2.04	1.07	0.79	20.50	12.12	7.56	6.40
All	100.00	100.00	100.00	100.00	100.00	100.00	100.00	100.00

Source: Agricultural Census Maharashtra State, relevant issues

A large majority of the landholders belonged to the small and marginal holding groups, and over the years, their number increased significantly among SCs and STs (Table 4.3).

The proportion of small and marginal holders together, which constituted around 76 per cent in 1980–81 among SCs, gradually went up to 83 per cent in 2010–11. Similarly, among STs, their number rose to 70 per cent in 2010–11 as opposed to 48 per cent in 1980–81. An increasing trend in the number and area of the marginal and small holders was accompanied by a corresponding decline in that of the large and medium holders. The number of large holders was negligible among both categories, though it was relatively higher among the STs. The increase in the number of small and marginal holders may be partly due to the division of large and medium holdings owing to family partitioning caused by population growth and to the allotment of small pieces of land through land reform measures. Marginalisation of holdings could also be attributed to the growing

Table 4.4 Percentage distribution of households without cultivated land

Year	All rural households		Rural labour households		Agricultural labour households	
	SC	ST	SC	ST	SC	ST
1983	–	–	70.01	67.44	69.88	66.86
1987–88	–	–	37.04	40.22	35.09	38.62
1993–94	59.70	52.02	65.54	68.84	64.52	67.79
2004–05	67.57	53.87	75.33	67.23	74.22	67.66

Source: (1) *Report on Employment and Unemployment of Rural Labour Households*, Rural Labour Enquiry, NSS-43rd Round (1987–88). (2) *Report on General Characteristics of Rural Labour Households*, Rural Labour Enquiry, NSS-50th Round (1993–94). (3) *Report on General Characteristics of Rural Labour Households*, Rural Labour Enquiry, NSS-61st Round (2004–05).

impoverishment of small and marginal peasants, as reported in other parts of the country (Jodhka 2000; Mohanty 2000b).

In addition, information on the distribution of rural households without cultivated lands (Table 4.4) as revealed by the successive Rural Labour Enquiry reports based on NSS data indicates that the proportion of all rural households without cultivated lands among SCs and STs increased significantly between 1993 to 1994 and 2004 to 2005.

A vast majority of them were without cultivated land. The proportion of landlessness was still higher among rural labour households and agricultural labour households. The magnitude of landlessness which declined in the late 1980s increased further after 1990s. Landlessness was more among the SCs than STs.

Nonetheless, Maharashtra was considered one of the lead states with regard to the improvement of landholding position of SCs and STs till 1990–91. The share of land of SCs and STs improved, and they acquired lands from other caste groups. The impact of urbanisation and the effects of economic development enabled a section of them to buy land in the land market (Omvedt 1994a; Mohanty 2001b). The landholding position of STs was better than that of SCs compared to their demographic strength. It is primarily because, unlike SCs, STs were not originally landless, and plenty of land was under their control. The improvement in the landholding position of SCs and STs was attributed to the impact of strong movements of the lower castes and tribes against the upper-caste-landed interests (Rajasekaran 1998; Mohanty 2001b). The *Satyasodhak* Movement led by Jyotiba Phule (which later extended its coverage to inequality in the distribution of land) and the subsequent activities of the Ambedkar and Republican Party of India challenged the dominance of the higher castes and helped the dalits to assert their land rights. Similarly, the well organised agitation of tribals, such as the *Shramik Sangathan* Movement of the *Bhils* in Dhule and the Gadchiroli districts against the alienation of tribal lands in the early '70s and the Warli Movement of *Kolis* in the Thane district put mounting pressure on the administration to promote and protect their land rights. However, after the 1990s, tribal movements in the state became weak and the dalit movements were largely confined to reservation issues.

Land use and cropping pattern

Along with the legislative and allied measures for the improvement of landholdings, the government of Maharashtra has made elaborate and continuous efforts to boost agricultural production by bringing more land under cultivation, improving the quality and extent of irrigation, inducing changes in the cropping pattern through expansion of cash crop cultivation, introduction of HYVs, adopting new agricultural machinery and implements, etc. Apart from the general programmes of agricultural development undertaken by both the central and state governments, the state government made special efforts through the Special Component Plan and Tribal Sub-plan for the modernisation of agriculture among the SCs and STs.[16]

Though irrigated land from categories increased over the years (Table 4.5), a large part of land belonging to these communities was unirrigated, and hence its use for agricultural purposes was dependent on rainfall.

Compared to STs, the area under irrigation was more with SCs at all points in time. This is mainly because SC lands are located in the plains adjacent to lands of the higher castes. While the ratio of irrigated land of SCs was only 6.48 per cent in 1980–81, it increased gradually to 17.04 per cent in 2010–11. But among the STs, it rose from 3.22 per cent to 12.32 per cent during this period. There are minimal differences in the ratio of irrigated land across the size classes of both of them. However, the growth is higher among the medium and large holders of both categories. Looking at the land-use pattern among SCs (Table 4.5), we find that the net cultivated area grew up to 96 per cent in 2010–11 as opposed to

Table 4.5 Area under different land uses (%)

Size classes	Scheduled castes					Scheduled tribes				
	Net cultivated area	Total uncultivated land	Land not available for cultivation	Net sown area	Irrigation ratio (% to net sown area)	Net cultivated area	Total uncultivated land	Land not available for cultivation	Net sown area	Irrigation ratio (% to net sown area)
1980–81										
Marginal	90.43	5.68	3.90	97.00	13.89	91.41	4.38	4.22	97.04	9.26
Small	90.31	6.39	3.30	96.20	8.88	91.43	5.22	3.34	96.01	5.82
Semi-medium	88.86	7.29	3.85	95.49	6.23	90.35	5.64	4.00	94.06	3.74
Medium	86.27	8.35	5.38	94.38	4.16	88.24	6.73	5.03	92.39	2.19
Large	80.93	12.34	6.72	91.00	2.92	82.44	9.39	8.17	87.81	1.63
All	87.46	7.93	4.60	94.91	6.48	88.03	6.76	5.21	92.53	3.22
1990–91										
Marginal	94.84	2.70	2.46	96.64	15.06	96.50	1.11	2.39	96.75	9.92
Small	94.87	2.93	2.20	95.87	9.85	95.93	1.55	2.52	96.23	6.79
Semi-medium	93.43	3.47	3.10	94.94	12.02	94.77	1.73	3.50	95.00	8.48
Medium	90.99	5.06	3.95	93.88	15.63	92.27	2.68	5.05	92.85	9.66
Large	89.78	6.51	3.71	86.54	19.62	89.58	3.55	6.87	85.14	9.83
All	93.12	3.84	3.04	94.60	13.17	93.61	2.21	4.18	93.48	8.77

(Continued)

Table 4.5 Continued

| | Scheduled castes | | | | | Scheduled tribes | | | | |
Size classes	Net cultivated area	Total uncultivated land	Land not available for cultivation	Net sown area	Irrigation ratio (% to net sown area)	Net cultivated area	Total uncultivated land	Land not available for cultivation	Net sown area	Net Irrigation ratio (% to net sown area)
2000–01										
Marginal	96.51	2.05	1.44	96.00	16.95	97.19	1.51	1.30	96.94	10.93
Small	96.24	2.47	1.29	95.86	11.06	96.56	2.06	1.38	96.26	7.36
Semi-medium	94.52	3.67	1.81	94.52	8.55	95.47	2.94	1.59	95.47	5.63
Medium	90.52	6.20	3.28	92.33	7.19	92.23	5.10	2.67	93.88	3.71
Large	83.52	10.32	6.16	91.45	4.34	88.51	7.01	4.48	89.75	3.53
All	94.42	3.62	1.96	94.80	10.67	94.45	3.52	2.04	94.96	5.91
2010–11										
Marginal	97.75	1.08	1.17	94.49	19.71	94.19	3.95	1.87	94.83	14.66
Small	97.52	1.09	1.39	94.49	16.05	93.52	3.63	2.84	95.53	13.34
Semi-medium	96.07	1.85	2.08	93.50	16.20	91.00	4.74	4.26	94.78	12.01
Medium	93.81	3.19	3.00	92.01	17.67	85.13	9.14	5.73	93.70	11.20
Large	88.36	5.20	6.44	90.48	14.74	75.79	16.54	7.68	90.28	8.66
All	96.29	1.76	1.95	93.72	17.04	89.56	6.23	4.21	94.49	12.32

Source: As in Table 4.2.

87 per cent in 1980–81. In the case of STs, it increased to 90 per cent in 2010–11 from 88 per cent in 1980–81. The proportion of net cultivated area was relatively higher among the marginal and small holders of both categories. More than 90 per cent of the cultivated area across the size classes came under the net sown area. However, in the case of the large holders, it is relatively less. The proportion of uncultivated land and the land not available for cultivation shows a declining trend, and the percentage of total area under these uses was negligible across the size classes. Landholders, especially those belonging to SCs, gradually expanded their cultivable area by making other categories of land cultivable. However, in 2010–11, the uncultivated land among the STs across size classes increased, particularly among the large holders. Probably the rising cost of cultivation and declining farm income compelled the tribal cultivators to keep their land uncultivated.

Data on cropping pattern available from the year 1990–91 broadly exhibits a similar pattern among SCs and STs (Table 4.6). The cultivators of these categories largely grow food grains. Of late there was a marginal shift towards high-valued crops like oilseeds, horticultural crops,[17] and vegetables. The area under food grains which covered more than 70 per cent of the gross cropped area in 1990–91 gradually declined to less than 60 per cent in 2010–11. The smaller is the land size, the greater is the area under food grains. Possibly due to a lack of irrigation facilities coupled with economic constraints and limited skill and knowledge, the farmers failed to expand their area under capital-intensive and profitable cash crops. However, the large and medium holders showed signs of change. Cropping intensity dramatically increased in 2010–11 across size classes of both categories. It was higher among SCs compared with STs. While the cropping intensity came to 140 among SCs, it was 133 in the case of STs in 2010–11. The low cropping intensity among the STs is due to relatively less irrigation facilities and rough terrain. In both categories, cropping intensity decreased when the size of holdings increased.

The NSS estimate of monthly income and expenditure of agricultural households for each size class indicated that the marginal and small holding households of both categories derived a major part of their income from wages, and net receipts from cultivation were far below their consumption expenditure (Table 4.7).

However, the average income of the large holders from cultivation was much higher for both SCs and STs. Though there were

Table 4.6 Area under crops (% to gross cropped area) and cropping intensity

SC/ST	Size classes	Total food grains	Oilseeds	Sugarcane	Fibre crops	Fruit crops	Vegetables	Cropping intensity
Schedule castes	*1990–91*							
	Marginal	77.03	5.25	2.62	10.92	0.49	0.37	115.01
	Small	70.68	5.43	1.46	18.61	0.55	0.44	110.89
	Semi-medium	68.46	6.14	2.21	17.73	1.81	0.32	110.14
	Medium	81.41	15.36	4.84	23.68	2.54	0.41	107.69
	Large	55.41	18.76	5.72	11.85	0.82	0.18	105.78
	All	72.01	8.32	2.77	17.77	1.34	0.37	110.12
	2000–01							
	Marginal	68.10	11.45	3.78	13.85	0.91	0.57	104.35
	Small	56.73	13.30	1.92	25.20	1.10	0.46	103.30
	Semi-med um	57.16	14.85	1.19	23.48	1.07	0.44	102.92
	Medium	54.42	16.73	0.91	23.87	1.56	0.39	102.82
	Large	62.89	12.75	0.58	20.15	2.07	0.26	101.87
	All	58.89	13.91	1.88	22.13	1.15	0.46	103.28
	2010–11							
	Marginal	59.67	14.97	3.99	12.91	2.30	1.21	138.78
	Small	55.27	17.15	2.46	18.57	1.64	1.18	140.72
	Semi-medium	55.84	17.27	2.19	18.05	2.06	1.36	140.62
	Medium	56.07	17.60	2.29	16.44	2.92	1.61	139.75
	Large	54.18	15.22	1.89	14.09	9.42	1.13	139.30
	All	56.41	16.76	2.65	16.85	2.30	1.30	140.12

Scheduled tribes							
1990–91							
Marginal	81.95	3.84	1.28	6.21	0.49	0.52	110.38
Small	75.22	4.85	1.05	12.79	0.37	0.38	107.41
Semi-medium	72.62	5.91	1.92	11.86	1.34	0.27	109.05
Medium	66.11	9.49	2.64	11.93	0.98	0.16	107.14
Large	65.56	11.13	2.74	7.68	0.42	0.12	105.82
All	70.86	7.29	2.03	11.26	0.87	0.26	107.82
2000–01							
Marginal	76.16	8.14	1.22	9.27	0.55	0.65	101.71
Small	66.39	9.87	0.85	17.31	0.62	0.60	101.62
Semi-medium	66.46	10.70	0.69	16.30	0.44	0.52	101.29
Medium	65.01	11.03	0.28	15.21	0.25	0.47	100.86
Large	63.36	11.27	0.38	12.37	0.34	0.36	100.38
All	66.79	10.37	0.65	15.30	0.44	0.53	101.23
2010–11							
Marginal	62.55	11.36	1.93	10.63	1.91	2.70	135.59
Small	57.58	13.24	1.62	16.92	1.55	2.09	136.60
Semi-medium	59.49	13.12	1.32	15.16	1.57	2.24	132.64
Medium	58.87	14.02	0.81	13.96	2.05	2.89	130.22
Large	56.24	12.39	0.77	13.68	4.68	2.33	128.37
All	58.99	13.12	1.33	14.79	1.87	2.41	133.30

Source: Same as in Table 4.2.

Table 4.7 Average monthly income and consumption expenditure (Rs.) of agricultural households for each size class of land possessed

Size class of land possessed (ha.)	Income from wages (Rs.)	Net receipt from cultivation (Rs.)	Net receipt from farming of animal (Rs.)	Net receipt from non-farm business (Rs.)	Total income (Rs.)	Total consumption expenditure (Rs.)	Net investment in productive asset (Rs.)
Schedule castes							
<0.01	3,706	345	898	65	5,014	4,170	-89
0.01–0.40	1,815	710	60	18	2,602	3,843	59
0.41–1.00	2,458	1,095	643	162	4,358	5,405	209
1.01–2.00	2,054	1,640	435	63	4,191	4,062	202
2.01–4.00	734	4,240	-361	180	4,793	5,529	581
4.01–10.00	5,291	6,207	3,920	1,490	16,908	8,503	1,480
10.00+	0	28,292	907	0	29,200	4,985	985
All size	2,388	1,774	641	192	4,995	4,859	271
Scheduled tribes							
<0.01	2,969	4	1847	116	4,936	3,481	259
0.01–0.40	3,768	819	216	410	5,212	4,083	703
0.41–1.00	2,495	797	101	65	3,458	4,701	208
1.01–2.00	4,010	1,741	163	1,086	7,000	5,073	940
2.01–4.00	1,713	3,698	407	97	5,915	4,901	204
4.01–10.00	388	7,693	-2	205	8,284	5,162	237
10.00+	2,679	12,812	-1,606	0	13,885	14,760	1,595
All size	2,991	1,926	186	467	5,570	4,881	536

Source: NSS Report No.569: Some Characteristics of Agricultural Households in India, 2012–13 (A-1253 and A-1254).

minimum variations in income from cultivation among the lower landowning households in both categories, the average income of an SC large holder was many times more than that of his ST counterpart. Despite their low-income, the tribal agricultural households relatively invested more in productive assets than SC households. Greater participation of SC households in rural non-farm activities could be one of the reasons for this. It is reported that while 9 per cent of the SC households in the state are self-employed in non-agriculture, only 3 per cent of the scheduled tribe households are engaged in such activities.[18] Though the non-farm activities in rural areas increased in Maharashtra in the post-reform period it is dominated by the non-SC/ST population, which comes to nearly 80 per cent (in 2009–10). Studies on Maharashtra report that less employment of SCs and STs is attributed to their lower educational attainment (Misra 2014). However, though the share of both SCs and STs increased over the years, the members of the SCs were ahead of their ST counterparts. While the share of SCs increased from 11 per cent in 1993–94 to 17 per cent in 2009–10, the proportion of STs employment during this period increased from 8 per cent to 13 per cent.

Analysis at the micro level provides more information on many aspects of the landholdings and agriculture among SCs and STs.

Analysis at the micro level

For examining empirically, the landholding and agricultural practices at the micro level, Latur and Dhule districts were selected, which are known for the highest concentration of SC and ST population, respectively, in the state.[19] One *taluka* from each district (Udgir from Latur and Akrani from Dhule) was chosen for drawing the sample households on the basis of the same procedure that was followed for the selection of districts. A sample of 100 households (50 from each *taluka*) was drawn from various categories of landowning households (marginal, small, semi-medium, medium, and large as defined by the Agricultural Census). However, there were no SC landholders belonging to the large farmers category. As there were only five landholders belonging to the medium group, all of them were included in the sample. Similarly, there were only three ST landholders from the large holdings group.

The landholding pattern and agricultural practices of the sampled SC and ST households are illustrated in Tables 4.8 and 4.9, respectively. It is found that the ST households acquired land mostly through inheritance. The allotment of land through land reforms was negligible. Only 3 per cent of their owned area could be attributed to these measures. Among the size classes, only small and semi-medium holders received around 10 per cent of their owned land as result of land reforms. Moreover, none of the ST landholders was able to buy land in the open market. Large holders of this category acquired a considerable portion of land through encroachment. It was reported that the encroachment of forestland was not uncommon in the tribal areas. Those who were relatively better-off encroached land and made it cultivable by entertaining the *talathis* and local revenue officials. As the tribals are mostly illiterate and unaware of the land laws, they accept the claims of the revenue officials. Sometimes, these officials demand regular payments either in cash or kind.

On the other hand, land reforms have contributed considerably to the improvement of the landholding position of SCs. More than 20 per cent of their land was provided through the distribution of either ceiling surplus or wasteland. Only 28 per cent of their land was inherited. The proportion of acquired land through inheritance was more among the marginal holders, mainly due to family partitioning. Landholders of all size classes, belonging to this category acquired land largely through purchase. They invariably bought land from the higher castes. More than 86 per cent of their purchased land belonged to other castes. Land reforms and development planning had a positive impact on the SCs, who earlier were mostly landless.

In addition, due to restrictions on land transfers from the SCs and STs to non-scheduled groups, the incidence of land sales was not recorded in the studied villages. However, in the tribal areas, though transfer of land through sale, mortgage, or gift from the STs is prohibited, many of the tribal landholders mortgaged their land with the non-tribal moneylenders with oral or informal agreements, with senior members of the village as witnesses who are often consulted in important matters. The moneylender leased out the land to some cultivators of the same village or cultivated it himself using the available cheap *adivasi* labour force. Land remained with the moneylender till the loan was repaid, and he cultivated it in lieu of interest payment. Over 10 per cent of the land of the sample holders was mortgaged with the moneylenders, and most of them were small and marginal holders. However, such informal land transfers were conspicuous by their absence among the SCs.

Analysis of the changes in landholding position of the sample households over the last ten years (1991–2001) indicates that the SCs have improved their position. The average holdings of these farmers increased across the size classes. While it increased by over 93 per cent for the marginal farmers, for small and medium farmers, it was up to 43 and 50 per cent respectively. Large farmers reported an increase of more than 18 per cent during this period. While excepting some improvement here and there, the overall landholding position of the ST farmers declined further. However, the reliability of these data is limited to the memory of the respondents. Lands of both the SCs and STs lacked irrigation facilities. Only 2 per cent of land of SCs and 5 per cent of land of STs was irrigated. Therefore, agricultural land use practices of the landholders of these communities were mostly dependent on the monsoon.

Among the SC holders, the area available for cultivation was more among the lower landowning cultivators. The higher the landholding size, the lower the proportion of cultivated area. While nearly 90–95 per cent land of the marginal, small, and semi-medium holders of this category came under the net cultivated area, among the medium holders, it was 79 per cent. The percentage of area not available for cultivation was negligible across the size classes. Almost the entire cultivated area of all the size classes among the SCs was sown, and the area under current fallow was insignificant. The net sown area was slightly less in the case of the medium holders. However, the cropping intensity was low for all the size classes and the variation was minimal. While the cropping intensity of the marginal holders was 105, for the small, semi-medium, and medium farmers it came to nearly 102. The low cropping intensity of these farmers was mostly due to the lack of irrigation facilities.

As regards the ST households, 86 per cent of their total land came under the plough. The marginal farmers of this category put 98 per cent of their land to agricultural use. Though the semi-medium and medium holders also had a relatively higher proportion of cultivated area, small and large holders had a considerable portion of area which was not cultivated or not available for cultivation. Land not available for cultivation for the small and large holders came to 16.57 and 12.39 per cent of their total land, respectively. This was because these lands were barren, uncultivable, and mostly covered by forests. The proportion of net sown area of these farmers was comparatively less than that of their SC counterparts. Though the percentage was higher in the case of the marginal and semi-medium holders, it was less among the large holders because 26 per cent of their cultivable land was fallow. The current fallow land of the small holders was also not insignificant (9 per cent). The primary reason for a relatively higher proportion of fallow land among the tribals was that they cultivated certain types of land every alternate year. Like the SC farmers, the cropping intensity among them was also low. However, a greater cropping intensity was noticed among the marginal farmers, which is attributable to a higher proportion of their land under irrigation.

The cropping pattern of landholders of SCs and STs indicates that they devoted their cropped area almost entirely to food crops. Their agriculture was essentially at the subsistence level, in which family consumption was the guiding factor. Differences across size classes were minimal. Lack of irrigation facilities and working funds compelled them not to go in for profitable capital-intensive crops, such as sugarcane or cotton. Even the higher landowning farmers were reluctant to cultivate these crops. Farmers predominantly cultivated *kharif* crops. As a result, agricultural activity was confined to a part of the year, and the remaining period did not provide enough farm work. Another important characteristic feature was that the cropping pattern of farmers of both categories showed diversity. It was the outcome of subsistence farming combined with hazardous agriculture. The farmers grew crops which were not viable in terms of yields and consequently of farm incomes.

The area under high yielding varieties of the farmers of both categories was largely confined to *jowar*, wheat, and tur. The area under HYVs was more among the SC farmers, which came to nearly 60 per cent. Farmers of both categories were poorly equipped to carry out agricultural production effectively. Modern agricultural machinery, such as tractors and pumpsets, were not seen even among the higher landowning groups. The ownership of minimum agricultural instruments, such as ploughs, was also inadequate across the size classes. Farmers of both categories found it difficult to apply high doses of agricultural inputs due to their poor economic position. Their application of chemical fertilisers was also low. ST cultivators mostly follow their traditional methods of cultivation based on the application of indigenously prepared manures and pesticides. Applications of inputs by SC farmers were higher. SC farmers applied pesticides for crops like cotton, tur, etc., which were largely under the HYVs and susceptible to pest attacks. On the other hand, the pesticide use was negligible among ST cultivators. Greater adoption of modern inputs by the SC farmers was largely due to the influence of the improved farming methods of the higher castes in the concerned locality. The ST farmers were disadvantaged in this respect due to their relative geographical isolation.

Lack of working capital was one of the major reasons to prevent these farmers from adopting new technologies and associating inputs extensively, which would have resulted in higher yields. The credit facilities provided by the formal credit institutions did not encourage the ST farmers to go to them for financing improved methods of cultivation and greater doses of agricultural inputs. However, formal credit agencies benefitted the SC cultivators. They took loans from them irrespective of their landholding sizes. A considerable proportion of their cost of cultivation was supported by formal credit facilities. Among the size classes, the marginal farmers appeared to be the major beneficiaries. On the contrary, although the ST farmers took loans from the formal credit institutions, their average as well as per acre quantum of loans were less. The per acre and average amounts of loans of marginal and small holders, which were relatively higher, covered only 38 per cent of their cost of cultivation. With exception of the large holders, farmers of remaining size classes were more or less indebted to moneylenders. Large farmers kept themselves free of debt because they cultivated mostly food grains which require low investment. Given the uncertainty involved in agriculture, they also did not like to invest more by borrowing from both formal and informal sources. Greater involvement of these farmers in the newly created rural non-farm activities could be one of the reasons for this. Compared to STs, the members of the SCs draw significant proportion of their income from non-farm activities like household and non-household manufacturing, handicrafts, processing, repairs, construction, etc.

Reasons like the lack of irrigation, working capital, appropriate agricultural machinery and instruments, etc., lead to the under-utilisation of agricultural land, which ultimately has a negative impact on agricultural income and productivity. While the overall per acre productivity of SC farmers was 1.80, it was 1.67 among ST farmers. Medium farmers of both categories appeared to be more productive than the farmers of other size classes. Overall, the productivity and per acre income are higher among the SC farmers compared with that of the ST farmers.

Table 4.8 Details of landholding and agricultural practices among scheduled castes

Sl. no.	Particulars	Marginal N = 20	Small N = 15	Semi-medium N = 10	Medium N = 5	All N = 50
1.	Average size of family	5.95	6.80	6.60	6.80	6.42
2.	Per cent of illiterate members	36.97	37.25	28.79	32.35	34.58
3.	Average size of family workers	3.30	3.33	3.50	3.60	3.38
4.	Average holding (in acres)	1.78	3.78	5.79	13.11	4.32
5.	Per cent of increase/decrease in average holding in the last ten years	+93.48	+43.18	+49.61	+18.42	+42.10
6.	Per cent of land inherited	38.25	31.37	15.98	30.51	28.12
7.	Per cent of land purchased	50.24	45.28	57.90	47.60	50.19
8.	Per cent of land received through land reform measures	11.51	17.62	26.12	21.89	20.19
9.	Per cent of land purchased from non-SC/ST to total land purchased	100.00	76.33	73.15	100.00	86.08
10.	Per cent of irrigated area to net sown area	2.98	3.85	0.00	0.00	1.61
11.	Per cent of net cultivated land to total land	94.72	93.13	90.00	79.33	88.36
12.	Per cent of uncultivated land to total land	2.47	3.24	8.95	19.45	9.57
13.	Per cent of land not available for cultivation to total land	2.81	3.63	1.05	1.22	2.07
14.	Per cent of net sown area to net cultivated area	99.41	98.41	98.08	96.10	97.87
15.	Per cent of current fallows to net cultivated area	0.59	1.59	1.92	3.90	2.13
16.	Cropping intensity	104.56	102.12	102.94	102.06	102.76

No.		Col1	Col2	Col3	Col4	Col5
17.	Area under food grains to gross cropped area	97.15	90.20	93.92	90.19	92.49
18.	Area under cash crops to gross cropped area	2.85	9.41	6.08	9.80	7.40
19.	Per acre use of chemical fertilisers (kg)	16.56	28.86	22.36	22.55	23.15
20.	Per acre use of chemical pesticides (lit.)	0.17	0.35	0.23	0.27	0.26
21.	Per acre cost of cultivation (Rs.)	2515.94	2224.63	1643.80	1530.94	1934.14
22.	Per acre loan (Rs.) from formal sources	2594.07	1054.61	760.46	686.27	1157.58
23.	Per acre loan from formal sources as the percentage of cost of cultivation	103.11	47.41	46.26	44.83	59.85
24.	Per acre loan (Rs.) from informal sources	0.00	18.83	0.00	0.00	5.21
25.	Productivity	1.78	1.64	1.99	1.87	1.80
26.	Per acre net income (Rs.)	1952.42	1425.08	1622.36	1335.73	1551.89

Source: Field Survey.

Table 4.9 Details of landholding and agricultural practices among scheduled tribes

Sl. no.	Particulars	Marginal N = 12	Small N = 12	Semi-medium N = 12	Medium N = 11	Large N = 3	All N = 50
1.	Average size of family	8.25	7.42	7.42	6.82	7.67	7.50
2.	Per cent of illiterate members	52.53	55.06	58.43	57.33	56.52	55.73
3.	Average size of family workers	2.33	2.58	2.33	2.45	5.33	2.60
4.	Average holding (in acres)	1.76	3.77	7.80	13.24	26.91	7.73
5.	Per cent of increase/decrease in average holding in the last ten years	−1.12	+15.64	+5.83	−7.86	−11.94	−3.01
6.	Per cent of land inherited	100.00	90.72	82.26	100.00	79.94	90.42
7.	Per cent of land purchased	0.00	0.00	0.00	0.00	0.00	0.00
8.	Per cent of land received through land reform measures	0.00	9.28	10.85	0.00	0.00	3.71
9.	Per cent of irrigated area to net sown area	21.50	0.00	6.13	4.79	0.00	4.91
10.	Per cent of net cultivated land to total land	97.87	76.23	88.19	90.09	77.04	85.70
11.	Per cent of uncultivated land to total land	2.13	7.20	8.68	5.75	10.58	7.44
12.	Per cent of land not available for cultivation to total land	0.00	16.57	3.13	4.15	12.39	6.85
13.	Per cent of net sown area to net cultivated area	100.00	91.30	98.79	95.43	73.95	92.09

14.	Per cent of current fallows to net cultivated area	0.00	8.70	1.21	4.57	7.91
15.	Cropping intensity	118.74	106.98	101.23	102.56	103.96
16.	Area under food grains to gross cropped area	98.78	93.47	88.12	89.57	91.26
17.	Area under cash crops to gross cropped area	1.22	6.53	11.64	6.39	7.04
18.	Per acre use of chemical fertilisers (kg.)	14.24	31.16	21.84	11.70	19.26
19.	Per acre use of pesticides (lit.)	0.01	0.00	0.01	0.00	0.003
20.	Per acre cost of cultivation (Rs.)	1,688.38	1,366.29	1,145.15	930.16	1,091.85
21.	Per acre loan (Rs.) from formal sources	655.00	510.26	206.67	238.32	259.16
22.	Per acre loan from formal sources as the percentage of cost of cultivation	38.79	37.35	18.05	25.62	23.74
23.	Per acre loan (Rs.) from informal sources	537.02	192.88	41.21	38.94	88.65
24.	Productivity	1.81	1.82	2.15	1.29	1.67
25.	Per acre net income (Rs.)	1,373.05	1,118.87	1,314.36	271.56	728.65

Source: Same as in Table 4.8.

Conclusion

Analysis at both the micro and macro levels indicates that though the SC and ST farmers continued to be disadvantaged with regard to land, there was improvement, especially among the SCs, as a result of land reforms. Improvement in the landholding position of these communities was also attributed to the effects of the lower castes and tribal movements in the state. The reason for the relatively higher improvement of SCs is not far to seek. Unlike STs, there is sustained mass mobilisation and rising assertion among the members of SCs throughout the state by organisations like the Republican Party of India. Moreover, STs did not have influential leaders (like Sushil Kumar Sindhe, who was the chief minister of the state and the union minister) to push up the tribal interests. Not only at the state level but also the lobby for SCs' welfare at the national level is stronger than that of STs (Xaxa 2001).

As regards the agricultural practices, the farmers belonging SCs as well as STs followed almost a similar pattern. Though the members of SCs were in a relatively advantageous position, the agriculture of the farmers of both categories of farmers was mostly at the subsistence level due to the lack of irrigation, credit facilities, and poor resource ownership. The cultivation of high-valued crops associated with application of high doses of modern inputs was beyond their reach. The effects of neoliberal policies slowed down the improvement of landholding position of SCs and STs. Neither the state-led nor the market-led modernising measures brought any substantial improvement in agricultural practices particularly among the STs. Therefore, a mere attempt to enhance the landholding position of these communities is naïve without a concomitant improvement in their resources, skills, and expertise for the effective use of their lands.

Notes

1 This is the revised part of the larger study titled "Land Holding and Use Pattern among Scheduled Castes and Scheduled Tribes in Maharashtra" undertaken by the author at the Gokhale Institute of Politics and Economics, Pune, and submitted to the Ministry of Agriculture, Government of India, New Delhi in 2001.
2 See *Draft National Land Reforms Policy*, Department of Land Resources, Ministry of Rural Development, Government of India. Accessed from http://rural.nic.in/sites/downloads/latest/Draft_National_Land_Reforms_Policy_July_2013.pdf on 13 October 2016, p. 2

3 The NDPLR while emphasising new land reforms in consonance with changing economic order also echoes the commitment to land rights of SCs and STs when it states, "The States shall review existing law and policies pertaining to alienation/transfer of land belonging to Scheduled Castes and Scheduled Tribes and take necessary steps for removing the constraints, if any, in protecting and restoring the allotted and other lands belonging to SCs and STs". See *Draft National Land Reforms Policy*, 18 July 2013, p. 8

4 For details, see (1) *Household Ownership Holdings in India, 2003*, NSS 59th Round, (January–December 2003), Report No. 491(59/18.1/4). (2) *Household Ownership and Operational Holdings in India, 2013*, NSS 70th Round, (January–December 2013), Report No. 571(70/18.1/1).

5 See (a) *Report on General Characteristics of Rural Labour Households, Rural Labour Enquiry*, NSS-50th Round (1993–94). (b) *Report on General Characteristics of Rural Labour Households, Rural Labour Enquiry*, NSS-61st Round (2004–05).

6 However, some studies on SCs and STs briefly touched on this issue while analyzing their socio-economic conditions. For example, see Omvedt (1994a), Mohanty (1997), and Chakrabarty (1998).

7 Though there are a few studies on SCs and STs in Maharashtra, they have only briefly or tangentially attended to the issues relating to land-holding and agricultural practices (Gare 1973; Nadkarni 1988). Studies having an analysis at both the macro and micro levels are rare.

8 A statewide comparative analysis of land reform measures made by Mohanty (2001b) shows that Maharashtra has better provisions for the protection and promotion of land rights of SCs and STs than many other states like Rajasthan, Odisha, Bihar, U.P, Punjab, Gujarat, and Andhra Pradesh.

9 Percentage distribution of SC and ST agricultural households engaged in agriculture and deriving income mostly from cultivation during the last year for each size class of land possessed in Maharashtra is given next:

Size class of land posses ses	Agricultural households engaged in cultivation		Agricultural households deriving income mostly from cultivation	
	SC	ST	SC	ST
< 0.01	36.5	1.7	0	0
0.01–0.40	96.4	96.0	74.8	53.7
0.41–1.00	98.1	100.0	38.2	62.9
1.01–2.00	100.0	100.0	81.8	79.5
2.01–4.00	99.7	100.0	98.2	98.7
4.01–10.00	100.0	100.0	89.2	100.0
10.00 +	100.0	100.0	100.0	100.0
All sizes	92.8	98.8	61.7	74.6

Source: *NSS Report No.569: Some Characteristics of Agricultural Households in India*, 2012–13 (A-120).

10 See *Maharashtra Development Report*, 2007 (Planning Commission). New Delhi: Academic Foundation, p. 209.

11 See *Rural Development Statistics 1999*, National Institute of Rural Development, Hyderabad.

12 Lok Sabha unstarred question no. 4009 dated 21 March 2013. Accessed from www.maharashtrastat.com/table/agriculture/2/agriculturalland-holdings/153/812097/data.aspx on 6 October 2016.

13 Maharashtra Agricultural Lands (Ceiling on Holdings) (Amendment) Ordinance was passed in 1993) to enable individuals, firms, companies, etc., to hold land on lease beyond the prescribed ceiling limit fixed earlier by the Maharashtra Agricultural Lands (Ceiling and Holdings) Act 1961 amended in 1975. However, the ordinance was not approved by the central government.

14 *Repot of the Commissioner for Scheduled Castes and Scheduled Tribes*, April 1980–March, 1981.

15 *Annual Report 1999–2000*, Ministry of Rural Development, Government of India, New Delhi.

16 The government of India introduced the Tribal Sub-Plan (TSP) in 1974–75 during the Fifth Five-Year Plan, and later in 1979–80 in the Sixth Five-Year Plan, the *Special Component Plan*, now called *Scheduled Caste Sub-Plan*, was made. This umbrella strategy was intended to direct plan resources across central, as well as state, departments, at least in proportion to the SC/ST population at the national level. These two approaches included special budgetary allocations for the development of agriculture of SCs and STs. In fact, Maharashtra's initiative in adopting the aforementioned approaches was considered a model for other states, and it was popularly known as the 'Maharashtra Model'.

17 The increase in area under horticultural crops was mainly due to the influence of the National Horticulture Mission, which was launched in 2005–06.

18 See *Employment and Unemployment Situation among Social groups in India*, 2015, NSSO Report 563.Government of India, pp. 66–7.

19 While more than 20 per cent of rural population in Latur belonged to SCs, STs constituted over 41 per cent of the rural population of Dhule district, as per the 2011 census. In Maharashtra as a whole, the SCs and STs form 12 per cent and 15 per cent of the total rural population.

Chapter 5

Neoliberal reforms, agrarian change, and rural women[1]

Neoliberalism is commonly understood as a set of policy prescriptions associated with the 'Washington Consensus', which emphasises 'market fundamentalism' (Cornwall *et al.* 2008). With robust optimism, the neoliberal economic agenda projects exemplary confidence in the efficiency of market mechanisms retrenching the state and follows the agenda that pursues fiscal restraint, open trade and capital accounts, and privatisation stressing sound finance, irrespective of social costs (Elson 2002: 3; Razavi 2002: 1). It strongly supports 'free enterprise' globalisation and a policy of 'grow first, redistribute after'. The policy shift called for substantial reforms in the agricultural sector, which included elimination of export subsidies, reduction in trade-distorting support, and other domestic support. Markets and the demand for agricultural commodities changed rapidly, especially for higher-value products and the advances in agricultural productive technology that accompanied the changes in the sector created an array of new choices for producers, altering what is produced, where it is produced, and how it is produced. Put succinctly, agriculture became technologically more sophisticated, commercially oriented, and globally integrated.

Although the rapid changes occurring in the agricultural sector created new sources of opportunities for livelihood, market participation and access to productive assets for both women and men, they also pose significant challenges and uncertainties for women in particular. Neoliberalism induced greater innovations and efficiencies in many cases but, at the same time, it favoured the producers who had more resources, information, and capacity to cope with increasingly stringent market demands (Joekes 1999; Young and Hoppe 2003). Poor and small producers, often women, were likely to be excluded because they were not able to compete with larger producers in terms of costs and prices. Ultimately, the emerging changes, as observed by the World Bank,[2] created conditions for a gender imbalance by increasing the vulnerability of individuals, especially poor women, who were historically placed in a disadvantageous position in relation to access and use of productive resources, and to crucial services and opportunities because of persistent cultural, social, and political biases. Additionally, neoliberal reforms undermined existing national policies, such as land reforms, right to food and livelihood, adopted to protect the interests of women and other weaker sections. To put it in the words of Razavi (2002: 2),

> Policies that were once considered to be crucial to rural development, such as redistributive land reforms, have simply disappeared from the development agenda, while the physical and social infrastructure in rural areas undergirding the rural economy and the welfare of its inhabitants has been left to oblivion. Rather than shifting the terms of trade toward agriculture, neoliberal policies have been, in effect, shifting the burdens of adjustment toward small farmers, and especially the women in rural households who often bear the double burden of farm (and off-farm) work and the care of human beings.

Even mainstream development agencies such as United States Agency for International Development (USAID) and World Bank which championed neoliberal policies acknowledged negative consequences for women (Rakowski 2000; Beneria 2003) and now many of these institutions laud the importance of giving women more roles in economic development.[3] Recently, the Food and Agricultural Organization (FAO) highlighted widespread gender gaps,

> Women in agriculture and rural areas have one thing in common across regions: they have less access than men to productive resources and opportunities. The gender gap is found for many assets, inputs and services – land, livestock, labour, education, extension and financial services, and technology – and it imposes costs on the agriculture sector, the broader economy and society as well as on women themselves.[4]

The *World Development Report* (2008) indicated that women represent a larger proportion of labourers than men in the agricultural sector, and their proportion in agricultural wages in the labour market increased because of globalisation. The *World Development Report* (2012: 201) on "Gender Equality and Development" noted,

Gender differences in productivity and earnings are systematic and persistent. Whether in agriculture or off the farm, among those self-employed or in wage employment, women exhibit lower average productivity and earn lower wages than men. These differences have been documented in both developed and developing countries, and although they have declined over-time . . . they remain significant.

Available literature at the global level on the impact of neoliberalism, however, generated a debate on feminisation of labour markets and gender gaps in work participation. One group of scholars argues that existing gender gaps may increase or shrink under the impact of neoliberal reforms because all sectors of agriculture were not equally affected by trade liberalisation on the one hand, and women and men were not evenly represented among various agricultural sectors, such as livestock, farming, or export (Garcia *et al.* 2006: 1). According to this line of argument, neoliberal policies have not been uniformly bad for women, and for some, women policies, such as market liberalisation, indeed contributed to their development. In some of the recent studies, it is demonstrated that rural women's work became more visible than ever, and they have moved from being unpaid family workers in agriculture to remunerated workers in both the agriculture and non-agricultural sectors (Lastarria-Cornhiel 2006; Deere 2009).

On the contrary, many studies indicate the markedly negative effects of neoliberal policies on women (Sparr 1995; Batliwala and Dhanraj 2004; Moghadam 2005). It is argued that neoliberal policies gave rise to the feminisation of labour accompanied by deterioration of working conditions – casualisation, low wages, and earnings, etc. Several case studies documented the way neoliberal policies forced poor women in poor countries to accept whatever paid work they could get, despite deteriorating pay and conditions in order to feed and clothe their families in the context of rising prices and falling male employment. Feminists also highlighted the systematic institutional bias against women in economic policy (Jaggar 2009; Young 2011). A new debate has emerged on land rights of women in the context of recent economic changes. This debate took place mainly in the pages of the *Journal of Agrarian Change*. While in the context of South Asia it is argued that the gender gap in the ownership and control of property furthers it in economic well-being, social status, and empowerment (Agarwal 1994, 2003), in sub-Sahara Africa, inadequate access to land is rarely a major factor constraining women's agricultural output and income (Whitehead and Lockwood 1999). In some areas of Africa where land is not abundant, the main constraint in farming and on income remains the lack of capital and labour (Whitehead and Tsikata 2001). Land as the single-most critical contributor to the gender gap has been questioned by others also (for example, Jackson 2003). It is argued that the interface between gender and land is context specific and cannot be adequately addressed through a uniform global prescription (Razavi 2003).

However, much of the discourse at the global level may not be relevant in India due to its typical socio-economic and historical features, as well as large regional variations. Moreover, it is argued that India's approach to liberalisation has been commended for its gradual implementation of reform measures and sympathy for poor and rural workers (Garikipati and Pfaffenzeller 2012; Ahluwalia 2002).

Agrarian change and rural women: the Indian debate

The impact of agrarian change on women in India emerged as an area of interest only in the 1980s[5] in response to the development of feminist science, the women's movement, and the shift in development policy from focusing on men as target groups for agricultural modernisation (based on the traditional 'male' image of agriculture) to making women an integral part of the development process. Studies of rural women in India in the context of agrarian change were missing both in early village studies and subsequent agrarian studies of the 1960s and 1970s. Though studies on agrarian change following the green revolution generated exciting debates on 'mode of production', 'social inequality', and 'agricultural growth and poverty' in the 1960s and 1970s, they were crying out for gender analysis. Nonetheless, there were a small number of studies which paid little attention to women (for example, Chakravorty 1975; Gulati 1975). However, they covered only a few aspects of agrarian change based on limited data. It is only during the 1980s that studies of women in agriculture generated scholarly debates,[6] mainly relating to their employment, wages, and earnings (Chattopadhyay 1982; Mencher and Saradamoni 1982; Agarwal 1984; Nayyar 1987; Acharya and Panwalkar 1988).[7]

Early literature on the gendered impact of agrarian change in India tends to subscribe to three diverse and conflicting lines of arguments by different groups of scholars: political economists, reformists, and gender analysts. The political economists contended that the agrarian change induced by capitalist development in the process of modernising agricultural practices impoverished the small farmers, landless labourers, and artisans, as a result of which women were compelled to take over agricultural wage labour to supplement their family income (Byres 1981; Mies 1986). To them, women were forced into the workforce at a faster rate than men, and their increasing participation did not indicate upward mobility but rather it deteriorated their working conditions and forced them to accept low wages. On the other hand, the reformists, who were mainly from large development research institutions like the World Bank and International Food Policy Research Institute, put forward the argument that the effects of agrarian change contributed to agricultural growth which eventually trickled down to male and female labourers through the creation of farm and off-farm employment opportunities. They are of the opinion that women's employment opportunities in agriculture grew more rapidly than men's. The rise in demand for non-agricultural goods and services led to male migration into non-farm work, thereby freeing up agricultural jobs for women. Putting precisely, according to the reformists, introduction of the green revolution package increased the demand for agricultural tasks that were primarily women's work, improved the working conditions, and reduced the male-female wage differentials (Mellor 1976; Walker and Ryan 1990; Hazell and Ramasamy 1991). Contrary to this, the gender analysts argue that modernisation of agriculture had differential effects on rural men and women. It widened the gender disparity in terms of workforce participation, wages, and access to productive assets (Agarwal 1984; Duvvury 1989).

Observations made by the three groups of scholars were based on evidences from the 1970s and 1980s, and after the 1990s, attempts were rarely made to substantiate this debate in the context of rapid economic development following neoliberal reforms that profoundly transformed the agricultural production systems with differential inclusionary and exclusionary implications for men and women.[8]

Nevertheless, few attempts have recently been made which directly or indirectly relate to the aforementioned three lines of arguments. While the studies of political economy orientation argued that there was an increasing trend of the feminisation of agriculture leading to a gender-based class division between men as cultivators and women as wage-labourers (Chaudhry 1994; da Corta and Venkateshwarlu 1999; Arun 2012), the studies of reformist orientation held that though female labour continued to be much less rewarding than male labour, the recognition of female labour improved and patriarchal relations weakened as the non-agricultural economy grew over the decades (Heyer 2016). The gender analysts demonstrated that given the pressure on profit margins as a result of rising input costs, women as cultivators as well as labourers were compelled to restrict bargaining for higher wages, which contributed to a widening gender wage gap. The effects of liberalisation resulted in further marginalising rural women into badly paid and socially degrading wage work (Vepa 2005; Garikipati 2009; Rao 2011; Usami 2011; Garikipati and Pfaffenzeller 2012). It is also argued that the lack of access to land and other productive resources keeps women always at a disadvantaged position as compared with men (Agarwal 2003; Rao 2008; Brara 2014).

It is, however, revealed that whether the early debate of the 1980s or more recent debate after 1990s in the aftermath of economic liberalisation, they were either confined mostly to the green revolution belt of traditional wheat-growing north-west states (like Punjab and Haryana) or to the rice-growing eastern and southern states (like Andhra Pradesh, Tamil Nadu, and Odisha) and states like Maharashtra were rarely a case in point. Additionally, in some cases, analysis was made at the national level without looking into state specific situations. Suffice it to say, a comprehensive analysis of the effects of agrarian change on rural women in the context of neoliberal reforms with reference to a state like Maharashtra (which is considered as the heartland of women movements in India) is missing in the current literature.[9] Maharashtra is one of the few states which have undertaken special measures for the welfare of women, and it is the first state in the country to formulate women's policy.[10] So far as the profile of rural women in the state is concerned (Table 5.1), about 2 million households were headed by females, with 69 per cent of female literacy, and a sex ratio of 952.

Hence, the present chapter makes an attempt to assess the impact of agrarian change on rural women in Maharashtra, mainly relating

Table 5.1 The profile of the rural women in Maharashtra – 2011

Sl. no.	Particulars	No.
1.	Total population	6,155,6074
2.	Female population	30,017,040 (48.76)
3.	Sex ratio	952
4.	Female population (SC)	3,669,766 (48.96)
5.	Female population (ST)	4,465,621 (49.58)
6.	No. of female-headed households	1,694,720*
7.	Literacy rate, 7+ years	18,091,286 (68.54)
8.	Workers	12,763,800 (42.52)
9.	Main workers	10,321,369 (34.39)
10.	Cultivators	4,231,645 (41.00)
11.	Agricultural labourers	4,998,031 (48.42)
12.	Household industry workers	180,790 (1.75)
13.	Other main workers	910,903 (8.83)
14.	Marginal workers	2,442,431 (8.14)
15.	Non-workers	17,253,240 (57.48)
16.	No. of female operational holders	2,052,519 (14.98)**
17.	Operated area (ha.)	2,585,253.43 (13.08)**
18.	Average size of holding (ha.)	1.26**

Source: *Primary Census Abstract 2011*, Maharashtra State.

Note: Figures in parentheses indicate percentage.
* Taken from the *Socio-Economic Caste Census 2011*.
** Taken from Agricultural Census, 2010–11

to workforce participation, landholding, wages, and earnings, and the conditions of female-headed households. The analysis is made at state, district, and village levels.

State level scenario

Workforce participation

The NSSO estimates reveal that the female worker population ratio (WPR) decreased over the years in rural India. The WPR for females in rural India (both principal and subsidiary status), which was 33 per cent in 1993–94 (50th round), declined to 25 per cent in

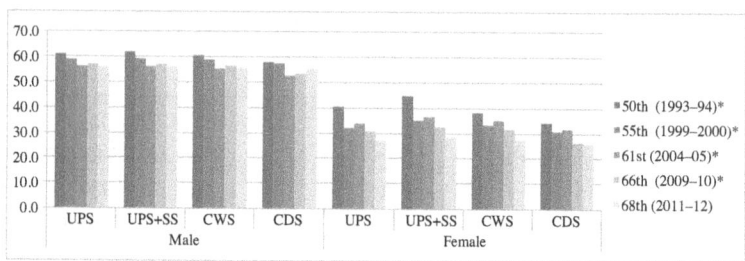

Figure 5.1 Worker population ratio (WPR) in rural Maharashtra

Source: *Employment and Unemployment Situation, NSSO 68th Round*, government of Maharashtra.

Note: (1) UPS: Usual Principal Status, SS: Subsidiary Status, CWS: Current Weekly Status, and CDS: Current Daily Status. (2) *Quinquennial surveys.

2011–12 (68th round), whereas the WPR for males declined marginally from 55 per cent to 54 per cent during this period. Though the NSSO estimate for Maharashtra reveals almost a similar trend, the rate of decline for males and particularly for females is relatively higher than that of all India level. The WPR in terms of both usual principal status (UPS) and subsidiary status (SS), which accounted for 45 per cent in 1993–94, went down to 29 per cent in 2011–12 for females, while in the case of males, it decreased from 62 per cent to 56 per cent (Figure 5.1).

It is worth noting that over the years the gender gap remains mostly unchanged. The Labour Force Participation Rate (LFPR) also shows the prevalence of a wide gender gap.[11] The current weekly as well as daily status reveals a similar pattern. Persistence of gender disparity in WPR and LFPR across social groups (SC, ST, and OBC) indicates a rising trend.[12] The decline in rural WPR may be attributed to the growing migration of workers from rural-to-urban areas.

It is evident that in terms of usual principal status, the proportion of rural female self-employed increased from 41 per cent in 1993–94 to 46 per cent in 2011–12, while for the self-employed males, it rose to 53 per cent from 47 per cent (Table 5.2).

In terms of both principal and subsidiary status, though rural female as well as male self-employment increased slowly, the proportion of males continues to be higher than of females. In regular wage/salaried employment, the males were also ahead of females,

Table 5.2 Percentage distribution of usually employed by broad category of employment for different NSSO Rounds

NSSO rounds	Principal status						Principal +subsidiary status					
	Self-employed		Regular wage/salaried		Casual labour		Self-employed		Regular wage/salaried		Casual labour	
	Male	Female	Male	Female	Male	Female	Male	Female	Male	Female	Male	Female
68 (Jul'11–Jun'12)	52.7	45.9	12.1	5.5	35.2	48.6	52.8	47.4	12.1	5.2	35.1	47.4
66 (Jul'09–Jun'10)*	51.2	44.8	11.1	3.6	37.7	51.6	51.3	45.6	11.1	3.4	37.6	51.0
61 (Jul'04–Jun'05)*	49.3	45.3	10.1	2.7	40.5	51.9	49.5	47.4	10.2	2.5	40.3	50.1
55 (Jul'99–Jun'00)*	48.3	40.0	8.4	1.4	43.3	58.6	48.6	44.7	8.4	1.2	43.0	54.1
50 (Jul'93–Jun'94)*	46.7	40.7	11.3	1.7	42.0	57.6	47.1	44.2	11.2	1.5	41.7	54.3

Source: *Employment and Unemployment Situation, NSSO 68th Round*, government of Maharashtra.

Note: * Quinquennial surveys

although both categories experienced a gradual growth. On the contrary, among the casual labourers, females had a greater representation, and the trend continued over the years. In short, while there was increasing feminisation of casual labour, male dominance continued in self-employment as well as regular wage and salaried categories in terms of both usual principal and subsidiary types of economic activities. As females had less access to land and other productive assets, self-employment was a difficult proposition for them, and they mostly worked as causal labourers with relatively lower wages. A very large proportion of rural female labourers (88 per cent) were engaged in agriculture and a majority of them worked as casual labourers, while the non-agricultural sector is dominated by the males.[13] The greater availability of women labourers in the rural areas kept the wage rates low as a result of which male labourers preferred to shift to the non-agricultural activities. In addition, the declining farm income accompanied by rising input costs, uncertainty, and risk due to dependence on the monsoon, as well as supply and price volatilities of the global market caused by trade liberalisation, made agriculture unviable. As a result, the male members moved out of agriculture, leaving females behind to carry the burden of agriculture leading to the feminisation of agriculture. This was observed by many micro- and macro-level studies (Ganguly and Talwar 2003; Garikipati 2006). However, the number of female cultivators which indicated an upward trend in the initial stage of economic reforms declined thereafter.[14]

The size-class-wise information[15] on female agricultural households employed in rural areas shows that while women belonging to small and medium landholding groups were mostly self-employed in agriculture, a higher proportion of women from the lowest as well as highest landholding size classes worked as an 'other' category of labourers. Greater representation of women in the 'other' category of workers may be attributed to the fact that while the marginal and near landless categories of women tended to leave agriculture due to lower wages and unviable holding size, women from higher landowning households preferred to move out of agriculture to the other emerging lucrative occupations, such as trading and business available in the rural areas. The casual wage labourers in agriculture were drawn mainly from the households with miniscule holdings. Changes in the sectoral distribution of employment in rural non-farm sector reveals[16] that though both male and female participation remained quite low, for males, it was 25 per cent in 1993–94 but increased to 29 per cent in 2009–10 as opposed to females, which was 8.8 per cent in 1993–94 and decreased to 7.8 per cent in 2009–10. Of the total rural non-farm workers, about 75 per cent were males, while 25 per cent were females in 1993–94; that increased to 79.2 per cent for males and decreased to 20.8 per cent for females in 2009–10.

Wages and earnings

It is a well-known fact that women represent the largest group of 'unpaid' workers, especially in rural areas. Their contribution to reproduction and range of activities within households, such as caring for the young and old, cooking, and other domestic chores, are not only unpaid, but hardly recognised in the system of national accounts or other economic statistics. When women are employed, they are usually paid less than men, even for the same task. According to the *World Development Report* (2008), in India, the average wage for agricultural casual work was 30 per cent lower for women than for men and 20 per cent lower for the same task. A recent review of wage data from different sources indicated an overall slowdown in the wages of male and female agricultural labour, as well as a widening of the gender gap, especially in operations such as transplanting, weeding, and harvesting that tend to be carried out largely by women (Chavan and Bedamatta 2006; Srivastava and Singh 2006; Rao 2011; Usami 2011). In rural Maharashtra too, an analysis of the growth of real wages suggests that the gap between male and female wages widened over the years (Table 5.3).

It may be noted that wage rates for ploughing and sowing, which were usually masculine works in rural areas, rose significantly between 2005 to 2006 and 2008 to 2009, whereas the wage rates for operations like transplantation and weeding remained stagnant or recorded significant negative growth, and for females, the negative growth was higher than for males. Compared with the period from1998 to 1999 through 2004 to 2005, between 2005 to 2006 and 2008 to 2009, the negative growth for wages in transplantation was higher. Though wage rates for transplantation grew rapidly for males between 1998 to 1999 and 2004 to 2005, for females, it witnessed a negative growth. However, the growth rate for female unskilled labour was marginally higher than that of males. In a

Table 5.3 Real wage rates for major occupations in rural Maharashtra

Particulars	Gender	1998–99	1999–2000	2000–01	2001–02	2002–03	2003–04	2004–05	2005–06	2006–07	2007–08	2008–09	Growth rate	
													1998–99 to 2004–05	2005–06 to 2008–09
Ploughing	Male	16.83	17.70	19.49	19.43	19.46	18.88	17.80	17.68	18.17	18.95	19.46	1.06	3.30**
	Female	—	—	—	—	—	—	—	—	—	—	—	—	—
Sowing	Male	15.87	16.08	17.56	18.29	18.19	18.05	16.98	16.60	16.86	17.37	18.13	1.68	2.93**
	Female	—	—	—	—	—	—	—	—	—	—	—	—	—
Transplanting	Male	16.13	16.36	20.11	20.40	19.55	19.80	19.14	18.08	17.64	17.06	16.32	3.10*	–3.41**
	Female	13.74	12.10	13.53	13.84	13.44	13.53	12.49	12.30	12.44	11.59	10.84	–0.25	–4.49*
Weeding	Male	—	—	—	—	—	—	—	—	—	—	—	—	—
	Female	10.64	9.70	10.86	10.80	10.55	10.15	9.51	9.81	9.96	10.21	10.02	–0.98	0.91
Harvesting	Male	14.36	15.38	17.32	17.69	17.88	17.17	16.64	16.17	16.73	17.35	17.62	2.47	2.94**
	Female	10.27	10.67	11.60	11.88	11.81	10.87	10.53	10.27	11.26	10.82	10.96	0.47	1.57
Unskilled labour	Male	14.28	14.20	14.61	15.21	14.95	14.78	14.45	14.10	14.01	14.10	14.67	0.49	1.24
	Female	9.95	9.70	9.63	9.79	9.72	9.48	9.22	8.97	8.84	9.21	9.48	–0.95**	2.04

Source: Usami (2011).

Notes: (1) The CPIA L and CPIRL (1986–87 = 100) are used as deflators. (2) * and ** stand for significance level at 10 and 5 per cent, respectively.

nutshell, effects of neoliberalism widened gender differentials in wages.

The data furnished by the *Rural Labour Inquiry Reports* indicated persistence of wide-ranging gender disparity in daily earnings of rural labour households (Table 5.4).

The earning differentials between male and female labour households were reflected in all agricultural operations, and with the march of time, they showed signs of growing inequality. Though women were usually more involved in weeding, harvesting, and transplanting vis-à-vis men, their average earnings from these operations were much lower than men. The average daily earnings from cultivation indicated a wide gender gap. Ploughing, which was essentially taboo for women, became more remunerative in recent times as opposed to other operations, and wherever women undertook this operation, they were underpaid. The gender gap was prevalent in the daily earnings of both rural labour and agricultural labour households (Table 5.5). The female average earning of rural labour households was not only less than their male counterparts in rural labour households but also much lower than the average earning of males belonging to agricultural labour households.

Landholding

It is widely recognised that the socio-economic disadvantages of women in households as well as in the local community are largely due to the denial of land rights to them (Rao 2008; Agarwal 2016). Therefore, strengthening women's access to and control over land is considered a key to raising their socio-economic status, power, and freedom within the households and village society. Though elaborate land reform measures were undertaken after India's independence to promote social justice, they were largely gender-blind, as they were based on the basic premise that land allocated to households mainly through male heads would benefit all the members equitably. Even though Maharashtra is one of the leading states in the successful implementation of land reforms in favour of weaker sections (Mohanty 2001b), women continue to be deprived of their rights over land. An analysis of changes in the share of women in operational holdings reveals the persistence of a striking gender gap. However, over the years, the share of women in operational holdings improved marginally across size classes both in terms of number and area (Table 5.6).

Table 5.4 Average daily earnings (Rs.) of all rural labour households in different agricultural occupations

Year	Ploughing		Sowing		Transplanting		Weeding		Harvesting		Cultivation	
	Male	Female	Male	Female	Male	Female	Male	Female	Male	Female	Male	Female
1993–94	19.93	15.35	23.35	13.06	21.95	14.95	17.72	11.70	19.63	12.72	18.19	11.26
1999–2000	40.65	29.74	38.96	24.29	32.45	25.85	36.24	24.27	38.07	23.75	35.99	23.87
2004–05	48.46	34.61	44.85	28.61	45.74	31.23	39.05	27.70	46.01	25.31	44.36	27.67

Source: Report on Wages and Earnings of Rural Labour Households (50th, 55th, and 61st round of NSSO).

Table 5.5 Average daily earnings (Rs.) of rural labour households and agricultural labour households in agricultural occupations

Year	Rural labour households		Agricultural labour households	
	Male	Female	Male	Female
1993–94	19.54	12.17	19.41	12.09
1999–2000	37.67	24.34	37.47	24.24
2004–05	44.56	27.41	43.98	27.18

Source: Report on Wages and Earnings of Rural Labour Households (50th, 55th, and 61st round of NSSO).

Table 5.6 Percentage share of female in operational holdings

Size class	No				Area			
	1995–96	2000–01	2005–06	2010–11	1995–96	2000–01	2005–06	2010–11
Marginal	14.11	16.91	16.9	16.21	14.03	17.23	17.23	16.24
Small	13.7	16.17	16.17	15.23	13.54	16.16	16.15	15.09
Semi-medium	11.59	13.44	13.44	12.62	11.33	13.23	13.23	12.35
Medium	8.85	10.57	10.53	10.16	8.69	10.39	10.36	10.04
Large	7.55	9.26	9.21	9.48	7.11	9.14	9.13	9.17

Source: Agricultural Census of India, relevant years.

While the number of holdings increased from 13 per cent in 1995–96 to 15 per cent in 2010–11, the area increased from 11 per cent to 13 per cent during the same period. It is interesting to note that the higher the size class, the lesser the women's access to land.

The improvement in landholding position may be attributed to the effects of women's movements in certain pockets that focused on the protection of their land rights. The *Shetkari Sanghatana Mahila Aghadi* led by Sharad Joshi, especially its *Lakshmi Mukti* campaign, focused on the restoration of women's right in agricultural land and other movable assets, particularly in developed parts of the state. The *Laxmi Mukti* programme was started in 1989, and its modus operandi involved voluntary transfers of land titles in the names of housewives on the lines of the Bhoodan Movement.[17] Land transfers under this campaign were treated as voluntary gifts and exempted from the payment of gift tax. The state government and its lower-level revenue bureaucracy cooperated in completing formalities of such transfers of documents and their registration in land records. The *Sanghatana* campaigned to ensure that each village would enlist at least 100 cases or about 50 per cent of the total *khatedars* (landowners) who would willingly sign documents of land transfer in the names of their housewives. It was claimed that as many as 0.2 million women received a 7/12 statement of land registration in their names through transfer by their husbands within a span of one-and-half years (Kale 2009, as cited in Dhanagare 2016a: 180). In a series of tours, women activists like Madhu Kishwar participated along with Sharad Joshi, and the *Laxmi Mukti* campaign was carried out in numerous villages. It was reported that farmers from Vidarbha and Marathwada regions were more willing to transfer land rights during the campaign than those in southern Maharashtra because of variations in land prices and the rental value of farmland (Dhanagare 2016a: 181). This was probably one of the reasons for the better landholding position of women in backward and most backward districts. However, with the decline of the *Sanghatana* from the 1990s, the campaign has more or less halted.

There were also women's movements like those fighting for the rights of single, deserted, and widowed women demanding housing and agricultural lands for them and also fighting for legal aid for widows to stake claims on their husbands' property. The *Bhumiheen Hakk Sanrakshan Samiti* (Association for Protection of Rights of Landless) raised the issue of 'joint pattas' for both men and women while demanding land for the landless. New campaigns beginning in 1987–88 to organise '*parityakta*' or abandoned women brought forward many aspects of fighting for women's access to property and employment. In WM, the *Stree Mukti Sangarsh Calval* (Struggle for Women's Liberation) centred its work on *parityakta* women getting their rights to house plots (of two *guntas* each) and along with other organisations began to formulate demands for giving control of at least a part of wastelands to women who would take up experimental sustainable farming. Recently, in five villages of WM, single and deserted women gained rights to over 1500 sq. ft. of housing land each with the support of *Stree Mukti Sangharsh* (Kulkarni *et al.* 2008). The latter initially spread its activities to Sangli, Satara, and parts of Kolhapur and Solapur districts. Its main task was to develop feminist consciousness among women and to challenge the existing sexual division of labour. The other notable effort in Maharashtra was the one led by Vaishali Patil of the *Adivasi Hakk Suraksha Sanghatana* (Organisation for the Protection of Tribal Rights) in the Raigad district of Maharashtra. In addition, *Mahila Arthik Vikas Mahamandal*, the women's economic empowerment organisation of the government of Maharashtra also undertook an awareness generation campaign to make women aware of their rights to their husbands' houses and properties. It undertook a unique initiative known as *Ghar Doghaanche Abhiyan* or the Home of Two campaign for the enforcement of women's property rights. The activities of *Maharashtra Rajya Shet-mazdoor Parishad* also contributed to the growing awareness on gender equality. It demanded equal wages for women and men, and in the course of time mobilised people using the slogans *'mageltyalakam'* and *'kamachayogya dam'* (work on demand with requisite wages), which were later included in the Employment Guarantee Act of 1977 (Chari 2006: 5143). Most of the female organisations were active in developed regions of the state. All these movements not only contributed to the women's landholding but also had also made an impact on other aspects of gender inequality.

It may be inferred from the earlier discussion that gender disparity is widely prevalent in rural Maharashtra. Though under changing conditions, the landholding and work participation, as well as wages and earnings, of women improved, but they hardly alter gender inequality deeply rooted in the patriarchal ideology of development. The following section illustrates the differential impact of agricultural transformation on gender disparity at the district level.

District-level analysis

There were 34 districts in Maharashtra, which were at different levels of agricultural development. In order to assess the differential impact of agrarian transformation on rural women, the districts were divided into three categories: advanced, backward, and most backward[18] in terms of the five major indicators of agricultural development such as irrigation intensity, fertiliser consumption (per hectare in kg.), area under HYV (percentage to gross cropped area), and use of tractors and pumpsets (nos/000'hc). As NSSO data were not available at the district level, the analysis is based on population census, agricultural census, and Labour Bureau reports.

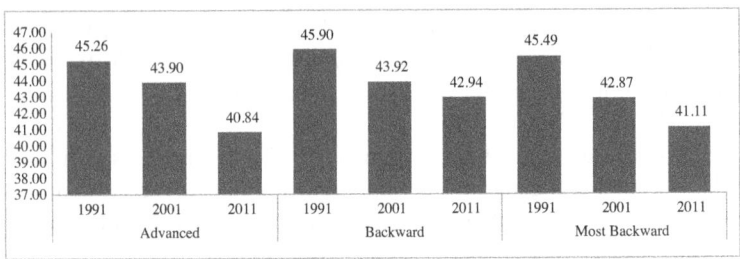

Figure 5.2 Percentage share of rural female workers

Source: *Primary Census Abstract* (Maharashtra State), relevant issues.

Workforce participation

Analysis of changes in the rural workforce showed that the share of rural female workers steadily declined across the districts between 1991 and 2011 (Figure 5.2).

Among three categories of districts, the shares of female workers were higher in backward and most backward districts, and the rate of decline was also relatively slow. It was evident that changes in rural economy following neoliberal reforms affected the workforce participation of rural women more adversely in developed pockets. The share of females in the agricultural workforce recorded considerable decline for all categories of districts accompanied by a rise in the non-agricultural workforce (Figure 5.3). A cursory glance at the district-wide information furnished by Labour Bureau revealed the widespread gender disparity with regard to LFPR and WPR in rural areas. Relative disparity was pronounced prominently in many advanced districts.[19]

Moreover, women in all the districts continued to work largely in the agricultural sector, though a marginal decline was noticed, accompanied by a rise in the share of non-agricultural workforce (Table 5.7).

As compared with the backward and most backward districts, the proportion of non-agricultural workers among women was higher in advanced districts. In all the categories of districts, male participation in the non-agricultural workforce was much higher than female counterpart. However, the proportion of male agricultural workers, which fell in 2001 as compared with 1991, accompanied by a corresponding growth in their share of non-agricultural workforce,

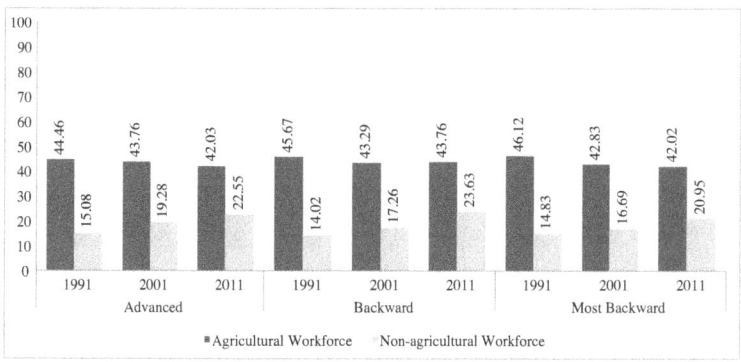

Figure 5.3 Percentage share of rural female agricultural and non-agricultural workforce

Source: *Primary Census Abstract* (Maharashtra State), relevant issues.

Table 5.7 Distribution of rural workers (%)

Categories of districts	Year	Cultivator		Agricultural labourer		Agricultural workers		Non-agricultural workers	
		Male	Female	Male	Female	Male	Female	Male	Female
Advanced	1991	48.96	49.16	23.83	43.18	72.79	92.34	27.21	7.66
	2001	46.96	51.38	23.08	37.01	70.04	88.39	29.96	11.61
	2011	46.10	44.62	27.38	42.73	73.48	87.34	26.52	12.66
Backward	1991	48.38	42.61	30.81	52.54	79.18	95.15	20.82	4.85
	2001	46.23	46.11	30.38	46.19	76.60	92.30	23.40	7.70
	2011	44.59	41.13	36.96	50.62	81.55	91.75	18.45	8.25
Most backward	1991	42.85	40.17	33.01	53.75	75.86	93.92	24.14	6.08
	2001	39.31	41.43	31.71	48.73	71.01	90.16	28.99	9.84
	2011	36.86	35.13	39.09	54.50	75.96	89.63	24.04	10.37

Source: *Primary Census Abstract* (Maharashtra State), relevant issues.

increased in 2011 across the districts, leading to a marginal decline in the gender gap in the participation in the agricultural workforce. Nevertheless, compared with males, the proportion of female agricultural workers was higher across the districts, and it was more in backward and most backward districts at all points of time. Agricultural labour,

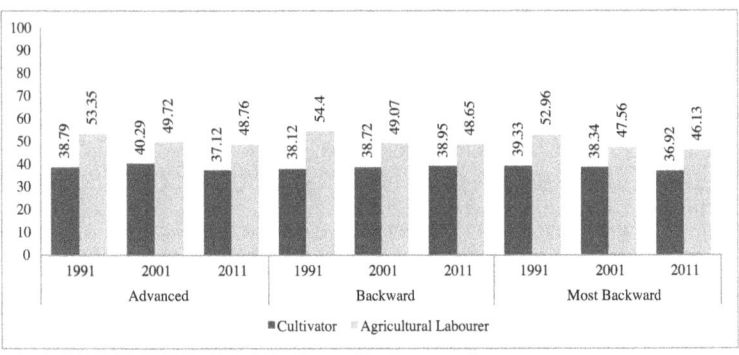

Figure 5.4 Percentage share of female cultivators and agricultural labourers

Source: *Primary Census Abstract* (Maharashtra State), relevant issues.

which carries low wages as compared with non-agricultural occupations and signifies a low status, absorbed a greater proportion of females than that of their male counterparts across the districts. For both males and females though, there was a fall in the proportion of agricultural labourers in 2001, as compared with 1991, across the districts; it rose again in 2011. The share of females as cultivators as well as agricultural labourers indicates a declining trend across the districts (Figure 5.4). The decline of farm income accompanied by low wages in agriculture under the emerging conditions of neoliberalism compelled women to gradually move to unskilled and low-paid jobs available in the rural non-farm sector.

Landholding

Size-class-wise changes in number and area of operational holdings of males and females across the three categories of districts are illustrated in Table 5.8.

We found that women remained greatly disadvantaged with respect to landholding, both in terms of area and number in advanced, backward as well as most backward districts. Compared with 2001, their position in 2011 declined further, particularly in backward and most backward districts. The developed districts experienced only a nominal improvement. Across categories of districts, women had insignificant representation in the higher size classes. The higher the size class,

Table 5.8 Size-class-wise operational holdings (%)

Category of districts	Size class	2001			2011			% change from 2001 to 2011	
		Male	Female	Gender gap	Male	Female	Gender gap	Male	Female
Number									
Advanced	Marginal	85.09	14.57	70.52	85.25	14.52	70.73	0.16	−0.05
	Small	85.84	13.99	71.85	85.74	14.07	71.67	−0.10	0.08
	Semi-medium	87.89	11.88	76.01	87.61	12.11	75.5	−0.28	0.23
	Medium	90.04	9.31	80.73	90.26	9.24	81.02	0.22	−0.07
	Large	88.03	6.35	81.68	89.55	7.92	81.63	1.52	1.57
	All	**85.97**	**13.7**	**72.27**	**85.88**	**13.87**	**72.00**	**−0.10**	**0.17**
Backward	Marginal	79.02	20.82	58.2	80.77	19.18	61.59	1.75	−1.64
	Small	81.85	18.03	63.82	83.18	16.74	66.44	1.33	−1.29
	Semi-medium	85.58	14.25	71.33	86.39	13.47	72.92	0.81	−0.78
	Medium	89.35	10.16	79.19	89.66	9.89	79.77	0.31	−0.27
	Large	86.54	7.8	78.74	87.19	8.18	79.01	0.65	0.38
	All	**82.11**	**17.70**	**64.41**	**83.02**	**16.87**	**66.15**	**0.91**	**−0.83**
Most backward	Marginal	81.87	17.87	64.00	84.28	15.57	68.71	2.41	−2.30
	Small	83.03	16.8	66.23	85.3	14.58	70.72	2.27	−2.22
	Semi-medium	85.41	14.33	71.08	87.76	12.03	75.73	2.35	−2.30
	Medium	87.81	11.67	76.14	88.6	10.94	77.66	0.79	−0.73
	Large	85.3	11.29	74.01	87.1	10.37	76.73	1.80	−0.92
	All	**83.55**	**16.15**	**67.4**	**85.58**	**14.23**	**71.36**	**2.03**	**−1.92**
Area									
Advanced	Marginal	85.15	14.61	70.54	85.31	14.5	70.81	0.16	−0.11
	Small	86.04	13.79	72.25	85.84	13.97	71.87	−0.20	0.18
	Semi-medium	87.94	11.82	76.12	87.82	11.9	75.92	−0.12	0.08
	Medium	90.21	9.05	81.16	90.39	9.06	81.33	0.18	0.01
	Large	80.93	5.64	75.29	88.08	7.31	80.77	7.15	1.67
	All	**86.99**	**11.87**	**75.12**	**87.17**	**12.27**	**74.90**	**0.18**	**0.40**

(Continued)

Table 5.8 Continued

Category of districts	Size class	2001			2011			% change from 2001 to 2011	
		Male	Female	Gender gap	Male	Female	Gender gap	Male	Female
Backward	Marginal	79.34	20.54	58.8	81.03	18.92	62.11	1.69	−1.62
	Small	81.97	17.92	64.05	83.37	16.55	66.82	1.40	−1.37
	Semi-medium	85.9	13.91	71.99	86.74	13.11	73.63	0.84	−0.80
	Medium	89.37	10.08	79.29	89.69	9.81	79.88	0.32	−0.27
	Large	78.58	7.18	71.4	83.88	7.5	76.38	5.30	0.32
	All	**84.36**	**14.82**	**69.55**	**85.13**	**14.42**	**70.71**	**0.76**	**−0.40**
Most Backward	Marginal	81.5	18.27	63.23	84.4	15.48	68.92	2.90	−2.79
	Small	82.85	16.98	65.87	85.42	14.46	70.96	2.57	−2.52
	Semi-medium	85.68	14.04	71.64	87.99	11.79	76.20	2.31	−2.25
	Medium	87.93	11.48	76.45	88.67	10.84	77.83	0.74	−0.64
	Large	82.37	10.75	71.62	85.34	9.94	75.4	2.97	−0.81
	All	**84.96**	**14.16**	**70.79**	**86.95**	**12.52**	**74.43**	**1.99**	**−1.65**

Source: *Agricultural Census of India*, relevant issues.

lesser the women's access to land. All the districts exhibit a striking gender gap both in number and area across size classes. While the gender gap reduced marginally in advanced districts between 2001 and 2011, it increased in backward and, particularly, in most backward districts. However, in terms of number as well as area of operational holdings, women remained in a better position in both backward and most backward districts.

Female-headed households

In all the categories of districts, female-headed households constituted a microscopic minority (Table 5.9).

The advanced districts had a higher proportion of female-headed households (13.16 per cent). For a majority of these households, manual casual labour was an important source of income, and it was more so in the case of backward and most backward districts. Over 59 per cent of female-headed households in backward

Table 5.9 Distribution of female-headed households by source of income

Categories of districts	Female-headed (%)	Percentage distribution of female-headed household by income source							Households with monthly income of highest-earning household member (%)		
		Cultivation	Manual casual labour	Domestic service	Foraging rag picking	Non-agricultural enterprise	Begging/charity/alms collection	Others	< 5000	5000–10000	> 10000
Advanced	13.16	30.49	50.23	4.07	0.27	1.50	0.44	12.99	79.59	13.74	6.65
Backward	10.05	25.30	59.46	3.23	0.38	1.02	0.64	9.88	83.53	11.05	5.32
Most backward	11.98	21.99	54.44	4.47	0.38	0.93	0.55	17.23	83.04	11.19	5.76
All	11.73	26.84	52.62	4.04	0.34	1.18	0.53	14.41	82.20	11.93	5.83

Source: Socio-Economic Caste Census 2011.

districts derived their income from manual casual labour services. The importance of agriculture as a source of income increased with agricultural development. About one-third of female-headed households in the advanced districts derived their income from cultivation, while in backward and most backward districts, it came to 25 and 22 per cent, respectively. A significant proportion of households in the three categories of districts, and especially in most backward districts, depend on 'other' sources of income. Across the districts, female-headed households had monthly income of less than Rs. 5000 from their highest-earning member, and the proportion was more in the backward districts.

To sum up, women continued to be the most disadvantaged section, though they witnessed marginal improvement in advanced districts, especially in terms of their access to land. The relative progress in the landholding position in advanced districts may be due to a better performance of districts like Pune, Satara, Sangli, and Kolhapur, which had better exposure to the women's movements since the early 1980s that highlighted their land rights and other issues.

Village-level analysis

The analysis is based on four villages of Maharashtra, two each from the Solapur and Akola districts, selected by ICRISAT for its village level studies during 2008–09.[20] While *Shirapur* and *Kalman* villages in the Solapur district were agriculturally advanced, *Kanzara* and *Kinkheda* villages in the Akola district were relatively backward. In *Shirapur*, more than 60 per cent gross cropped area was irrigated, and it had 6 tractors, 5 threshers, 450 bore wells, 2 irrigation pump houses, and good infrastructure. It grew mainly sugarcane, sorghum, and wheat. Similarly, *Kalman* had good infrastructure and was well connected by roads to important cities and towns. There were 45 sprinkler sets and 30 drift sets. Twenty-four per cent of its gross cropped area was irrigated. It had 11 tractors, 7 threshers, and 190 bore wells, with an agricultural extension centre. On the other hand, *Kinkheda* and *Kanzara* in Akola district had less area under irrigation. While the gross irrigated area in *Kinkheda* was 17 per cent, it was only 3 per cent in *Kanzara*. Both of the villages mostly grew soybeans, cotton, and pigeon peas. However, between the two villages, infrastructure was better in *Kanzara*. It had 10 tractors, 8 threshers, 4 flour mills, 32 bore wells, 12 irrigation pump houses, and 20 self-help groups.

Resource ownership and decision-making

Village-wise information on gender-based ownership of resources (land, livestock, credit, and machinery), crop inputs (seed, fertiliser, pesticides, etc.), main crop and by-products, and quantity sold is presented in Table 5.10. Female members hardly owned resources in advanced and backward villages. Resources like land, machinery, livestock, and agricultural inputs were mostly owned by male members. Joint ownership was also negligible in all the villages. Compared with advanced villages, the gender disparity in ownership of resources was relatively less in the two backward villages, particularly in *Kinkheda*. In this village, the percentage share of ownership was better for all varieties of resources. The gender division on making decisions in using resources and other matters like household maintenance, education, and marriage of children also indicates male dominance across villages, though marginal differences do exist (Table 5.11).

However, in significant numbers of households, both males and females take decisions jointly on matters like investment, hired labour, children's marriages, and education in all the four villages. Among all in the most backward village of *Kinkheda*, the female members had a greater say in the decision-making process.

Household gender division of major crop cultivation activities

In all four villages, activities which are considered technical and skilled like selection of seeds and crops, marketing, irrigation, etc., were mostly undertaken by male members (Table 5.12).

Except operations like weeding, there was hardly any job which was primarily done by female members. Activities like harvesting, threshing, and seed storage were done by both males and females. In *Shirapur*, activities like selection of seeds and crops, land preparation, marketing, and plant protection were mostly done by male members, while threshing, harvesting, seed selection and storage, fertiliser application, transportation of manure, etc., were taken care of by both male and female members. In *Kalman*, crop selection, land preparation, marketing, plant protection, selection of seeds, etc., were considered masculine jobs. Agricultural operations like hand weeding, sowing, fodder harvesting and stacking, watching, threshing, harvesting main crop, seed selection and storage,

Table 5.10 Percentage share in resource ownership

Particulars	Solapur						Akola					
	Shirapur			Kalman			Kanzara			Kinkheda		
	Male	Female	Both	Male	Female	Both	Male	Female	Both	Male	Female	Both
Land	91.5	5.7	2.8	97.6	1.2	1.2	89.2	1.2	9.6	93.1	6.9	0.0
Livestock	91.1	5.0	4.0	100.0	0.0	0.0	88.2	2.9	8.8	84.1	5.8	10.1
Machinery	90.0	0.0	10.0	100.0	0.0	0.0	97.8	0.0	2.2	91.1	6.7	2.2
Pesticides	95.4	3.1	1.5	98.8	1.2	0.0	95.2	1.2	3.6	90.7	7.0	2.3
Fertilisers	94.0	2.0	4.0	98.8	1.2	0.0	94.0	1.2	4.8	90.7	7.0	2.3
Seed	95.0	2.0	3.0	98.8	1.2	0.0	94.0	1.2	4.8	90.7	7.0	2.3
Crop main production	96.0	2.0	2.0	97.6	1.2	1.2	97.6	1.2	1.2	81.4	8.1	10.5
Sale quantity	96.0	2.0	2.0	97.6	1.2	1.2	96.4	2.4	1.2	79.1	8.1	12.8
Fodder production and use	92.0	2.0	6.0	44.3	1.3	54.4	96.4	1.2	2.4	79.1	7.0	14.0

Source: Authors own calculation from ICRISAT Village Level Primary Data (2008).

Note: Households reporting not applicable (NA) have been excluded.

Table 5.11 Percentage share in decision-making

Particulars	Solapur						Akola					
	Shirapur			Kalman			Kanzara			Kinkheda		
	Male	Female	Both	Male	Female	Both	Male	Female	Both	Male	Female	Both
Investment	62.6	5.0	32.4	64.2	3.8	32.1	77.1	6.3	16.7	73.3	6.7	20.0
Land	86.7	5.7	7.6	95.3	1.2	3.5	94.0	1.2	4.8	71.3	6.9	21.8
Hired labour	45.5	9.1	45.5	31.0	1.1	67.8	74.7	3.6	21.7	47.7	7.0	45.3
Pesticides	95.4	3.1	1.5	96.4	1.2	2.4	95.2	2.4	2.4	77.9	7.0	15.1
Fertilisers	92.9	2.0	5.1	92.9	1.2	5.9	92.8	3.6	3.6	74.4	7.0	18.6
Seed	93.9	2.0	4.0	94.1	1.2	4.7	92.8	3.6	3.6	74.4	7.0	18.6
Migration	62.7	4.9	32.4	43.3	0.0	56.7	92.0	3.0	5.0	47.3	7.7	45.1
Machinery	89.7	0.0	10.3	88.9	0.0	11.1	97.8	0.0	2.2	76.7	7.8	15.6
Livestock	88.1	3.0	8.9	95.4	0.0	4.6	82.4	2.9	14.7	56.5	5.8	37.7
Credit	70.3	3.6	26.1	74.5	3.8	21.7	85.6	4.1	10.3	76.7	6.7	16.7
Crop main production	91.9	2.0	6.1	90.6	1.2	8.2	95.2	1.2	3.6	74.4	8.1	17.4
Sale quantity	89.9	2.0	8.1	85.9	1.2	12.9	94.0	1.2	4.8	66.3	8.1	25.6
Fodder production and use	88.9	3.0	8.1	38.0	1.3	60.8	95.2	1.2	3.6	62.8	7.0	30.2
Household maintenance	69.5	7.1	23.4	2.8	3.8	93.4	2.0	89.0	9.0	26.4	20.9	52.7
Children marriage	58.6	7.1	34.3	2.9	4.4	92.6	68.0	2.1	29.9	33.0	7.7	59.3
Education of children	59.3	8.6	32.1	2.9	2.9	94.2	62.9	8.2	28.9	28.6	16.5	54.9

Source: Authors own calculation from ICRISAT Village Level Primary Data (2008).

Note: Households reporting not applicable (NA) have been excluded.

Table 5.12 Gender-based household division of crop cultivation activities (%)

Crop cultivation activity	Solapur						Akola					
	Shirapur			Kalman			Kanzara			Kinkheda		
	Male	Female	Both	Male	Female	Both	Male	Female	Both	Male	Female	Both
Hand weeding	0.0	26.3	73.7	3.5	4.7	91.8	1.1	59.6	39.3	0.0	29.2	70.8
Sowing seed	52.0	0.0	48.0	7.1	1.2	91.8	2.2	23.6	74.2	1.1	36.0	62.9
Selection of crop	97.0	2.0	1.0	91.8	2.4	5.9	92.1	1.1	6.7	76.4	5.6	18.0
Fodder harvesting and stacking	25.0	0.0	75.0	4.7	1.2	94.1	87.6	0.0	12.4	67.4	6.7	25.8
Watching	17.0	4.0	79.0	3.5	1.2	95.3	100.0	0.0	0.0	100.0	0.0	0.0
Land preparation	93.0	0.0	7.0	89.4	2.4	8.2	69.7	0.0	30.3	80.9	0.0	19.1
Threshing	0.0	1.0	99.0	3.5	1.2	95.3	85.4	0.0	14.6	98.9	0.0	1.1
Harvesting main crop	0.0	1.0	99.0	3.5	1.2	95.3	1.1	0.0	98.9	9.0	4.5	86.5
Interculture/mech. weeding	0.0	30.0	70.0	34.5	1.2	64.3	100.0	0.0	0.0	95.5	0.0	4.5
Marketing	99.0	1.0	0.0	96.5	2.4	1.2	98.9	1.1	0.0	95.5	2.2	2.2
Seed selection and storage	13.0	2.0	85.0	3.5	1.2	95.3	97.8	1.1	1.1	79.8	10.1	10.1
Irrigation	45.0	4.0	51.0	51.3	1.3	47.4	100.0	0.0	0.0	100.0	0.0	0.0
Plant protection measures	96.0	0.0	4.0	98.8	1.2	0.0	100.0	0.0	0.0	98.9	0.0	1.1
Selection of variety	98.0	2.0	0.0	96.5	2.4	1.2	98.9	0.0	1.1	84.3	5.6	10.1
Chemical fertiliser appl.	20.0	4.0	76.0	3.5	1.2	95.3	2.2	67.4	30.3	6.7	33.7	59.6
Transport of manure and appl.	7.0	1.0	92.0	42.4	1.2	56.5	100.0	0.0	0.0	96.6	0.0	3.4

Source: Authors own calculation from ICRISAT Village Level Primary Data (2008).

Note: Households reporting not applicable (NA) have been excluded.

and fertiliser application were done by both males and females. In both the backward villages of *Kanzara* and *Kinkheda*, male members were predominant in the selection of crops, fodder harvesting and stacking, watching, threshing, interculture, marketing, seed selection and storage, irrigation, plant protection measures, selection of seeds, and transport of manure application. In these villages women had a greater participation in operations like hand weeding, sowing, and chemical fertiliser applications. Except important operations, the participation of male members in advanced villages was comparatively less, as they were partly engaged in salaried jobs, petty business, etc., due to the proximity of villages to urban areas. Moreover, some of them also went for temporary outmigration labour earnings.

Wage differentials

The gender disparity in the wage rate was common across the villages (Table 5.13).

Though the average daily wages for both farm and non-farm work were higher in advanced villages, gender disparity was more compared with the backward villages. In advanced villages, the female workers received less than half the wages of male workers for both farm and non-farm activities. However, the wage rates for females in one of the backward villages (*Kanzara*) for non-farm work were also lower. Compared with advanced villages, opportunities for non-farm work in backward villages were lower.

In short, striking gender disparities exists in the four villages in terms of ownership and use of resources, decision-making, distribution of crop production activities, and daily wages. Disparities are lesser in backward villages as compared with advanced villages.

Conclusion

Agrarian changes brought about by neoliberal reforms widened the gender disparities. While women's concentration in agriculture declined, it was still highest among all the sectors. The emerging changes in rural economy reduced the employment opportunities for women in agricultural as well as in specialised non-agricultural sectors. They gradually pushed them from somewhat secure self-employment to the casual labour category. The striking gender disparity was all pervasive throughout rural Maharashtra in terms of

Table 5.13 Daily wage rates (Rs.)

| Particulars | Solapur | | | | | | Akola | | | | | |
| | Shirapur | | | Kalman | | | Kanzara | | | Kinkheda | | |
	Male	Female	%*	Male	Female	%*	Male	Female	%*	Male	Female	%*
Farm work	97	48	49.5	86	40	46.5	67	37	55.2	56	33	58.9
Non-farm work	176	77	43.8	108	50	46.3	111	38	34.2	86	46	53.5

Source: Authors own calculation from ICRISAT Village Level Primary Data.

Note: * Percentage of female wage to male wage.

employment, access to land, wages, and earnings. Variations across the districts were marginal. The village-level analysis also indicates the persistence of disparity not only in ownership of resources and earnings but also in distribution of productive activities and decision-making on utilisation of resources. It may be concluded that the changes in rural economy and conditions of agricultural production under neoliberal regime further marginalised women and widened gender disparity. The declining trend in feminisation of agriculture is not accompanied by the corresponding growth of viable employment opportunities in the non-farm sector. The better-paid jobs available in non-agricultural sectors are dominated by the male members, whereas women remained with low-paid insecure jobs. Hence, the arguments made by the gender analysts hold true in the context of Maharashtra. The marginal improvement of women in certain respects, like access to land, is largely due to the effects of movements that focused on land rights and requisite wages for women. The remark made by Ghosh (2013: 15) is worth quoting here:

> The economic history of India (as elsewhere) suggests that greater social recognition of women's work, as well as better remuneration and conditions of such work, come about through struggles and social movements that give voice to such demands and force both governments and the public to respond. The fact that such voices have been getting louder recently is therefore at least one possible source for optimism.

Notes

1 This is a revised version of the paper presented at Department of Sociology, University of Pune in November 2017.
2 See *Gender in Agriculture: Source Book*, 2009, Washington, DC: World Bank.
3 While the Department for International Development (DFID) considered including women as a weapon in the fight against poverty, the World Bank proclaimed that investing in women entrepreneurs was 'smart economics'. See "Smart Economics", by Mayra Buvinic and Elizabeth M. King, 2007, *Finance and Development*, IMF, Vol. 44, No. 2.
4 See "The State of Food And Agriculture: Women in Agriculture-Closing the Gender Gap for Development", 2011, *Rome: Food and Agriculture Organization of the United Nations*, p. 5.

5 Earlier studies mostly focused on urban women. The understanding of rural women was mostly based on oral and local histories. Even at the global level, the study of women in agriculture received due attention only since the 1970s, following Ester Boserup's (1970) seminal work on *Women's Role in Economic Development*.

6 A brief summary of these debates is available in Duvvury (1989: WS 96–112).

7 Besides, the *Indian Journal of Agricultural Economics* brought out a special edition (Vol.40, No.3), which published a number of studies on these issues.

8 During the 1990s, agrarian studies in general lost its central position in the academic agenda, let alone studies on women in agriculture, and a decline in the interest became conspicuous in the last two decades, mainly due to the effect of neoliberal reforms. The new studies focused largely on violence against women, their participation in local governance, etc. The response of rural women to agrarian change and emerging gender relations in agriculture received rare attention. It is surprising to note that the *Economic and Political Weekly*, which brings out special issues on women studies regularly two times in a year, hardly published any articles on gender dimensions of agrarian change between 1991 and 2016 (till April), barring a single paper by Rao (2005), which looked into gender equality and rural poverty reduction through women's control over land.

9 The studies in Maharashtra were either based on some selected districts or regions relating to some aspects of impact of agricultural modernisation on rural women (for example, Gadre and Mahalle1985; Joshi and Alshi 1985). Although the later study made by Acharya and Panwalkar (1988) examines the temporal and spatial variation in the structure of female labour force in the context of land systems, cropping pattern, and commercialisation of agriculture, the analysis was based on NSSO and census data of early 1970s and 1980s.

10 The women's policy in the state was declared in 1994 and was revisited in 2001. A new women's policy was announced in 2013. Recently, several women-specific schemes were undertaken by the state to empower women. See *Economic Survey of Maharashtra, 2015–16*. Directorate of Economics and Statistics, Planning Department, Mumbai: Government of Maharashtra, p. 162.

11 The Labour Force Participation Rate (the number of persons/person-days in the labour force per 1,000 persons/person-days) in 2011–12 for females was 274 as opposed to 568 for males in UPS terms. In terms of both UPS and SS, the LFPR for females was 288, whereas for males it was 571. For details, see report on "Employment and Unemployment Situation" based on data collected in the state sample of the 68th round of the National Sample Survey (July 2011–June 2012), Directorate of Economics and Statistics, Planning Department, Government of Maharashtra, Mumbai, pp. 10.

12 LFPR and WPR (per 1,000) according to usual status (ps+ss) in rural Maharashtra is as follows:

Year	Social group	Male		Female	
		LFPR	WPR	LFPR	WPR
2004–05	SC	566	550	492	490
	ST	570	562	530	529
	OBC	577	569	476	474
2011–12	SC	570	565	379	378
	ST	583	580	458	458
	OBC	569	562	359	357

Source: *Employment and Unemployment Situation among Social Groups in India*, 68th round of NSSO. Report No. 563 (July 2011–June 2012), p. 37.

13 The following distribution of labour force (%) in the age group of 15–59 according to current weekly status (2009–10) provides information on employment of male and female labourers in agriculture and non-agriculture in rural Maharashtra:

Sex	Agriculture				Non-agriculture			
	Self-employed	Regular wage/ salaried	Casual labour	Total	Self-employed	Regular wage/ salaried	Casual labour	Total
Female	40.4	0	47.5	87.9	4.5	5.0	2.6	12.1
Male	36.3	0.2	28.5	65.0	12.0	14.7	8.3	35.0

Source: *NSSO*, 66th Round, State Sample Maharashtra.

14 As per census reports, the number of female cultivators increased from 45 per cent in 1991 to 47 per cent in 2001 and declined to 41 per cent in 2011. However, the male cultivators indicate a gradually declining trend.

15 Per 1,000 distribution of persons among agricultural households by principal status during the agricultural year of 2012–13 for each size class of land possessed (female-Maharashtra) as given by NSSO is presented next:

Size class of land possessed (hectares)	Self-employed		Regular employment		Casual wage labour		Other
	Agriculture	Non-agriculture	Agriculture	Non-agriculture	Agriculture	Non-agriculture	
<0.01	92	7	0	1	273	47	579
0.01–0.40	271	16	0	18	128	12	556

Size class of land possessed (hectares)	Self-employed		Regular employment		Casual wage labour		Other
	Agriculture	Non-agriculture	Agriculture	Non-agriculture	Agriculture	Non-agriculture	
0.41–1.00	406	4	0	5	98	6	480
1.01–2.00	478	19	0	5	61	6	432
2.01–4.00	545	5	0	2	23	1	422
4.01–10.00	493	4	0	3	7	0	493
10.00+	371	0	0	0	0	0	630
All Size	442	10	0	6	69	6	467

Source: *Some Characteristics of Agricultural Households in India (2012–13)*, NSS 70th Round, Report No.569, January–December 2013, p. A-875.

16 For details on sectoral distribution on employment (ps+ss) in rural non-farm sector of Maharashtra, see Misra (2014).
17 Bhoodan Movement was launched by Acharya Vinoba Bhave in 1951, motivating people to donate land to the landless. Within a few years, more than four million acres of land was donated, and the movement reached its peak in late 1960s.
18 Of the 29 undivided districts (excluding Greater Mumbai), 8districts (Bhandara, Jalgaon, Kolhapur, Nashik, Pune, Sangli, Satara, and Solapur) are catagorised as agriculturally advanced states, 9 districts (Ahmednagar, Aurangabad, Buldhana, Dhule, Jalna, Latur, Nagpur, Nanded, and Osmanabad) as backward districts, and the remaining 12 districts (Akola, Amravati, Beed, Chandrapur, Gadchiroli, Parbhani, Raigarh, Ratnagiri, Sindhudurg, Thane, Yavatmal, and Wardha) as most backward districts. The newly created districts after 1991 have been merged with their respective parent districts as data were not available on these districts for previous years.
19 In advanced districts like Satara, Sangli, Pune, Kolhapur, and Nashik, females lag far behind males in terms of both LFPR and WPR. In backward and most backward districts, the disparity is relatively less. For more details, see the *Report on District Level Estimates for the State of Maharashtra 2013–14*, Government of India, Ministry of Labour and Employment, Labour Bureau, Chandigarh, pp. 15–16.
20 The data used here are drawn from the sample households of four villages of Maharashtra selected by ICRISAT for its Second Generation Village Level Study in 2008. A total of 444 sample households was drawn from these 4 villages (Shirapur-144, Kalman-106, Kanzara-101, and Kinkheda-93).

Chapter 6

Rural poverty and labour migration

In a country like India where disparity in economic development within and across states is growing rapidly in the context of recent growth patterns and policies, labour migration assumes more importance than ever before. Arguably, under the hegemony of neoliberal capitalism, migration has resurfaced as an agency for development in recent years. However, in the literature on development studies in India, migration, especially of rural labourers, has not received due attention. While national-level data highlight that migration is selective with opportunities biased against the poorer,[1] the available limited micro-level studies often report very high rates of migration among the poor, and migration is considered a part of livelihood strategy of the labouring poor (Connell *et al.* 1976; Breman 1996; Rogaly *et al.* 2001; de Haan 2002; Deshingkar and Start 2003). In view of this micro-macro paradox,[2] analysis of patterns of rural labour migration with reference to a developed state like Maharashtra, where studies on labour migration rarely exist,[3] deserves high research priority.

In the available literature, Maharashtra is viewed as the destination for migrant labourers from labour-surplus poor states like Bihar, Odisha, and Uttar Pradesh (Deshingkar *et al.* 2006; Rodgers and Rodgers 2010; de Haan 2011). To put it somewhat differently, Maharashtra is generally considered a labour-scarce state.[4] However, it is reported that in recent decades, the number of agricultural labourers in the state grew considerably. As per the census estimate, the number of rural agricultural labourers increased from 7.8 million in 1991 to 10 million in 2011. Moreover, it is estimated that the proportion of landless families in the state increased from 43 per cent in 1993–94 to 50 per cent in 2011–12 (Rawal 2013). According to a recent (2009–10) estimate of the government of

India, as per the Tendulkar committee approach, one out of three people in rural Maharashtra, amounting to 18 million were poor. The highlight of the *World Development Report* (2009: 163) is that the largest flows of migrants – within districts, across districts, and across states – were from lagging rural to leading rural areas is important in the context of Maharashtra, given its glaring regional disparity in agricultural development.[5] Large tracts of cultivated area in Vidarbha and Marathwada are without irrigation and crop failure and drought were regular visitors in these regions (Mohanty 2001a). In such a context, the present chapter makes an attempt to analyse the emerging pattern of migration and to what extent the labouring poor opt for migration as a coping strategy as found in poor states. The chapter is divided into four sections. Section II provides an account of the debates on labour migration. Section III describes the characteristics of rural labour force. Section IV analyses poverty and migration trends at the district level, and Section V documents incidences of seasonal migration and analyses the impact of the National Rural Employment Guarantee Act (NREGA) on labour migration. The last section presents concluding remarks.

Debate on labour migration

The early debate on labour migration in India was based on two main approaches: *Marxian* and *neoclassical*. The Marxist scholars viewed migration as fulfilling an essential function in capitalist development. With the introduction of capitalistic production into agriculture, the demand for an agricultural labouring population fell as a result of which the agricultural population was constantly on the point of passing over into the urban or manufacturing proletariat. They argued that the capitalist agrarian transition following the green revolution would result in the impoverishment of small peasantry, leading to a reserve army of labour which would eventually be forced to join the urban-industrial labour force (for example, Breman 1985). This line of argument, which was linked with the debate over the mode of production in Indian agriculture, however, died down in due course, as there was little evidence to show that pre-capitalist peasant social formation in India was getting transformed through commercialisation, depeasantisation, and proletarianisation under capitalist farming.

On the other hand, the neoclassical approach based on the models of Lewis[6] (1954), and Todaro[7] (1969) explains migration on the geographical differences in the supply and demand for labour. The resulting differentials in wages caused workers to move from low-wage, labour-surplus regions to high-wage, labour-scarce region. It was based on the assumption that migration decisions were made on the rational cost-benefit calculation of income-maximising labourers operating in well-functioning market. Neoclassical migration perspective viewed rural-to-urban migration as an integral component of the development process by which surplus labour in the rural sector supplied the workforce for the urban-industrial sector (Lewis 1954; Todaro 1969). However, evidences accumulated over several decades from the early village studies, and subsequent analysis on labour migration did not support the arguments of the neoclassical approach (Connell et al. 1976; Nayyar 1978; Bharadwaj 1989).

Stated precisely, the debate between neoclassical and Marxian scholars tapered off as migration in the1970s, the 1980s, and the 1990s from rural-to-urban areas showed a declining trend (Mohanty 2016: 21). However, with the advent of neoliberal globalisation, which pursued rapid transformation of agrarian rural economy, the emerging pattern of migration generated a new debate.

Based on his analysis of the declining importance of agriculture and village economy, Gupta (2005) indicated the sign of the 'vanishing village' and the rising trend of rural-to-urban migration. To him, "the villager is as bloodless as the rural economy is lifeless. From rich to poor, the trend is to leave the village" (2005: 757). Similar arguments were also put forth by others. It was estimated that the number of dispossessed and displaced peasants in India who would migrate from rural-to-urban areas by 2015 was expected to be equal to twice the combined population of the United Kingdom, France, and Germany (Sharma 2007, as cited in Araghi 2009).

Taking Sanyal's (2007) lead, Chatterjee (2008) challenged the hypothesis on the vanishing trend of the village and agriculture in India, and argued that such a transition is unlikely to take place due to state interventions in response to the demand politics of electoral democracy. According to him, with continuing rapid growth of global capitalism, more and more primary rural producers lost their means of production, the government welfare policies and programmes which were launched under the conditions of democratic politics tended to reverse the effect of primitive accumulation. In Chatterjee's (2008: 62) words,

> there will be more and more primary producers, i.e., peasants, artisans and petty manufacturers, who will lose their means of production. But most of these victims of primitive accumulation are unlikely to be absorbed in the new growth sectors of the economy . . . the passive revolution under conditions of electoral democracy makes it unacceptable and illegitimate for the government to leave these marginalised populations without the means of labour to simply fend for themselves. . . . Hence, a whole series of governmental policies are being, and will be, devised to reverse the effects of primitive accumulation.

On the other hand, a group of scholars of neoliberal orientation subscribed to the view that though the effects of neoliberal reforms created opportunities for employment in the urban-industrial sector, the surplus labouring poor in rural India were unable to take up these employment opportunities due to a lack of necessary human capital. In their view, those who migrated from rural-to-urban areas were mainly the richer ones (Dubey *et al.* 2006; Srivastava 2011; Kundu and Saraswati 2012).

Yet a number of micro-level studies indicated circular or seasonal labour migration due to poverty and deprivation (Reddy 1990; Breman 1996; de Haan 2002; Deshingkar 2006). It is argued that migrant labourers remained footloose, a phenomenon which has led to the continual circulation rather than the permanent out-migration of workers from the countryside. Based on a series of studies in western India, Breman (2010: 3) writes,

> A large number of people who leave the villages nowadays do not "arrive" in the cities. To the extent their mobility is intra-rural in nature, it is usually only for the duration of a season; they return to their place of origin when their presence is no longer required. Thus, labour migration usually labour circulation. My field work for the last half century has focused on the coming and going of these people in south Gujarat.

All these arguments, however, are either based on national-level analysis or with reference to selected states like Gujarat and Andhra Pradesh, and experiences of states like Maharashtra were largely overlooked. In some cases (for example, Chatterjee 2008), the arguments were hypothetical and lacked substantial empirical evidences. Rare attempts have been made to substantiate this debate further.

Characteristics of the rural labour force in Maharashtra

As per the 2011 census, there were 27 million rural main workers in Maharashtra as opposed to 21 million in the 1991 census. However, in terms of percentage, there was a marginal decline in the size of rural main workers from 44 per cent in 1991 to 43 per cent in 2011 (Table 6.1).

While developed districts like Satara, Sangli, and Solapur experienced considerable increase in the size of rural main workers, most of the agriculturally backward districts like Gadchiroli, Raigarh, Chandrapur, and Bhandara reported a noticeable decline. This decline in backward districts may be attributed to the migration of workers to the developed districts. A large section of the rural workforce was engaged in agriculture across the districts. However, in districts like Raigarh, Thane, and Pune, which have large urban centres in proximity to the industrial establishments, the size of agricultural workers declined, indicating a shift towards non-agricultural occupations.[8] In the state as a whole, the proportion of agricultural labourers went up between 1991 and 2011. The growth was significant in agriculturally backward districts. In most of these districts, a vast majority of the rural workers were agricultural labourers. In districts like Amravati, Akola, Yavatmal, and Wardha, each cultivator gets two to three agricultural labourers on an average. The percentage of agricultural labourers to cultivators increased significantly between 1991 and 2011 in most of the districts.

According to the recent estimates of the Labour Bureau though, a majority of the rural workers in most of the districts were self-employed, a significant proportion of them worked as casual labourers (Table 6.2).

The higher proportion of casual labourers belonged mostly to the backward districts like Nanded, Parbhani, and Latur. Wage and salaried employees were fewer across the districts, barring Thane, and contract workers were negligible in number. Though in some districts like Kolhapur, a large portion of the rural workers worked for the whole year in a number of districts like Gadchiroli,

Table 6.1 Rural main workers and agricultural labourers

District	Rural main workers as percentage to rural population		Agricultural workers as percentage to main workers		Agricultural labourers as percentage to rural main workers		Agricultural labourers as percentage to cultivators	
	1991	2011	1991	2011	1991	2011	1991	2011
Ahmednagar	44.5	47.9	81.4	82.1	28.9	27.5	55.1	50.4
Akola	46.9	43.8	88.9	90.2	58.1	61.5	188.1	213.7
Amravati	45.7	40.8	88.0	88.6	61.8	66.3	235.9	296.7
Aurangabad	45.8	46.3	87.1	87.5	34.9	34.3	66.9	64.4
Beed	44.8	48.5	87.4	88.4	32.7	31.2	59.9	54.6
Bhandara	48.1	34.7	78.2	77.4	34.7	44.8	79.7	137.0
Buldhana	48.6	46.4	89.9	90.6	46.4	53.4	106.4	143.4
Chandrapur	47.4	40.7	81.4	82.5	42.1	50.1	107.3	155.0
Dhule	43.0	43.8	87.2	88.4	44.3	55.8	103.1	171.2
Gadchiroli	46.7	37.5	88.1	84.5	29.0	35.2	49.2	71.4
Jalgaon	44.6	43.6	86.0	87.6	50.6	60.1	142.6	219.0
Jalna	47.1	46.5	90.5	88.6	38.9	34.8	75.2	64.8
Kolhapur	42.1	41.6	76.5	70.7	20.0	17.8	35.5	33.6
Latur	42.6	43.4	87.4	85.1	43.0	46.7	96.8	121.5
Nagpur	46.8	42.9	78.0	76.4	43.6	48.5	126.6	174.0
Nanded	43.3	43.4	87.4	86.5	45.6	48.1	109.0	125.1
Nashik	47.7	48.1	86.4	88.3	30.3	36.4	54.0	70.2
Osmanabad	43.9	45.3	87.0	85.6	43.0	40.8	97.7	91.1
Parbhani	46.3	46.9	89.8	90.2	45.4	43.2	102.2	91.7
Pune	43.3	45.6	76.5	68.9	23.1	20.8	43.4	43.1
Raigarh	41.1	32.4	71.7	50.6	20.0	20.5	38.8	68.2
Ratnagiri	38.1	35.3	77.9	68.3	10.1	20.0	14.8	41.3
Sangli	38.6	40.2	78.0	76.0	27.1	26.8	53.2	54.3
Satara	37.7	41.2	76.8	74.1	23.1	23.3	43.2	45.9
Sindhudurg	39.0	26.2	76.5	60.7	12.7	18.5	19.9	43.7
Solapur	42.1	44.4	81.8	83.3	39.4-	37.5	92.9	81.7
Thane	43.9	34.8	70.0	55.7	22.9	27.3	48.7	96.1
Wardha	47.4	45.6	84.9	84.5	52.5	54.5	161.7	181.5
Yavatmal	47.7	46.5	87.6	90.0	57.9	60.0	195.5	200.6
Maharashtra	44.2	43.1	82.9	81.6	36.6	39.4	79.1	93.5

Source: *Primary Census Abstract* (Maharashtra State), relevant issues.

Note: The newly created districts have been merged with the older districts, as data for 1991 census is not available for these new districts.

Table 6.2 Distribution of rural workers by activity and duration of work (2013–14)

Districts	Distribution of workers according to activity (%)					Distribution of workers available for 12 months but actually work done according to ps + ss status (%)			
	Self-employed	Wage/salaried employee	Contract worker	Casual labour		Worked for 12 months	Worked 6–11 months	Worked 1–5 months	Did not get any work
Ahmednagar	63.5	8.3	1.9	26.3		77.3	21.7	0.5	0.5
Akola	56.0	11.0	0.4	32.5		39.6	55.0	1.4	4.1
Amravati	56.6	7.5	0	35.9		65.4	33.0	0.3	1.3
Aurangabad	62.9	2.9	3.0	31.2		50.5	46.0	0.1	3.3
Bhandara	76.2	6.1	3.1	14.5		50.9	46.5	0.3	2.3
Bid	66.6	4.2	0.6	28.7		55.4	43.4	0.2	1.0
Buldhana	65.0	6.0	0.1	28.8		29.3	68.7	0.6	1.4
Chandrapur	71.1	10.4	1.7	16.9		72.1	27.2	0.3	0.4
Dhule	74.1	3.3	0.2	22.4		47.5	51.1	0.3	1.1
Gadchiroli	55.2	3.4	1.9	39.5		10.7	88.8	–	0.5
Gondia	66.8	7.4	3.1	22.7		59.5	36.9	0.1	3.4
Hingoli	55.5	6.0	0.5	38.1		68.2	30.2	0.5	1.1
Jalgaon	65.3	5.2	0.5	29.0		43.6	53.8	0.2	2.3
Jalna	61.7	3.1	2.8	32.4		42.9	53.8	–	3.3
Kolhapur	65.7	21.5	1.8	11.0		90.3	8.4	0.2	1.1
Latur	53.9	4.1	0.3	41.8		63.5	36.4	–	0.1
Mumbai	–	–	–	–		–	–	–	–

(Continued)

Table 6.2 Continued

Districts	Distribution of workers according to activity (%)				Distribution of workers available for 12 months but actually work done according to ps + ss status (%)			
	Self-employed	Wage/salaried employee	Contract worker	Casual labour	Worked for 12 months	Worked 6–11 months	Worked 1–5 months	Did not get any work
Nagpur	58.4	8.6	1.3	31.8	59.7	38.5	0.2	1.6
Nanded	57.3	2.7	0.1	40.0	41.2	56.5	0.6	1.8
Nandurbar	63.4	1.8	6.1	28.7	19.3	80.5	0.2	–
Nashik	52.3	4.1	2.4	41.1	25.1	73.1	–	1.8
Osmanabad	62.1	7.3	0.6	29.9	35.1	63.2	0.9	0.7
Parbhani	48.5	8.8	0.4	42.3	69.0	29.8	0.2	1.0
Pune	57.1	9.9	8.7	24.3	68.5	29.2	0.6	1.7
Raigarh	41.8	19.9	6.9	31.5	52	39.2	0.6	8.1
Ratnagiri	64.3	17.0	1.5	17.1	80.4	12.9	1.5	5.1
Sangli	67.9	12.5	1.0	18.6	74.9	23.9	0.6	0.7
Satara	61.2	11.5	1.6	25.6	56.2	42.4	0.3	1.1
Sindhudurg	64.7	14.5	1.7	19.0	38.8	45.0	8.0	8.1
Solapur	49.7	7.6	0.4	42.2	80.2	18.9	–	1.0
Thane	35.2	31.1	1.8	31.9	65.4	31.5	0	3.1
Wardha	62.5	9.2	4.7	23.6	51.1	48.2	0.6	0.1
Washim	64.0	5.2	0.5	30.2	32.2	66.6	0.3	1.0
Yavatmal	73.6	1.8	2.9	21.7	28.4	70.6	0.2	0.8
Overall	60.5	8.3	2.0	29.1	55.3	42.7	0.4	1.7

Source: Report on District-Level Estimates for the State of Maharashtra 2013–14; Labour Bureau; Ministry of Labour and Empowerment; Government of India, Chandigarh.

Table 6.3 Profile of rural labour households (2004–05)

Sl. no.	Particulars	No.
1.	No. of rural households (000)	11,955
2.	Percentage of rural labour households	45.8
3.	Percentage of rural SC labour households to total rural labour households	21.9
4.	Percentage of rural ST labour households to total rural labour households	16.7
5.	Percentage of agricultural labour households to total rural labour households	79.3
6.	Percentage of SC agricultural labour households to total rural agricultural labour households	22.6
7.	Percentage of ST agricultural labour households to total rural agricultural labour households	17.8
8.	Percentage of rural labour households without land	65.1
9.	Percentage of SC rural labour households without land	75.3
10.	Percentage of ST rural labour households without land	67.2
11.	Average no. of agricultural labourers per rural labour households	1.8
12.	Average no. of non–agricultural labourers per rural labour households	0.3
13.	Average daily earning of rural labour (male) households from agricultural occupations (Rs.)	44.6
14.	Average daily earning of rural labour (female) households from agricultural occupations (Rs.)	26.2
15.	Average daily earning of rural labour (male) households from non-agricultural occupations (Rs.)	73.9
16.	Average daily earning of rural labour (female) households from non-agricultural occupations (Rs.)	39.7

Source: (1) *Report on General Characteristics of Rural Labour Households*, Rural Labour Enquiry (61st Round of NSSO) 2004–05. (2) *Report on Wages and Earnings of Rural Labour Households*, Rural Labour Enquiry (61st Round of NSSO) 2004–05.

Nandurbar, and Yavatmal; the workers worked mostly for a period of 6–11 months. The proportion of rural workers who did not get any work or worked only for a period of one to five months was negligible across the district. Nevertheless, in few districts like Raigarh and Sindhudurg, around 8 per cent of the rural workers did not get any work in terms of principal and subsidiary status.

Of the total number of rural households in Maharashtra, labour households constituted 46 per cent in 2004–05 and nearly 80 per cent of them were agricultural labour households (Table 6.3). While 22 per cent of these households belonged to SCs, 17 per cent were from STs. A large majority of rural labour households were asset poor. Over 65 per cent of them were without land, and the proportion of landless labour households was still higher among the SCs and STs. Regarding the average daily earning of rural labour households, while male labourers received Rs. 44 from agricultural occupations and Rs. 74 from non-agricultural occupations, the women on the other hand earned between Rs. 26 and Rs. 40 in 2004–05.

Rural poverty and migration pattern

Though Maharashtra is one of the developed states, except WM and a few districts like Nagpur from other regions, the rest of the state was economically backward and poor. According to the estimate from the Kelkar Committee (2013), in most of the districts of the Marathwada and Vidarbha regions, the per capita income was much lower than that of the state average (Table 6.4).

In districts like Nadurbar, Gadchiroli, and Hingoli, the per capita income was almost half of the state average and almost all of these regions were backward in agricultural development.[9] In terms of rural poverty ratio (as per the planning commission estimate based on conventional approach), the state was above the national average during the 1970s and 1980s. Only after the 1990s, it came close to the national average (Table 6.5).

However, the estimate based on the Tendulkar Committee approach reported higher proportion of rural poor in the state in 2004–05, which comes to 48 per cent. The estimate shows that nearly 19 million people were poor in rural Maharashtra in 2009–10, amounting to 30 per cent of the total rural population. The proposed methodology of the recent Rangarajan Committee, which was set up in response to the criticism of Tendulkar's methodology, however, lowered the poverty ratio considerably. Though the planning commission estimates indicate the downward trend of rural poverty over the years, the comparability of the recent estimate with the earlier ones was questioned by several scholars due to change in the recall period in the NSSO survey.[10] None of the alternative estimates considerably changed Maharashtra's position

Table 6.4 District-wise per capita income (2008–09 to 2011–12) at current prices

Sl. no.	Districts/regions	Per capita income (Rs.)	Sl. no.	Districts/regions	Per capita income (Rs.)	Sl. no.	Districts/regions	Per capita income (Rs.)
1.	Aurangabad	74,769	13	Yavatmal	45,755	25	Dhule	52,155
2.	Jalna	45,644	14	Wardha	56,511	26	Nandurbar	35,587
3.	Parbhani	46,465	15	Nagpur	86,273	27	Jalgaon	62,575
4.	Hingoli	36,746	16	Bhandara	50,803	28	Ahmednagar	59,675
5.	Beed	44,059	17	Gondia	44,302	29	Pune	112,247
6.	Nanded	41,581	18	Chandrapur	62,511	30	Satara	63,420
7.	Osmanabad	43,886	19	Gadchiroli	38,067	31	Sangli	65,147
8.	Latur	45,971		**Vidarbha**	**57,079**	32	Solapur	59,573
	Marathwada	**49,653**	20	Thane	113,483	33	Kolhapur	78,887
9.	Buldhana	41,974	21	Raigad	95,083		**Rest of Maharashtra**	**81,719**
10.	Akola	52,098	22	Ratnagiri	63,916		**Maharashtra**	**69,953**
11.	Washim	43,588	23	Sindhudurg	66,849			
12.	Amravati	52,892	24	Nashik	72,162			

Source: GoM (2013: 446).

Table 6.5 Incidence of rural poverty in Maharashtra and India (percentage of poor)

Assessment year	Lakdawala methodology		Tendulkar methodology		Estimate based on the Rangarajan Committee's proposed methodology	
	Maharashtra	India	Maharashtra	India	Maharashtra	India
1973–74	57.7	56.4	–	–	–	–
1977–78	64.0	53.1	–	–	–	–
1983	45.2	45.7	–	–	–	–
1987–88	40.9	39.1	–	–	–	–
1993–94	37.9	37.3	–	–	–	–
1999–2000	23.7	27.1	–	–	–	–
2004–05	29.6	28.3	47.9	41.8	–	–
2009–10	–	–	29.5	33.8	27.6	39.6
2011–12	–	–	24.2	25.7	22.5	30.9

Source: *Report of the Expert Group to Review the Methodology for Measurement of Poverty*, Planning Commission, Government of India, June 2014.

in relation to national averages (Mishra and Panda 2006). Even the estimate made by Deaton and Dreze (2002) indicated higher incidence of poverty in rural Maharashtra as compared to the national average by as much as 5 per cent. To put it precisely, all the available evidences on poverty estimates (barring the Rangarajan Committee) showed that the proportion of poor in rural Maharashtra was about the same as that in the national average.

Though poverty ratio in the state as a whole declined the district-level estimates[11] showed a significant upward trend in a number of districts between 1993 to 1994 and 2004 to 2005 (Table 6.6).

Higher incidence of rural poverty was noticed in agriculturally backward districts like Gadchiroli, Parbhani, Beed, and Yavatmal. Poverty ratio in the backward district of Gadchiroli was as high as 52 per cent in 2004–05. Almost all the agriculturally advanced districts[12] experienced decline in poverty over the years. It is to be noted that the higher the incidence of poverty, the greater the concentration of agricultural labourers and vice versa. In the context of the high poverty ratio and greater incidences of agricultural labourers, the agriculturally backward districts like Beed, Gadchiroli,

Table 6.6 District-wise rural poverty in Maharashtra

District	1993–94		2004–05	
	Average per capita consumption per month (rupees at current prices)	Incidence of poverty (percentage)	Average per capita consumption per month (rupees at current price)	Incidence of poverty (percentage)
Ahmednagar	269.2	33.4	622.4	10.8
Akola	260.3	30.8	548.4	22.6
Amravati	282	24.2	475.7	29.4
Aurangabad	283.2	34.9	450.4	34.1
Beed	251.8	34.2	469.6	36.4
Bhandara	293.4	23.3	526.6	30.1
Buldhana	245.2	37.8	581.1	21.3
Chandrapur	284.7	32	623.1	26.3
Dhule	232.3	45.6	489.7	27.1
Gadchiroli	301.4	28.6	413.4	51.9
Gondia	–	–	484	34.6
Hingoli	–	–	613.9	24.8
Jalgaon	239.8	37.2	507.9	24.9
Jalna	280.8	31.3	638	24.3
Kolhapur	375	7	626.7	10.6
Latur	343.6	24.7	556.6	36.8
Nagpur	310.2	24.2	507.7	30.3
Nanded	287.3	34.2	502.1	35.2
Nandurbar	–	–	470.9	44
Nashik	270.8	36	468.5	37.5
Osmanabad	369.9	16.7	667.3	16.1
Parbhani	301.8	28.6	478.6	38.7
Pune	318.7	17.9	824.9	7.1
Raigarh	398.8	4.9	651.1	21.1
Ratnagiri	311.8	24.9	638.2	16.6
Sangli	401.6	14.4	583.9	12.1
Satara	300.1	22.6	637.7	9
Sindhudurg	286.8	20.5	565.5	10.7
Solapur	278.3	31.2	638.2	10.9
Thane	430.9	8.3	627.8	35.2

(Continued)

Table 6.6 Continued

District	1993–94		2004–05	
	Average per capita consumption per month (rupees at current prices)	Incidence of poverty (percentage)	Average per capita consumption per month (rupees at current price)	Incidence of poverty (percentage)
Wardha	308.8	27.4	575.3	11.5
Washim	–	–	516.5	23.2
Yavatmal	270.4	25.7	485.9	33.4
Mumbai	–	–	–	–
Maharashtra	**302.9**	**26.6**	**570.4**	**24.2**

Sources: *Maharashtra Human Development Report 2012* (Appendix C, p. 155).

Note: Estimates of poverty for 1993–94 and 2004–05 correspond to the official state-level rural poverty lines of Rs. 194.9 and Rs. 362.3 per capita per month, respectively.

Yavatmal, and Parbhani experienced regular migration as a coping strategy of labouring poor. It is reported that labourers from these districts migrated mostly to the agriculturally advanced regions of Maharashtra and the adjoining state of Gujarat (Breman 1985; Neelamma 2010; Bansode 2011; Nandy 2013). As a result of the large influx of labourers from backward areas to the developed districts who offered to work for less wages, the native labourers were compelled to migrate to the urban areas of Mumbai, Pune, and other cities in search of better employment. The *Mumbai Human Development Report* (2010) revealed that nearly two million migrants in Mumbai were from Maharashtra in 2001, and most of them were from Satara, Sangli, Ratnagiri, and Sindhudurg. According to the recent estimate of the NSSO (64th Round), a large proportion of the migrants of Mumbai came from the rural areas of Maharashtra. Of every 1,000 migrants in urban Maharashtra, 370 came from villages within the state.[13] An analysis of intra-state migration stream of 1991 and 2001 population census indicates that nearly 40 per cent of the rural-to-urban migrants in the state went to three major cities: Mumbai, Pune, and Thane. Nashik received about 8 per cent of the rural migrants from various parts of Maharashtra. However, the rate of rural migration to Mumbai, which was the major urban destination, declined to 15 per cent in 2001 as opposed to the 20 per cent in 1991.

Table 6.7 Migration from Maharashtra to other states

States	Rural to rural		Rural to urban		Urban to rural		Urban to urban	
	1991	2001	1991	2001	1991	2001	1991	2001
Andhra Pradesh	11.3	10.7	7.9	5.3	6.3	6.6	7.4	6.4
Bihar	0.1	3.0	0.3	1.1	0.5	2.2	0.5	0.9
Goa	2.1	1.8	1.9	2.0	6.6	4.5	3.2	3.9
Gujarat	30.5	28.5	47.4	56.4	18.6	12.6	27.9	26.2
Haryana	0.2	0.5	0.5	1.0	0.7	1.8	1.2	1.5
Karnataka	25.1	29.1	9.1	8.0	18.9	19.4	13.4	14.5
Kerala	0.6	0.7	0.5	0.6	7.0	6.1	2.5	2.7
Madhya Pradesh	25.1	18.8	24.4	14.8	17.9	15.2	21.1	19.3
Odisha	0.4	0.2	0.2	0.3	0.7	1.0	0.4	0.7
Punjab	0.4	0.6	0.8	1.3	2.0	2.3	1.3	1.4
Rajasthan	0.5	0.7	1.5	1.4	5.8	8.7	3.4	3.7
Tamil Nadu	0.2	0.5	0.8	1.3	2.7	2.8	3.8	3.7
Uttar Pradesh	1.4	0.8	1.6	1.0	7.1	6.0	3.8	3.3
West Bengal	0.2	0.5	0.6	0.4	1.0	2.8	1.3	1.6
Other States and UTs	1.8	3.7	2.4	5.0	4.2	7.8	8.8	10.2
Total	100.0 (195,601)	100.0 (235,222)	100.0 (164,231)	100.0 (215,078)	100.0 (107,844)	100.0 (109,060)	100.0 (300,019)	100.0 (319,581)

Source: Migration Tables (D-2), Census of India, relevant years.

Note: Figures in parenthesis denote total number of out migrants from Maharashtra to other states.

Rural-rural, rural-urban, and urban-rural migration from Maharashtra to other states is presented in Table 6.7.

It is evident that rural-to-rural migration in Maharashtra was more compared with other streams. In 2001, the number of rural-to-rural out migrants from Maharashtra was 0.235 million as opposed to 0.196 million in 1991. Urban-to-rural migration decreased, and the increase between 1991 and 2001 was negligible. Maximum number of migrants from rural Maharashtra went to the rural and urban areas of the adjoining states of Gujarat, Karnataka, and Madhya Pradesh. Among the states Gujarat received the highest number of migrants in terms of rural-to-rural and rural-to-urban migration. For more than half of the rural-to-urban and nearly one-third of the rural-to-rural migrants, Gujarat was the place of destination. Several micro-level studies corroborate this trend (for example, see Breman 1985, 2010).

Looking at the intra-district migration stream (Table 6.8), it is found that rural-to-rural migration is predominant, and it shows a rising trend in most of the districts.

Among the districts, higher rates of rural-to-rural migration were experienced in Gadchiroli, Beed, Bhandara, Dhule, and Yavatmal. As regards the rural-urban stream, higher rates of migration were reported in Aurangabad, Nashik, Pune, Jalgaon, and Amravati districts, which had bigger urban centres. The rural-to-urban migration was invariably less compared to the rural-to-rural migration, and in the majority of the districts, it declined between 1991 and 2001. In districts like Sindhudurg, Ratnagiri, and Gadchiroli, people rarely migrated from rural-to-urban areas within the district. Urban-to-rural migration, though negligible, was relatively higher in few districts like Sindhudurg, Ratnagiri, Amravati, and Akola. In many districts, this stream of migration showed a marginal increase indicating a pattern of increasing circular migration. In a nutshell, it is evident that while rural-to-rural intra-district migration was more common in the backward and poverty-ridden districts, rural-to-urban migration was prominent in districts which were better off and comparatively advanced.

Although census data indicate streams of migration at the district level, it does not provide information on other important aspects, such as temporary and household migration, economic class of the migrants, and their activity status before and after migration. Hence, for more details, it is essential to analyse NSSO data, though it does not provide district-level information.

Table 6.8 Intra-district migration stream

Districts	Total migrants		Percentage to total intra-district migrants					
			Rural to rural		Rural to urban		Urban to rural	
	1991	2001	1991	2001	1991	2001	1991	2001
Ahmednagar	253,677	427,101	73.8	76.1	13.1	13.8	8.7	5.9
Akola	200,124	192,610	61.4	60.4	20.2	17.3	10.3	12.9
Amravati	155,734	227,003	59.9	54.8	21.7	22.7	11.6	13.8
Aurangabad	138,396	219,551	67.0	61.5	22.6	26.9	5.8	6.4
Bhandara	111,851	193,649	76.3	78.8	13.6	9.7	6.0	8.6
Beed	122,815	156,632	66.9	73.2	22.3	14.8	5.1	7.0
Buldhana	201,533	206,427	67.1	66.2	17.4	17.2	9.2	9.9
Chandrapur	127,364	238,924	76.8	61.2	9.3	19.8	10.9	9.6
Gadchiroli	42,873	96,408	77.5	87.9	15.7	6.8	4.4	4.2
Jalgaon	325,507	379,898	63.2	61.9	18.4	18.2	9.3	8.4
Jalna	62,345	115,424	73.4	73.0	18.7	12.9	4.3	6.3
Kolhapur	258,179	380,076	67.8	66.1	15.2	14.5	10.9	11.8
Latur	119,262	157,717	61.1	62.4	29.9	25.0	4.9	6.1
Nagpur	90,433	227,474	57.2	50.1	20.1	23.8	14.8	13.5
Nanded	191,267	200,002	64.2	67.4	22.5	21.4	6.2	5.7
Dhule	95,377	258,627	49.0	77.5	35.8	12.8	3.7	5.4

(Continued)

Table 6.8 Continued

Districts	Total migrants		Percentage to total intra-district migrants					
			Rural to rural		Rural to urban		Urban to rural	
	1991	2001	1991	2001	1991	2001	1991	2001
Nashik	326,895	580,319	68.0	67.2	16.4	19.5	7.9	6.3
Osmanabad	88,354	85,064	67.9	70.2	20.7	17.9	5.5	5.3
Parbhani	192,611	169,572	63.5	67.5	23.7	20.2	4.6	4.5
Pune	250,977	562,464	52.0	51.8	20.5	20.4	16.3	7.5
Raigarh	144,715	202,268	69.8	62.6	14.5	17.3	9.2	10.1
Ratnagiri	133,950	183,105	77.4	72.2	7.6	8.8	12.6	14.9
Sangli	155,647	239,750	64.7	67.3	19.7	16.4	9.3	9.3
Satara	258,906	358,251	77.9	71.4	10.9	10.7	8.8	11.9
Sindhudurg	71,441	89,546	71.0	70.0	6.6	6.1	19.4	20.6
Solapur	287,316	339,733	67.2	67.9	12.5	13.3	15.0	10.7
Thane	138,270	413,617	59.4	44.4	15.5	16.1	6.8	4.2
Wardha	101,793	114,918	63.0	59.4	19.1	16.2	12.0	16.9
Yavatmal	212,864	269,675	71.2	71.0	15.3	14.7	8.9	9.1
Maharashtra	4,860,476	7,285,805	66.5	64.8	17.5	16.8	9.4	8.8

Source: Migration Tables (D-2), Census of India, relevant years.

According to the 64th round of NSSO, around 34 per cent of the rural population in Maharashtra were out migrants, and the percentage of migrants was less among the SCs and STs.[14] Of the total rural out migrants, while 55 per cent migrated within the same district, 36 per cent migrated within the state, but to other districts. Taken together, 94 per cent of the rural out migrants moved within the state.[15] Large sections of rural out migrants migrated only in the recent years.[16] Rural-to-rural migration was more compared to rural-to-urban migration.[17] Of the total rural migrants, nearly 70 per cent came from rural areas of the same district, other districts, and other states. Internal rural migration was predominant within the state and migration from other states was negligible (6 per cent). However, a significant number of urban people also migrated to rural areas within the district and to other districts in the state. Data on household migration also exhibits a similar pattern (Table 6.9).

Table 6.9 Percentage distribution of migrant households by location of the last usual place of residence

Location of the last usual place of residence		*Area*	
		Rural	*Urban*
Same district	Rural	28.7	12.5
	Urban	11.9	7.9
	All	40.6	20.4
Same state but another district	Rural	41.8	21.9
	Urban	17.1	26.0
	All	58.9	47.9
Same state	Rural	70.5	34.4
	Urban	29.0	33.9
	All	99.4	68.3
Another state	Rural	0.4	17.4
	Urban	0.2	13.2
	All	0.6	30.6
Another country		0.0	1.1
All		100.0	100.0

Source: *A Report on Migration Particulars (Maharashtra State)*, 64th Round of NSSO (July 2007–June 2008).

Over 70 per cent of the migrant households of rural Maharashtra were previously residing in rural areas, whereas only 34 per cent of the households in the rural areas moved to urban areas within the state. Migrant households which returned from urban-to-rural areas within the state constitute 29 per cent. It is reported that the households that migrated in search of employment or for better employment opportunities were largely from the lower income group. The well-off section in the rural areas migrated primarily for reasons like studies and transfers of services.[18]

Analysis of the economic activity of rural migrants before and after migration shows that the migrants tended to retain their pre-migration sectors of work (Table 6.10).

Those who were working in the agricultural sector before migration mostly remained in the same sector after migration, though their status changed considerably, particularly in the case of self-employed migrants. The shift from non-agriculture to agriculture was more compared to the shift from agriculture to non-agricultural sector. A significant number of rural people who were self-employed in agriculture, as well as in non-agriculture before migration, became agricultural casual labourers after migration. Similarly, casual and waged labourers retained their same economic status after migration. A majority of the rural migrants had low educational attainments, because of which they found it difficult to get employment in the urban areas.[19] The migration pattern reveals a clear-cut trend towards rural-rural migration as agriculture continued to be the dominant sector of employment before as well as after migration. Though the unemployment rate was negligible among the migrants, a significant number of regular waged non-agricultural labourers (11 per cent) remained unemployed after migration.

Rural-to-rural household migration whether within the district or within the state largely took place for a temporary period (Table 6.11). While within the district, rural-to-rural temporary migration constituted 71 per cent, within the state it was 75 per cent. Rural-to-urban temporary migration came to 66 per cent within the district. Similarly, rural-rural temporary migration from one district to another was 78 per cent as opposed to 37 per cent in the rural-urban stream. On the contrary, migration from urban-to-rural areas was more permanent in nature. While 34 per cent of the rural migrant households moved to urban areas permanently within the

Table 6.10 Distribution of rural migrants by usual principal activity before/after migration (%)

Usual activity before migration		Usual activity after migration											
Status	Industry	Self-employed		Regular wage/salaried		Casual labour		Total employed		Unemployed	Not in labour force		All
		Agri	Non-agri	Agri	Non-agri	Agri	Non-agri	Agri	Non-agri		Student	Others	
Self-employed	Agriculture	12.4	0.3	0	1.2	13.4	0.3	25.8	1.8	0.3	5.4	66.7	100
	Non-agriculture	4.2	25.4	0	3.9	12.6	1.9	16.7	31.2	1.7	15.3	35.1	100
Regular wage/ salaried	Agriculture	0	0	42.3	0.6	11.6	0	53.9	0.6	0	0.6	44.8	100
	Non-agriculture	8.7	1.1	0.2	37.8	11.1	1.9	20	40.8	11.6	13.7	13.9	100
Casual labour	Agriculture	2.6	0	0	0.3	41.4	0.4	44	0.8	0.8	2.4	52.6	100
	Non-agriculture	1.5	1.8	0	0.4	14.8	20.3	16.3	22.5	1	6.9	53.4	100

Source: A Report on Migration Particulars (Maharashtra State), 64th Round of NSSO (July 2007–June 2008).

Table 6.11 Percentage distribution of migrant households by pattern of migration

Location of last usual place of residence		Rural			Urban		
		Temporary	Permanent	All	Temporary	Permanent	All
Same district	Rural	71.1	28.9	100.0	65.9	34.1	100.0
	Urban	45.4	54.6	100.0	47.3	52.7	100.0
	All	63.6	36.4	100.0	58.7	41.3	100.0
Same state but another district	Rural	77.7	22.3	100.0	37.2	62.8	100.0
	Urban	43.4	56.6	100.0	28.2	71.8	100.0
	All	67.8	32.2	100.0	32.3	67.7	100.0
Same state	Rural	75.0	25.0	100.0	47.7	52.3	100.0
	Urban	44.3	55.7	100.0	32.7	67.3	100.0
	All	66.1	33.9	100.0	40.2	59.8	100.0

Source: *A Report on Migration Particulars (Maharashtra State)*, 64th Round of NSSO (July 2007–June 2008).

district, the migration to other districts on a permanent basis was much higher (63 per cent).

The NSSO (64th round) report reveals that the temporary migrants in rural areas were mostly engaged in agriculture, indicating seasonal migration of workers for activities like sowing and harvesting. Over 57 per cent of them were employed in agriculture. Besides agriculture, construction activity in rural areas also provided sizable employment to temporary migrants. Nearly 15 per cent of them in rural areas were employed in construction activities. It is also reported that in terms of the longest duration of work, agriculture accounted for more than 75 per cent of migrants. It is interesting to note that the lower the quintile group of Monthly Per Capita Consumer Expenditure (MPCE) is, greater the concentration of temporary migrants in agriculture.[20] Thus, temporary migrants in the rural areas were mostly the labouring poor.

However, both NSSO and census data misrepresented the relationship between poverty and migration, as they suffered from major shortcomings.[21] They missed out on circular or seasonal migration, and seriously underestimated the mobility of a significant portion of people for whom migration was an integral part of livelihood

(Deshingkar and Farrington 2009; Srivastava 2011). Studies based on these official statistics, therefore, indicates that migration was higher among the better-off groups than among the poor (Dubey *et al.* 2006; Kundu and Sarangi 2007; Srivastava 2011). Analysis of seasonal migration patterns provides firm evidence on the impact of rural poverty on migration.

Seasonal labour migration

Based on national level official statistics, it is reported that in the last few decades, the number of circular and short-duration migrants increased considerably in the country (Srivastava 2011). Temporary migration in rural areas was mostly seasonal in nature. Higher temporary and seasonal migration were reported in a number of village studies (Deshingkar *et al.* 2008; Mosse *et al.* 1997). According to the estimate by Keshri and Bhagat (2012), based on the 64th Round of NSSO, nearly 73 million people were temporary migrants for all ages in 2007–08 in Maharashtra. In their earlier estimate, based on the 55th Round, Maharashtra was one of the top-ranking states in seasonal and temporary migration. It was next to Uttar Pradesh and Madhya Pradesh with 92 million migrants (Keshri and Bhagat 2010). A survey[22] conducted during 2005–06 in Maharashtra covering 115 villages of as many as 20 districts reported that 50 per cent of the surveyed villages and 54 per cent of the households of these villages reported seasonal migration. A recent study undertaken on circular migration based on ICRISAT longitudinal village-level data by Badiani and Safir (2009) in the four villages of Maharashtra, two each from the Akola and Shola-pur districts, revealed that a sizable fraction of households in all villages was seasonal migrants. They also found that in addition to having experienced an increase in the number of circular migrants, the villages witnessed a gradual increase in the number of individuals permanently leaving the villages.

During sugarcane-crushing season (November to April/May), labourers from resource-scarce areas of the central Maharashtra and Marathwada region migrated to the lush sugarcane belt to find employment in sugarcane farm activities. A majority of these labourers came specially from Beed, Osmanabad, Jalgaon, Jalna, Parbhani, Aurangabad, and Latur districts (Chithelen 1980; Hatekar 2003; Smita 2008; Nandy 2013). The sugar belt consisted of the districts in WM-Ahmednagar, Pune, Satara, Sangli, Kolhapur, and Sholapur extending into Surat (Gujarat) in the north and Belgaum (Karnataka) in the south. A study commissioned by Janarth, an Aurangabad-based NGO, estimated that about 0.65 million labourers migrated from central to WM for sugarcane cutting each year (Hatekar 2003). The migrants largely come from Beed. In a recent study by Bansode (2011), it was found that nearly 60 per cent of sugarcane cutters in WM were from Beed. Similar estimates were reported by others.[23]

Labourers from the backward districts also migrated to Gujarat and Karnataka to work in sugarcane field. According to SETU, an NGO based in Ahmedabad, an estimated 0.15 million labourers were received in the Surat district of Gujarat for sugarcane cutting every season, of which 75 per cent were from the Khandesh region of Maharashtra. Teerink (1995) reported compelling conditions of seasonal labour migrants from Khandesh to sugar cooperatives in Gujarat and documented their distressful experiences. As estimated by Janarth, migration into Belgaum was in the range of 25,000 labourers per season. These labourers usually camped in clearings earmarked by sugar factories in the vicinity of sugarcane fields. Each family lived in a small conical hut or *kopi* made of a bamboo mat and poles, and *kopis* are often cramped together, with humans and livestock living in close proximity (Smita 2008).

Migration starts to begin around October to November, with migrant families spending the next six to eight months on the worksite and then returning to their villages before the next monsoon. The labourers were recruited in *tolis* or groups (consisting of 30–100) through agents, popularly known as *Mukadams* appointed by the sugar factory managements. The factory management made advance payments to the *Mukadams*, who booked these labourers well in advance by offering the *Uchal* (advance payment). The *Mukadams* received a commission from the wages of the labourers. The migrant labourers took advances from the *Mukadams* mainly for agricultural activities in their native villages during the monsoon period. They were normally drawn from the poorest strata of the society and were mostly landless or marginal farmers belonging to OBC, SC, and ST categories (Chithelen 1980; Teerink 1995; Hatekar 2003; Smita 2008; Bansode 2011; Nandy 2013). Scarcity of resources and debt burden accompanied with dry spells of the monsoon compelled them to migrate to sugarcane factories in search of a livelihood. Of the two kinds of circular migration among the poor, *viz.* coping and accumulative migration, as classified by Deshingkar and Farrington (2009: 18–19), these seasonal migrants belonged to the coping type, as they hardly accumulated any assets, savings, and investments. Chithelen (1980) categorised these migrants as bonded labourers because whatever they earned could not be used in its entirety to repay the advance due to the *mukadam's* cut – i.e. the commission consumed part of the earnings. It is reported that these labourers took no money back home with them (Breman 1996; Marius-Gnanou 2008). A similar kind of neo-bondage of sugarcane cutters was described elsewhere (Srivastava 2005; Breman 2007; Guérin *et al.* 2009).

Besides sugarcane cutting, large-scale rural-to-rural seasonal migration also took place for other activities like brick making, stone quarrying, and rural construction (roads and houses). A recent study by Neelamma (2010), based on 1,200 seasonal migrant families covering 11 districts, noted that agricultural labourers who were illiterate and mostly belonged to SC, ST, and OBC groups left their villages under distressful conditions to work as labourers in four major industries: brick making, stone quarrying, cotton ginning, and construction. Of these, the largest number of migrant workers was found in construction activity. Compared to agricultural work, these labourers got better wages, though they experienced poor living and working conditions. Navale *et al.* (2014) also documented the migration of seasonal labourers from drought-prone and poverty-ridden areas of Sholapur, Osmanabad, Latur, and Beed to the stone-quarrying industry in Pune. The study undertaken by Jesim (2006) reported that a significant proportion (18 per cent) of the temporary migrant workers in Dharavi's leather accessories manufactures were from rural Maharashtra. It showed that these migrant workers belonged to the poor and landless classes, and they eventually returned to their place of origin in the rural sector, a process termed 'circular migration' by Breman (1996, 1997). A number of recent village studies from other parts of India also revealed a sharp increase in temporary migration, particularly from poverty-prone locations (e.g. Rogaly *et al.* 2001 in West Bengal; Karan 2003 in Bihar).

Impact of NREGA

Maharashtra was a pioneering state to provide employment to the rural poor through Employment Guarantee Scheme (EGS) during the drought years. The scheme was started on a pilot basis in 1965 in Sangli district and was subsequently implemented across the state in 1972, following a severe drought in the state. A statutory basis was provided to the scheme with the enactment of Maharashtra Employment Guarantee Act of 1977. Its primary objective was to create employment opportunities in rural areas for all adults above 18 years of age willing to provide unskilled manual labour on a price-rate basis. However, after an initial success for a few years (Dutt and Ravallion 1992; Dev 1995; Gaiha 1996), EGS coverage declined sharply, particularly in the 1990s. It was reported that in many cases, EGS work was commenced after people migrated. Once the majority of the families of a village migrated, there was not sufficient strength to start any work, which could be completed before the monsoon began. It was observed in one village that the *sarpanch* himself wanted work but did not know how to avail the scheme (Datar 2007). Moreover, the poor, especially the landless and chronically unemployed, were rarely covered under the scheme. The following illustration was a classic example:

> The very poor of Dahanu – the landless labour – migrate to work in brick kilns and other industries in neighbouring Kanwel and Boisar attracted by better wages and immediate payment. They return to Dahanu at Holi (mid-March) and having exhausted their earnings. As EGS earnings begin to get exhausted during the rainy season, recruiters from industries outside offer advances to tide over the rainy season. At the end of the rainy season (October) these recruiters come back to the villages to pick up those who owe them money. They are then made to work in the brick kilns/ factories, being provided with just enough money to survive, in near bonded labour conditions. They are allowed to go back to their villages just before Holi with a pittance as payment. This vicious circle then continues. When EGS work begins in earnest after the rainy season, in December-January, many of the poorest have already migrated.
> *(Krishnaraj et al. 2004: 1742)*

It was widely viewed that EGS functioned more as a relief programme during years/seasons of crisis rather than as an employment guarantee scheme (Dutt and Ravallion 1992; Echeverri-Gent 1988; Dev and Ranade 2001; Ganesh-Kumar *et al.* 2004). In a nutshell, EGS did not have a substantial effect in checking labour migration from the rural areas.

The EGS in its new form as National Rural Employment Guarantee Act (NREGA) came into being by legislation on 25 August 2005. It was implemented on 2 February 2006 in a phased manner, and by 2008, it was expanded to all districts of India. The law was initially called the National Rural Employment Guarantee Act but was renamed on 2 October 2009, as Mahatma Gandhi National Rural Employment Guarantee Act (MGNREGA). One of the major objectives of NREGA was to reduce distress migration by providing employment opportunities on demand. It provided a legal guarantee for at least 100 days of employment in every financial year to adult members of any rural household willing to do public work. In the first phase, the Act was implemented in the most backward districts (12 in number), and in the second phase it was extended to 6 more districts, and the remaining 15 districts were brought under the Act in the third phase.[24] It was believed that the NREGA would have a significant positive impact on seasonal rural-urban migration by providing employment to rural workers during the lean season.

Evaluation of the performance of NREGA, however, revealed that the number of employment days generated by the scheme was grossly inadequate to prevent labour migration in the rural areas (Kajale and Shroff 2011; Shah 2012). There was continuous decline in employment generation particularly during 2008–09 and 2010–11. The number of households provided with employment came down to 451,000 in 2010–11 as opposed to 906,000 in 2008–09.[25] The number of households completing 100 days of employment decreased from 32,000 to 23,000, amounting to a 28 per cent decline during this period (Kajale and Shroff 2012: 3). Moreover, wide deviation was reported in terms of actual employment generation. The actual employment generation was quite less than 100 days in a year in almost all the districts of Maharashtra.[26] A detailed study made by Kajale and Shroff (2011) in five districts revealed that out of 205 households, only 30 per cent could get employment for 100 days, and a significant number of households migrated due to non-availability of work under NREGA. It showed that there was an absolute decline in the number of households provided with employment, man days of employment generated, the amount spent, and the number of works completed. In their analysis, employment under NREGA fell short of creating the level of employment that matched with that created under EGS before 2006.

Recently, it was reported that the MGNREGA's failure prompted large-scale migration of farm labourers from villages of Marathwada region to Mumbai under distressful conditions.[27] A villager from Marathwada narrated,

> Around 100–150 people have left for Pune from our village, another 90 have gone to Andhra Pradesh. No sugar factory, not even in Karnataka, where many of our people go to, is offering advance payments. And, there is no work under the employment guarantee scheme (EGS) here. Villagers have been forced to sell their buffaloes to abattoirs; the next step will be to leave.[28]

According to a recent survey on migrants in Mumbai, only 45 per cent of the total surveyed migrants had the MGNREGA job card, and while 25 per cent worked under the scheme, only 8 per cent had benefitted from the scheme.[29] The Kelkar Committee (2013) also noted that though MGNREGA held promise, the implementation was not upto the desired level, as a result of which seasonal migration continued. It says, in Maharashtra, 0.833 million tribal families with job cards received employment of only 13.39 million labour days in 2011–12 (an average of 16 days per family). Kumar and Chakraborty (2016), in a statewide analysis of the performance of MGNREGA, indicated Maharashtra's relatively poor performance. In their estimate, Maharashtra provided only 40–50 days of employment as opposed to the provision for 100 days. It was reported that out of 19,516 drought-affected gram panchayats in 21 districts in the state, as many as 6,352-gram panchayats did not show any expenditure under MGNREGA.[30]

Conclusion

The pattern of migration among rural labourers was quite complex in Maharashtra and took multiple directions. In the last few decades, the size of labourers, particularly the agricultural labourers, increased considerably, accompanied by a rise in the poverty ratio in many districts. However, unlike many backward states, out-state migration of rural labourers in Maharashtra was negligible. Migration took place largely within the state. The limited out migrants mostly went to the rural and urban areas of Gujarat and Karnataka. Rural-to-rural labour migration was predominant than the other streams. While labourers from the backward regions migrated to rural areas of developed pockets, labourers from the advanced regions migrated to the urban areas. Rural-to-rural migration was mostly temporary and seasonal in nature. The service sector, which emerged in a big way in the urban areas in the recent decades, was unable to attract rural labourers as it required a high-level of skill. The low level of education and skill of rural population was one of the most important reasons combined with the high cost of living in cities, for low rate of rural-urban migration.

The census or NSSO data do not capture adequately the relationship between poverty and migration due to their inherent shortcomings. The available independent surveys, reports, and micro level studies clearly show migration, particularly seasonal and temporary migration as coping strategies of the rural labouring poor to escape from poverty and rising incidences of unemployment, as well as agricultural crisis caused by drought conditions. Neither the argument that vanishing sign of villages and agriculture would lead to rapid rural-urban migration nor does the argument that government intervention would reverse the trend of distress migration holds true in the context of rural Maharashtra. Similarly, the argument that the rich tend to migrate more than the poor is also far from truth. Agriculture and rural economy continue to absorb a vast section of rural labour force. Labour mobility in the rural areas largely took place within the agricultural sector, from backward to developed pockets, and within the different sectors of rural economy. Labour migration was mainly poverty induced. Government intervention through programmes like MGNREGA was not adequate enough to reverse the trend of large-scale distress migration. The findings of the study come closer to Breman's thesis on labour circulation and footloose labour. He rightly observed,

> The periodic drift or workers in an uneven rhythm to inclusion exclusion ranges from short to long distances, extending far beyond the borders of their own state. In an employment arena that is more fluid and much wider in scale than ever before, the rural hinterland functions as an almost inexhaustible reservoir of informal sector workers who leave their homes and after some period of time, mostly return again to their place of origin.
>
> *(Breman 1996: 11)*

Elsewhere Breman *et al.* (2009: 3) also noted,

> This proletariat constitutes a huge reserve army of labour hired and fired according to the need of the movement, in agriculture

but increasingly also in other economic sectors. The extension in the scale of rural labour market gave rise to new patterns of both intra-rural and rural-urban wage labour circulation. Men, women and children became footloose, going off but also coming back again while in many cases their temporary exodus to other destinations was accompanied by an influx of migrants from the more remote hinterland contracted for work for which local labour used to be hired.

Notes

1 The recent NSSO report (2007–08), which shows the incidence of migration among different income groups, indicates higher propensity of migration in the top income deciles than in the lower ones. The analysis of NSS data by de Haan (2011) also suggests that migrants are relatively better off than the non-migrants.

2 The micro-macro paradox with regard to the relationship between poverty and migration has also been indicated by de Haan (2011).

3 In the literature on migration in India, studies on Maharashtra are conspicuously absent. The recently released four important volumes on migration in India (Rajan 2011; Breman *et al.* 2009; Deshingkar and Farrington 2009; Mishra 2016) do not contain a single paper on Maharashtra. *The Indian Journal of Labour Economics*, which brought out a special issue on migration in 2011, also did not include any article on Maharashtra.

4 The recent *World Development Report* (2009: 163) says that internal migrants in India flow to prospering states like Maharashtra and workers from backward states like Bihar, Odisha, Rajasthan, and Uttar Pradesh routinely move to the developed green revolution pockets of advanced states to work on farms. The report also notes that about three million people moved in the second half of the 1990s from the lagging Indian states of Bihar and Uttar Pradesh to prosperous states like Maharashtra and Punjab. Many studies report labour migration to Maharashtra from other states (Breman 1985; Bhagat 2016; Sekhar 2012), and one rarely comes across a study showing a reverse trend.

5 See (1) *Report of the Fact Committee on Regional Imbalance* (Dandekar Committee), 1984, Bombay: Planning Department, Government of Maharashtra; (2) *Report of the High-Level Committee on Balanced Regional Development: Issues in Maharashtra*, 2013, Bombay: Planning Department, Government of Maharashtra. Also see Mohanty (2009).

6 The Lewis model viewed that migration from rural areas would primarily be driven by the existence of surplus labour, along with expanding opportunities of employment for such labour in urban areas. As the industrial sector expands, it would attract unemployed worker from the rural areas who would migrate to urban areas in the expectation of higher urban earnings relative to their earnings in the rural areas. Stated precisely, the Lewis model is founded on the assumption that migration

takes place from the rural sector with low wages and where the marginal product of labour is below zero to the modern capitalist sector in which the wage rate is much higher. Eventually, surplus labour in the traditional sector would disappear being absorbed by the modern capitalist sector.

7 Todaro's (1969) basic two-sector model of rural-to-urban labour migration also considers the role of internal migration in which the urban sector draws labour force from the rural sector.

8 According to NSSO estimates, the number of rural people engaged in non-agriculture in the state as a whole increased from 17 per cent in 1993–94 to 21 per cent in 2009–10 in terms of both principal and subsidiary status. For details, see Misra (2014: 142).

9 In terms of irrigation, mechanisation, area under HYV, and chemical fertiliser consumption, these districts were ranked far below the other districts. For details, see Mohanty (2000a).

10 For details, see Sen and Himanshu (2004) who have examined the alternative poverty estimates.

11 The estimate was made based on the central sample of the NSS for the corresponding rounds; hence the estimates for Maharashtra are not strictly comparable with district-level estimates, which are based on pooled state and central samples.

12 Districts like Satara, Sangli, Pune, Solapur, and Kolhapur are considered to be agriculturally the most advanced districts of Maharashtra (Mohanty 2000a). These districts reported low incidence of rural poverty.

13 See "70% Migrants to Mumbai Are from Maharashtra", *Times of India*, 17 September 2012 (Mumbai Edition). Accessed from http://timesofindia.indiatimes.com/city/mumbai/70-migrants-to-Mumbai-are-from-Maharashtra/articleshow/16428301.cms on 18 August 2016.

14 Among the social groups, 72 per cent from scheduled tribes, 67 per cent from scheduled castes, and 67 per cent from other backward castes were non-migrants. See NSSO, 64th Round, 2007–08.

15 For details, see *A Report on Migration Particulars (Maharashtra State)*, 64th Round of NSSO (July 2007–June 2008).

16 As per NSSO (2007–08) estimate, over 61 per cent of rural out migrants migrated during the last ten years. For details, see *A Report on Migration Particulars (Maharashtra State)*, 64th Round of NSSO (July 2007–June 2008), p. 11.

17 The percentage distribution of persons who stayed away from village/town by the location of the last residence is given next:

Sector	Same district		Other district of same state		Other state		Other country	All
	Rural	Urban	Rural	Urban	Rural	Urban		
Rural	31.3	12.2	33.0	15.9	5.6	2.1	0.0	100.0
Urban	14.1	26.7	13.9	28.8	6.8	7.4	2.3	100.0

Source: *A Report on Migration Particulars (Maharashtra State)*, 64th Round of NSSO (July 2007–June 2008)

18 Distribution of migrant households in rural Maharashtra by reason for each quintile group on MPCE reveals that while 51 per cent of the households belonging to the quintile group of 0–20 migrated in search of employment opportunities more than 36 per cent of households of 20–40 quintile group migrated for the same reason. A significant proportion of households (about 25 per cent) from these quintile groups also migrated in search of better employment opportunities. On the other hand, the migration of households from higher quintile groups (60–80 and 80–100) were mostly caused by reasons such as studies and transfer of services. For details, see *A Report on Migration Particulars (Maharashtra State)*, 64th Round of NSSO (July 2007–June 2008), p. T-20.

19 As per the NSSO estimate, while 45 per cent of the rural migrants are not literates, about 40 per cent have education up to primary and middle level. For details, see *A Report on Migration Particulars (Maharashtra State)*, 64th Round of NSSO (July 2007–June 2008), p. 5.

20 For details, see *A Report on Migration Particulars (Maharashtra State)*, 64th Round of NSSO (July 2007–June 2008), p. 11.

21 Deshingkar and Farrington (2009) listed out the shortcomings like underestimation of seasonal or circular migration, women migration, illegal migration, and movement of members of scheduled castes and scheduled tribes.

22 It was part a of nationwide large-scale survey which was conducted during 2005–06, covering 1,501 villages across 289 districts in the country, by the National Council for Applied Economic Research, New Delhi and University of Maryland, USA, for analysing human development in India. See Desai *et al.* (2010).

23 Recently, it has been reported that every year, about 0.6 million people from Beed district alone temporarily migrate to Karnataka, Telengana, and Western Maharashtra to work on sugarcane fields. See Neerad Pandharipande, "Marathwada's Drought: History of the State's Sugarcane Addiction Long Precedes the Water Crisis". Accessed from www.firstpost.com/india/marathwadas-drought-history-of-states-sugarcane-addiction-long-precedes-water-crisis-2739574.html on 11 July 2016. According to a study done by Dilasa Janvikas Pratishthan, an NGO working with farmers, around 0.8 million people from Beed migrated because of drought to western Maharashtra, other parts of the state, and neighbouring states to work on sugarcane fields and transport sugarcane to factories. See G. Seetharaman, "Beed in Marathwada Bone Dry After Two Years of Drought", *Economic Times*, 10 April 2016. Accessed from http://articles.economictimes.indiatimes.com/2016-04-10/news/72209793_1_marathwada-fodder-and-water-much-water on 11 July 2016.

24 Various districts came under the purview of NREGA in three different phases: Phase I (2006) – Ahmednagar, Amravati, Aurangabad, Bhandara, Chandrapur, Dhule, Gadchiroli, Gondia, Hingoli, Nanded, Nandurbar, and Yavatmal; Phase II (2007) – Thane, Wardha, Buldhana, Osmanabad, Akola, and Washim; and Phase III (2008) – Raigad,

Ratnagiri, Sindhudurg, Nashik, Jalgaon, Pune, Satara, Sangli, Solapur, Kolhapur, Jalna, Parbhani, Beed, Latur, and Nagpur.

25 See *Economic Survey 2011–12*, Government of Maharashtra, p. 188.

26 An analysis of employment generation under NREGA in Maharashtra between 2006 to 2007 and 2009 to 2010 revealed that among various districts, only Nandurbar and Thane reported more than 100 average man days only during 2008–09. In all other districts, the average man days per household generated through NREGA was far below 100 during the entire period. Even in the year 2008–09 that recorded the highest man days created under the scheme, except few backward districts which showed about 40–50 average man days, employment generation was markedly lower in the remaining districts. For details, see Shah (2012)

27 For example, see *India Spend Primary Survey – Migrants in Mumbai*, May 2016, Mumbai.

28 See "Maharashtra: Drought Driving Labourers Out of Sugarcane Fields to Cities". *Hindustan Times*, 13 September 2015. Accessed from www. hindustantimes.com/mumbai/maharashtra-drought-driving-labourers-out-of-sugarcane-fields-to-cities/story-M1D0wqgbsuAQTQohi7uDzM .html on 11 July 2016.

29 For details, see *India Spend Primary Survey – Migrants In Mumbai*, May 2016, Mumbai

30 See the *Hindu*, 10 April 2016, Mumbai, Accessed from www.the-hindu.com/news/cities/mumbai/news/state-ignored-our-warnings-on-drought-yogendra-yadav/article8457273.ece on 22 July 2016.

Chapter 7

Changing response to agrarian crisis

From rebellion to suicides[1]

The agrarian crisis in India and the nature of the peasant response to such a crisis have assumed increasing importance in research and policy debates, particularly in the context of recent widespread agrarian distress manifested in the wave of peasant suicides in alarming numbers.[2] The available literature views peasant suicides as a response to agrarian crisis caused by the rising cost of cultivation, indebtedness, and declining farm income following neoliberal reforms (See, for example, Shiva and Jafri 1998; Vasavi 1999; Deshpande 2002; Mohanty 2005; Mishra 2006b; Singh *et al.* 2016). However, a review of the historical account of the agrarian economy and society reveals that though exploitative land revenue policies and agricultural modernising measures during the colonial period led to the agrarian crisis, as evidenced from widespread indebtedness, land alienation, and impoverishment of peasantry, they did not result in any such suicide waves. Rather, the peasants responded to this crisis through a series of revolts,[3] particularly in the later part of the nineteenth century (Dhanagare 1983; Guha 1983; Hardiman 1992). Moreover, after independence, the peasants in states like Maharashtra, Punjab, and Karnataka also joined organised movements in the 1970s and 1980s, demanding better pricing, subsidy, and export policy in response to the growing crisis in agriculture (Dhanagare 1994; Gill 1994; Omvedt 1994b). It is widely agreed that the achievements of these revolts and movements were not unimpressive; they were followed by various kinds of ameliorative measures (Dhanagare 1983; Radhakrishnan 1989; Mohanty 2001b). In this context, it is pertinent to examine why the peasants in India who showed rebellious spirit during the second half of the nineteenth century and in the aftermath of the green

revolution during the1970s and 1980s resorted to killing themselves in the post-reform period.

This chapter is confined to the state of Maharashtra, which experienced peasant rebellions during the colonial period (like the Deccan
Riots in 1875) against the moneylenders (Kumar 1968; Catanach
1970), organised peasant movements in the post-independence
period (Omvedt 1980; Dhanagare 1994), and emerged as one of the
top-ranking states in terms of incidences of peasant suicides in the
country in the post-reform period (Mohanty 2001a; Mishra 2006a;
Mitra and Shroff 2007; Nagaraj 2008; Sainath 2009). The purpose
of the present study is not to analyse the causes of the Deccan Riots,
the agrarian movements of the 1970s and 1980s, and the current
wave of peasant suicides, but to examine the conditions that led the
peasants to respond to the crisis in varied ways. In this context, it
makes a twofold attempt. First, it tries to compare changes in agrarian conditions and the consequent agrarian crisis[4] in Maharashtra
during the colonial period (particularly in the 1860s and 1870s),
post-independence period (the 1970s and 1980s), and post-reform
period (after the 1990s). Second, it analyses why the peasants[5]
rebelled at one point in time and joined protest movements and
committed suicides at other points.

Theoretical debate

The process of integration of peasant economy into global economy
and the response of peasants to this change has been the subject of
enduring scholarly debate since the 1960s.[6] This debate revolved
around two competing models of collective action of the peasants –
namely, 'moral economy' and 'political economy' (Redfield 1960;
Moore 1966; Hobsbawm 1969; Wolf 1969; Thompson 1971;
Migdal 1974). It has subsequently been refined by the arguments
advanced by James C. Scott (1976), representing the 'moral economy' and the counterarguments made by Samuel L. Popkin (1979)
on behalf of 'political economy', who adumbrated the central tenets
of each position based on the studies of peasant behaviour in the
context of South-East Asia.

According to Scott (1976), the peasants are likely to rebel when the
'ethics of subsistence'[7] and the 'ethics of reciprocity'[8] are violated
through the penetration of capitalism and modern state. He argues
that the effects of colonialism in the process of commercialising

agriculture and enhancing tax[9] increased the risk of the peasants that threatened the ethics of subsistence and reciprocity, and the resulting moral outrage of the peasants led to rebellions. He believes that a peasant always opts for the safety-first principle and follows the principle of risk aversion. Rather than maximising his average return, he prefers to avoid the likelihood of having a disaster. Scott writes,

> Living close to the subsistence margin and subject to the vagaries of weather and the claims of outsiders, the peasant household has little scope for the profit maximization calculus of traditional neoclassical economics. Typically, the peasant cultivator seeks to avoid the failure that will ruin him rather than attempting a big, but risky, killing. In decision-making parlance his behavior is risk-averse; he minimizes the subjective probability of the maximum loss.
>
> *(Scott 1976: 4)*

Scott's account on peasant society reveals that the pre-capitalist peasant society is characterised by a variety of social arrangements which include labour exchanges, rent reduction at the time of crop failure, gifts by patrons at the birth of a child or the death of a peasant, and the like that insure against a subsistence crisis where "all should have a place, a living not that all should be equal" (Scott 1976: 40). As long as the elites respect this principle by protecting the peasants against ruin in bad years, their position is considered legitimate, and they are treated as the leaders of the moral community. To quote Scott (1976: 192) again,

> They (peasants) take up arms less often to destroy elites than to compel them to meet their moral obligations. . . . Regardless of the particular form it takes, collective peasant violence is structured in part by a moral vision, derived from experience and tradition, of the mutual obligations of classes in society.

He argues that peasants defend not only the narrow margin of subsistence but also their social space. A peasant needs a minimum level of resources to discharge his necessary ceremonial and social obligations, and falling below this level is not only to risk starvation but also to suffer profound loss of standing within the community. Violation of these standards provokes resentment and resistance not only because needs were unmet, but because rights were violated.

On the other hand, Popkin (1979), the ardent supporter of the political economy approach contests Scott's line of argument. He looks at the peasant as a self-interested and rational decision-maker. Although peasants live close to the margin of existence, they have some surpluses and make risky investments, political as well as economic, on the basis of rational logic. To Popkin, peasants are individual cost-benefit calculators, and they are not very different from people in small-scale business in western countries. He asserts that the peasants continuously try not merely to protect but also raise their subsistence level. It is viewed that peasant involvement in commercial agriculture is not a last-grasp alternative to declining income or threatening situations, but a response to new opportunities. Popkin (1979: 21) notes,

> Peasants often are willing to gamble on innovations when their position is secure against the loss and when a success could measurably improve their position. There are times when a small loss would mean a big fall but there are also times when a small loss would mean little and a win would move the peasant up one level.

He rejects Scott's view that a peasant rebellion is a defensive response against threats to subsistence guarantees and the erosion of previously beneficial institutions and relationships. For him, rebellions are positive attempts to gain something better based on "cost-benefit calculations about the expected returns on his own inputs". Thus, Popkin argues that peasants rebel not to preserve their traditional village life, but because they seek upward mobility. He adds that "throughout the world, peasants have fought for access to markets, not as a last gasp when all else has failed them, but when they were secure enough to want to raise their economic level and 'redefine' cultural standards!" (Popkin 1979: 80). They do not protest in response to a crisis, imbalance, and temporary decline of their income or to restore the traditional system; rather, they protest to check the markets. To quote him again,

> Moral economists . . . use terms like *decline, crisis, imbalances, decay, loss of legitimacy*, or the *erosion of traditional bonds* to describe the conditions under which protests will occur. On the contrary, I argue that short-term declines or drops are neither necessary nor sufficient for protests, that even without any

drops in welfare peasants seek individual or collective means to improve their situation. Protests are collective actions and depend on the ability of a groups or class to organize and make demands. . . . Peasant struggles are frequently battles to tame markets and bureaucracies, not movements to restore "traditional" systems.

(Popkin 1979: 35)

In contrast to Scott, he contends that peasants join political movements on the basis of the same kind of rational investment logic that they follow in their daily affairs; they must be convinced that the leader can affect individual benefits that exceed the individual cost and effectively deal with freeriders. For Popkin, peasants are household utility maximisers who, although they have the same safety first goals as Scott assumes, behave according to their self, rather than the collective interest (Feeny 1983: 779). They behave rationally in both market and non-market situations, seeking to maximise the potential benefits for themselves and their families. Unlike Scott, he notes that peasants are competitive and conniving individuals who often see other villagers as competitors. Viewing thus, he treats a peasant as a universalising economic man who acts within varying sets of constraints, including those that characterised pre-capitalist agrarian societies and societies incorporated into the global economy (Keyes 1983: 756).

The arguments made by Popkin led to a dispute with Scott,[10] which generated an exciting debate known as the Scott-Popkin debate.[11] The debate served as a benchmark for later discussions on both the nature of peasant societies and causes of peasant rebellions in the context of the third world countries of Asia and Africa. While some of the scholars agree with Scott's *moral economy* (Moise 1982; Evans 1987), others subscribe to Popkin's *rational economy* approach (Feeny 1983; Keyes 1983). Yet many of them also hold that Scott and Popkin are complimentary rather than contradictory to each other (for example, Colburn 1982; Brocheux 1983; Huang 1985; Little 1989). Though this debate is based on experiences of many Asian, African, and Latin American countries, it is rarely extended to countries like India,[12] which experienced a series of peasant resistance movements from the colonial to the contemporary period. It may be due to the orthodox assumption that Scott is inappropriate for the modern or highly individualised context where capitalist transition is consolidated, and Popkin is irrelevant where simple individual cost-benefit calculation is an inaccurate model of peasant decision-making process (Kurtz 2000: 104), while India does not confirm to either of the contexts.

Hence, an attempt is made here to examine whether the principles of moral economy and/or rational economy govern the peasants' behaviour in Maharashtra and to what extent they explain the varied responses of peasants to agrarian crisis at different phases in time.

Agrarian crisis of the Bombay Deccan: The Deccan Riots, 1875

The Bombay Deccan consisted of Khandesh,[13] Poona, Ahmednagar, Nashik, Sholapur, and Satara districts. Prior to the British advent, a major part of the region was under the rule of Muslim oligarchy and Hindu rulers. The political regime during this period was dependent on land and agricultural surplus. Unlike the feudal lords in Europe, the overlordship in Bombay Deccan in the seventeenth and eighteenth centuries mainly related to land revenue rights. However, both the Hindu and Muslim rulers were least concerned about agricultural development. A major part of the agricultural surplus, which was extracted mostly through harsh methods, was spent to meet the rising military and administrative expenses. Due to the scarcity of cultivators relative to arable land, there was hardly any competition among the cultivators for land and landlessness was rare (Hatekar 1996: 452). The governments did not have information about the crops grown by each cultivator, and the relative qualities of soil and land tax were paid by the village as a collective unit, the exact sum being negotiated between the village *patil* and the *mamlatdars*.

Nevertheless, the Peshwa's tax collectors, the *mamlatdars*, were known for their rigid, unfeeling, and corrupt methods of tax collection. Moreover, the crisis of *Maratha* polity that started in the late eighteenth century, drastically affected the agrarian economy, followed by three major wars with the British (1775–82, 1803–05, and 1817–19). In order to raise more revenue to increase the strength and efficiency of his army, following the acceptance of subsidiary alliance in 1803, the Peshwa introduced a new system of revenue which had a disruptive effect on peasants (Guha 1985: 12). Large tracts of Khandesh and Ahmednagar were utterly desolated by the Holkar's army and Pindaris, and in some areas, the marauding armies extracted money from the peasants.[14] Besides, the famine of 1789–91 in the Krishna Valley in the south and of 1802–03 in the northern Deccan led to a significant decline of rural population.[15] Peasants responded to this crisis by migrating to other areas. There were instances where the whole village declared that it could not pay the rates in force and threatened to desert the land and settle elsewhere. Nonetheless, the agrarian economy had autonomy, and the agricultural population could secure minimal survival in the normal years, except in years of famines and military raids (Brahme and Upadhyaya 1979: 18). Landholdings were occupied by two classes of peasants: landed proprietors called *mirasis*, *mirasidars*, or *thalkaris* and farmers or tenants called *uparis* (occupants on annual basis). The village power structure revolved around the *patil* and *panchayat* (village council for deciding judicial disputes). Agricultural services were organised around the patron-client relationship inherent in the *Vatandari* and *Balutedari* system,[16] and most needs of the village population relating to production and consumption were met from within the village or a group of villages. Though rural indebtedness was widespread,[17] peasants were hardly under any threat from moneylenders due to the numerical preponderance of the *kunbi* peasants who dominated the village *panchayat* (where disputes relating to recovery of debts were resolved).[18] In a nutshell, inter-dependence among different castes and groups, as well as the dominance of the *ryots*, united different groups within a village and created a social climate in which collective action, rather than individual behaviour, formed the basis of social order (Kumar 1965: 634).

The British took control of Bombay Deccan in 1818 from Peshwa Baji Rao II and introduced the *ryotwari* system, following the footsteps of the earlier *Maratha* rulers. However, the East India Company gave up the system of farming land revenue introduced by the Peshwa and went back to the earlier system of Nana Fadnavis.[19] Though assessments were made high by the *Marathas*, they rarely succeeded to collect them in full. The new land revenue assessment introduced in 1818–19 was too high, and it was difficult to save the *ryots* from ruin.[20] Within a few years of the commencement of the British rule, Bombay Deccan was plunged into an agricultural depression that lasted for over a quarter of a century due to an increase in population accompanied by a decline in the area under cultivation (Guha 1985: 15). There was a sharp decline of prices of agricultural produce after 1821, leading to agricultural stagnation.

The effects of low prices were aggravated by heavy taxation. A class of moneylenders (*sowkars*) consisting of *Gujaratis*, *Marwaris*, and *Chitpavan* Brahmins emerged who advanced credit to the peasants at exorbitant rates of interest for the payment of land tax.

In due course, the method of revenue collection introduced in 1818–19 was found to be faulty as the full amount of assessment could not be recovered. Subsequently, the Pringle system was introduced to bring the various methods of assessment under one system. However, this system also resulted in an uneven and heavy assessment which ultimately led to the introduction of the Revenue Survey and Assessment in 1836 by Goldsmid, which was settled for a period of 30 years. The land was divided according to its quality, the assessment was imposed according to the classification and the amount of demand was determined with the experience of former years. The introduction of the Goldsmid rates marked a new era for the Deccan *ryots*, as it lowered the rental of the *talukas*, resulting in better collection of land revenue. Large tracts of land which were lying waste and all the cultivable land were brought under the plough. Within five years, the cultivated area increased by 60,000 acres, and there was a proportionate increase in land revenue too (Kumar1965: 22). Condition of the cultivators improved, and they could pay their rent and cultivate land with reasonable efficiency (Choksey 1955; Kumar 1965). However, this resulted in a substantial increase in loan activity and debt. By 1851, about 95 per cent of the peasants in Poona district were indebted to moneylenders and the interest paid by the villagers exceeded the state revenue demands (Rao 2009: 56). Captain Anderson's writing of the *talukas* Newasa in Ahmednagar collectorate in 1854 reveals

> I am led to believe from enquiries made by me at the time of settlement that about two-thirds of the *ryots* are in the hands of *Marwaris*, and the average debt of each individual is not less than Rs. 100. This under any circumstances would be a very heavy burden on the agricultural population, and owing to the harsh and usurious proceedings of many of the *Marwaris*, the system has endangered so much bad feeling.
>
> *(Report of the Deccan Riots Commission,*
> *hereafter RDRC 1878: 159)*

The economic condition of peasants improved significantly in the period of the American Civil War caused by a rise in the price of cotton as India was called upon to supply cotton to the English market. Peasants growing food crops switched over to cotton cultivation. For example, though Indapur was not a cotton growing district, *ryots* of this region started immediately planting this crop as soon as reports of the profits from cotton production reached them. The poor farming families also ploughed marginal land with bad soil in areas of low rainfall to cultivate cotton (Ludden 1999: 199–200). By 1867–68, 30,000 acres of land had been planted with cotton which ultimately boosted the price of food crops, and *jowri* skyrocketed to twice its normal price (Kumar 1965:622). When unfavourable season commenced in 1867 and prices fell rapidly, the burden of debt was severely felt. With the close of the American War in 1865, the inflow of capital ceased. Prices did not at once fall, as 1866–67 was a season of severe drought, 1867–68 of partial failure and 1870–71 of serious failure. The effect of the sudden fall in prices between 1871 to 1872 and 1873 to 1874 reduced the landholder's power of paying and motivated creditors to enhance their security by turning personal debt into land mortgage and to check further advances to peasants. Simultaneously, large expenditure on public works, which from the time of the construction of the railways down to 1871 had subsidised the district with large sums in wages, was suddenly contracted.[21] Instead of reducing revenue in times of distress, the government enhanced it further through the revised settlement between 1869 and1872 owing to the expiry of the term of 30 years of the earlier settlement.[22] Moneylenders also did not spare the peasants at this time of crisis.

Peasants began to mortgage and sell their lands more than before. In Bhimthari *talukas* alone there were 757 registered transactions, by mortgages and sales amounting to Rs. 206,000 between1865 and 1875.[23] The government's Resolution 726, 5 February 1875 in the Revenue Department, which was framed to prevent the sale of land, directed that the process to recover land revenue should be the first against the movable property of the occupant and that land should not be sold until after the sale of the movable property. The moneylenders turned this order to their own advantage at the expense of the landholders.[24] The district deputy collector of Khandesh observed,

> It appears from the circumstances which came now and then to my knowledge connected with the realisation of land revenue, that the condition of some of the landholders has been

actually worse than that of day labourers. Though they with all their family members work hard throughout the year, they are obliged to take the whole produce to their creditors, and to stand at their mercy for livelihood and even for the assessment of land.[25]

Peasants paying the government Rs. 10–20 owed the moneylenders rupees 1,000–2,000 (Charlesworth 1972: 402). The debt continued for generations and the moneylender often manipulated debt records to take possession of the debtor's land. Talegaon subjudge's letter reveals

> the indebtedness of the *kunbi* is increasing, and he is sinking deeper and deeper into the mire. A few rupees borrowed from a creditor in an evil hour often serve to deprive him of his immoveable and movables and all, and oblige him to depend on his creditor for his very bread.[26]

Sholapur sub-judge's letter substantiates the same point more forcefully:

> Money-lenders generally buy such land, and it is seen that land is fast passing into the hand *sowkars* from those of cultivators. The aim of the creditor is generally to possess himself of everything the debtor has; so he never stops till his debtor has lost everything.[27]

In addition to the volume of land revenue, the manner of its collection added to the woes of cultivators. When difficulties arose in the collection of revenue, crops were not allowed to be removed from the village stackyard until the revenue had been paid or some satisfactory security proffered.[28] Prior to the arrival of the British, the usual method for recovery of debt was for the *sowkar* to send a *mohosu*, a servant whose maintenance had to be paid daily by the debtor, or to place a servant in *dharna*[29] at his door, or to confine the debtor in his house, or otherwise subject him to restrain. However, such methods could not be put in force against anyone because the dealings of the *sowkar* and the *ryot* were based on mutual faith. Moreover, the creditor received little or no assistance from the state in recovering debts (RDRC 1878: 13–14). Consequent upon the introduction of the first regular procedure for the administration of civil justice in the Bombay Presidency in 1827, new courts came into operation which provided security to the *sowkar* and afforded the means for speedy realisation of debt. Moreover, the passing of civil procedure code in 1859 swept away the discretion which the court had possessed under the regulations of 1827. It abolished the exemption of the *ryot's* necessary cattle and implements from seizure for debt and regulated the process of imprisonment so as to make it a most effective instrument for bringing pressure to bear on the debtor. It was observed that compared with any other country, the law in India was much more unfavourable to the debtor and even the Roman law was not so severe.[30] The new law gave the creditor the power to compel the debtor to do whatever would be less grievous than imprisonment and usually cultivators preferred to remain in their native village and work on their ancestral lands at all costs rather than going to a distant jail which had all the terrors of the unknown to them (RDRC 1878: 40). The issue of arrest warrant was used as a threat for repayment and settlement of debt.[31] Failure to pay the interest brought the debtors into court. The number of suits for debt in the courts of Poona doubled between 1867 and 1873, and applications for the execution of decrees increased from 12,000 in 1868 to 28,000 in 1873 (Keatinge1912: 85). A large number of peasants were arrested under decrees and imprisoned across districts in 1871 and 1872.[32] Many of the cases appended to RDRC narrating the debt bondage of the imprisoned peasants illustrate the gravity of the situation.[33]

The cumulative impact of the excess land revenue, the rapacious attitude of the moneylenders accompanied by the oppressive state laws made the conditions of peasants miserable. At this point in time, the *Poona Sarvajanik Sabha*, which emerged in 1870 as a reorganised form of Poona Association founded in 1867, took an active interest in agrarian problems and educated the peasants about the adverse impact of British land revenue policies, particularly the resettlement of the Deccan by Francis (Charlesworth 1972: 120n). The *Sabha* initially consisted of 95 representatives who were elected by over 6,000 persons representing all castes, creeds, and interests and a number of affiliated bodies were founded with the *Sabha* in Satara, Wai, Solapur, Nashik, etc. (Kumar 1965: 627). The *Sabha* strongly defended the interests of the peasants under the leadership of Brahmin intellectuals like Ganesh Vasudeo Joshi and Madhav Gobinda Ranade. It sent agitators to the villages to whip up the *Kunbis'* opposition to rates imposed under the new survey and to explain the reasons for their poverty. The *Sabha*'s success in arousing the peasants against the new settlement was spectacular, which became immediately apparent to the British authorities. The activities of the *Sabha* became more aggressive with the joining of Vasudeo Balwant Phadake who successfully motivated and organised the peasants against the moneylenders. The new elites of the *Sabha*, as observed by Kumar (1965: 628), stood as *deshmukhs*, defending the interest of their rural clients through the use of Ricardian and Utilitarian values.

The first sign of open hostility to the *Marwari* moneylenders among the villagers of Poona began in Karde in Sirur, where a *deshmukh* named Babasaheb mobilised the villagers and persuaded them not to borrow and work for them or buy from them in December 1874. The villagers opened a grocer's shop at which all the village purchases were made. In addition, the villagers annoyed the *Marwaris* by throwing dead dogs and other filth into their houses. As a result, the *Marwaris* were put to inconvenience. They were scared, and they sought police protection and appealed to the magistrate.[34] As noted by Fukazawa (1976: 19–20), there were three types of boycotts of moneylenders by the villagers: (a) opposition to new moneylenders who intended to settle down in their villages, (b) refusal to cultivate lands whose occupancy right was obtained by the moneylenders, and (c) total boycott of moneylenders by the villagers as a whole. However, such boycotts did not last long due to non-cooperation of the neighbouring villages and intervention of the police. Nevertheless, hatred towards the moneylenders, which had been fuming for some time, became bitterer.

The first actual outbreak against the *Gujarati* moneylenders occurred at Supa, a large village in the Bhimthari *Talukas* of Poona on 12 May 1875. Their houses and shops were attacked by a large mob, and one house was burnt down. Within 24 hours of it, the leading *Marwari* moneylender of Kedgaon, about 14 miles to the north of Supa, was attacked, his stacks were burnt down, and his house was set on fire. Within a few days, riots occurred in four other villages of Bhimthari, which spread to the neighbouring sub-divisions of Indapur and Purandhar. Riots also occurred in Sirur about the same time. The first act of violence was committed at Navra where a *Marwari* who had left the village for safety was mobbed and prevented from moving his property. Soon other villages in Sirur followed the example of Navra. In 15 of Sirur and 3 of Haveli villages, riots either broke out or were threatened.[35]

The regiment of Poona Horse at Sirur supplied parties to help the Magistrate and police in restoring order. Though riots were mostly against the *Marwaris*, in some villages where they were not found, Brahmans were attacked. Similar outbreaks occurred in the neighbouring *talukas* of Ahmednagar, Shrigonda, Parner, Nagar, and Karjat. In Poona, serious disturbances took place in five villages of Bhimthari *talukas* and six villages of Sirur *talukas*. Disturbances were threatened but averted by the arrival of police in 17 villages of Bhimthari, 10 of Sirur, 1 of Indapur, and 3 of Haveli. In Ahmednagar disturbances took place in 6 villages of Parner *talukas*, 11 of Shrigonda, 4 of Nagar *talukas*, and 1 of Karjat. In Poona, 559 persons were arrested of whom 301 were convicted and 258 discharged, and in Ahmednagar, 392 persons were arrested of whom 200 were convicted and 192 discharged.

Scholars who worked on the colonial agrarian history of Bombay Deccan were divided on the issue of intensity and magnitude of these riots. While Charlesworth (1972: 614–615) characterised them as an insignificant demonstration and a momentous non-event, others described them as a noteworthy peasant rebellion where the range of participation was far and wide, involving rich, poor, and middle peasants, indebted and debt-free peasants, village servants and agricultural labourers (Catanach1970; Fukazawa 1976; Nand 1980; Rodrigues 1998). In the words of Fukazawa (1976: 1),

> The Deccan riots of 1875 were, however, very extraordinary in the sense that they were the only massive uprisings of peasants against moneylenders in the long run history of western India during the British period. They were epoch-making in the agrarian history of western India.

To summarise, discontentment among peasants in Bombay Deccan during the 1870s following the agrarian crisis caused primarily by excessive land revenue, commercialisation of agriculture, unscrupulous moneylending practices and mobilisation by *Poona Sarvajanik Sabha*, ultimately led to the Deccan Riots.

Drought and agrarian crisis (1970–73): rise of the farmers' movement

The state of Maharashtra was formed on 1 May 1960, prior to which it was a part of the bilingual Bombay state along with Gujarat. After independence, the then Bombay state introduced many measures for rural and agricultural development, following the broad guidelines of the union government. It began with the formulation and implementation of land reforms which mainly included abolition of intermediaries and tenancy reforms.[36] To curtail the dominance of moneylenders, the Bombay Moneylenders Act of 1946 was brought into operation which required the moneylenders to obtain licenses for carrying on their business and put a ceiling to their lending rates. This Act was amended in 1954–55 to make moneylending without license a cognisable offence.

With the introduction of the First Five-Year Plan in 1952, packages of state-directed programmes were launched, including the Community Development Programme and National Extension Scheme, which were implemented through the newly created Panchayati raj institutions. The erstwhile Bombay state and the subsequently formed Maharashtra were the first of few states in the country to introduce Panchayati raj institutions in accordance with the recommendations of Balwantrai Mehta Committee Report, which was subsequently revised in the light of suggestions made by Naik and Patil Committees. In addition, the cooperative movement in Maharashtra gained momentum in the 1960s as a result of which many cooperative sugar factories together with a network of cooperative credit societies came into being under the state support to provide a strong base to the peasants to improve their economic conditions. Cultivation of crops like sugarcane that required high investments compared with other crops spread revolving around cooperative sugar factories. Keeping in view the malpractices in the marketing of agricultural produce, which resulted in the exploitation of cultivators by the middlemen and thereby not receiving a fair share of the value of their produce, the state introduced the Maharashtra Produce Marketing (Regulation) Act 1963.[37] The Agriculture Produce Market Committees (APMCs) were established under this Act, which provided for the setting up of Maharashtra Agriculture Marketing Board to ensure that the produce was weighed correctly and the peasant got a fair price for his goods (Baviskar 1980).

The plan investments in irrigation works, electrification of roads, market, and extension of credit significantly changed the agrarian structure. In tune with the government of India's strategy to augment food production, the state government launched high yielding varieties (HYVs) in 1960–67 as a result of which new varieties of *jowar*, wheat, and rice were adopted extensively. To meet the rising credit needs of the peasants, primary agricultural cooperative credit societies were expanded in terms of number as well as membership.[38] By 1960, their network reached 96 per cent of the villages of the state. Elaborate arrangements were made to bring improved varieties of seeds, fertilisers and pesticides, energised irrigation, and agricultural credit to the doorstep of the peasants at highly subsidised rates. Peasants eagerly awaited a trickle from the ongoing agricultural revolution against the background of devastating experience during the colonial period (Dhanagare 1990: 353). This apart, the election promises of socialism and eradication of poverty, gave rise to new hopes among the rural poor (Mies 1976a: 174).

Though the new arrangements following independence improved the agrarian conditions, agricultural production remained by and large stagnant, except the few good-crop years like 1960–61. Agricultural production declined for about a decade and a number of major and minor droughts reduced food production significantly, accompanied by a fall in the production of most commercial crops despite a rise in cultivated area, improved varieties of seeds and other supporting measures (Ladejinsky 1973: 383). Though the area under irrigation witnessed growth, it was only 8 per cent of net sown area during the TE 1970–71 (Mohanty 2009: 66).[39] In the early '70s, for three successive years (1970–71 to 1972–73), the state was in the grip of a severe drought. It was estimated that out of about 35,000 villages, 23,000 villages in 21 districts (out of 26 districts) in 1970–71, and 15,000 villages in 20 districts in 1971–72 were affected by scarcity. Drought conditions in 1972–73 were more widespread and acute, which hit 25 districts and nearly 31,000 villages.[40] As per the official estimate, loss in production was 55 per cent of the *kharif jowar*, 65 per cent of *rabi jowar*, 75 per cent of *bajra*, 50 per cent of paddy, 50 per cent of *tur*, and other pulses (Brahme 1973: 49). Over 70 per cent of groundnut crop and 40 per cent of cotton crop also failed.

A comparison of the area, production, and yield of major food crops between TE 1968–69 and TE 1970–73 (Table 7.1) as well as assessment of the conditions in 1970–71, 1971–72, and 1972–73 in relation to the situation in 1968–69, which is considered a normal year (Brahme 1983), indicates the severity of crop losses.

The estimate made by Dyson and Maharatna (1992) shows that cereal production during 1972–73 failed to less than half its levels in 1967–68. Oughton (1982) reveals that from 1970 to 1971 through 1972 to 1973, the Index of the Value of Output per hectare for major crops drastically went down compared with the

Table 7.1 Area, production, and yield (1970–71, 1971–72, and 1972–73)

Particulars	Reference year	Bajra	Jowar	Cereals	Food grains
Area	1970–71% to 1968–69	102.01	91.95	95.68	97.45
	1971–72% to 1968–69	60.02	98.07	92.61	91.83
	1972–73% to 1968–69	67.61	79.4	79.29	78.97
	TE 1972–73% to TE 1968–69	79.83	90.79	90.43	90.67
Production	1970–71% to 1968–69	131.77	44.84	76.61	78.1
	1971–72% to 1968–69	40.00	54.43	68.59	69.2
	1972–73% to 1968–69	34.16	35.1	41.37	42.62
	TE 1972–73% to TE 1968–69	82.31	46.98	66.63	67.86
Average yield	1970–71	408.45	275.11	461.55	432.63
	1971–72	210.72	313.1	426.9	406.78
	1972–73	159.76	249.42	300.75	291.35

Source: *Agricultural Statistics at a Glance*, relevant issues.

preceding and succeeding years,[41] accompanied by a rise in rural mortality rate from 13 per 1,000 in 1970 to 16 per 1,000 in 1973. The per capita production of cereals fell drastically during the three years of 1970–71 to 1972–73 and the shortfall was not made up by increased trading. Though the net inter-state imports of grains doubling in 1970 over 1969 and almost trebling by 1971, the per head availability of food grains was insufficient to maintain the levels of the late 1960s (Oughton 1982: 179–181). Due to continuous drought conditions, prices of fodder mounted way beyond the capacity of the peasants to sustain them. In the affected regions, peasants were unable to buy cattle and were trying to sell them to slaughter houses. As a result, half of the cattle perished by starvation or sent to the slaughter houses (Ladejinsky 1973: 385).

Due to the agrarian crisis during 1971–72, only 30 per cent of the total outstanding loans in primary agricultural cooperative Societies (PACSs) were recovered. In its review of agricultural development and cooperative credit in Maharashtra, the Reserve Bank of India noted the poor financial position of the majority of the PACSs in 1973–74 and recommended closure of non-viable credit societies either by amalgamation or liquidation.[42] The number of agricultural labourers increased from 4.5 million in 1961 to 5.4 million in 1971 and land distribution became skewed (Brahme and Upadhyaya 1979: 123). There were instances of distress sale of assets by peasants (Oughton1982: 189–190). The economic conditions of agricultural labourers and village artisans deteriorated remarkably during these years.[43] Increased migration of the scarcity-ridden rural population from Nashik, Solapur, Bhir, Osmanabad, Aurangabad, Ahmednagar, and Poona towards cities like Mumbai and Pune made the situation alarming (Brahme 1983: 88–89). On the whole, there was a widespread distress among the peasants during the drought years.

In response to the persistent agrarian crisis, distressed peasants started organising themselves, demanding relief measures which ultimately gave birth to powerful statewide movements. The slogan "drought is not from nature but is man-made and people responsible for creating the drought" was first put forward by the peasants. Though it is argued by Dhanagare (1990) that for nearly three decades (1945–46 and the 1970s), there were no peasant movements in Maharashtra, the available information provides evidence of peasant agitation in the early 1970s. Following the deteriorating economic conditions of the peasants, many activists and intellectuals formed a committee named *Duskal Nivaran and Nirmoolan Samiti* (Drought Relief and Eradication Committee), demanding a comprehensive agrarian development programme. Started in 1969, the *Yuvak Kranti Dal* (Yukrant), which in the beginning focused on the problems of the deprived groups and unequal land distribution, took up the issue of drought conditions and became part of *Duskal Nivaran and Nirmoolan Samiti*. Protesting against galloping prices, demonstrations were held under the leadership of communist parties in the district of Chandrapur where four persons were killed (Ladejinsky 1973: 394). Political parties like CPI, CPI (M), Lal Nisan Party, the PWP, and the Socialist Party joined together and organised agitations all over the state with the active participation of peasants demanding relief work, increase in wages, supply of food grains through public distribution system, etc. (Brahme and Upadhyaya 1979: 167–168). A huge rally of famine-affected workers and peasants was held in March 1973 in Ahmednagar. In April 1973, an anti-hoarding campaign was launched. *Morchas* and *gheraos* were a daily phenomena at one or the other of the 5,400 scarcity relief works. The leftist parties jointly organised the landless agricultural workers and peasants to fight for their rights to work and subsistence wages. The

Maharashtra Rajya Shet-Mazdoor Parishad, established by Lal Nisan Party leaders, organised demonstrations of around 25,000 rural people with the slogan '*ala re ala Shetkari ala*' (here comes the peasants).[44] It spread its agitation to many parts of Maharashtra which continued for about a decade but became weak in the late 1980s. During 1972–73, *Shetkari Shet-Mazdoor Panchayat*, which was set up by the socialist party in 1969, was also active and took up the issues relating to drought conditions. Omvedt's (1975) account indicates the magnitude of peasant resistance in the state in the early 1970s.[45] Details of the agitations, issues involved and their coverage are given in Table 7.2.

These agitations were made up primarily of poor peasants and agricultural labourers who began to assert themselves, whereas rich peasants showed least interest (Omvedt 1975: 43–4). Besides, there was a movement in Shahada in Dhule district under *Shramik Sangatana* (Labour Association). Begun in 1972, it mobilised tribal landless labourers and poor peasants against the outrageous practices of local landlords, who were mostly non-tribals. The drought of 1972–73 and the consequent famine led to a rapid growth of the movement. The *Shramik Sangathna* adopted the 'mass line' approach, following non-violent and democratic means like strike, boycott, *satyagraha, gherao*, processions, *bandhs*, etc., as opposed to the annihilation line of the Naxalite movement which took place in Andhra Pradesh and West Bengal during the 1970s.[46] Similarly, in Thane the tribal peasants revolted against the local landlords, moneylenders, police, and forest department officials in 1975 under the banner of a local non-party militant organisation called *Bhoomi Sena* (Land Army).

Though drought-related protest movements became weak following the onset of the monsoon in 1973, agitations were launched on other issues relating to employment guarantee scheme, supply of credit and inputs, and fair price for agricultural produce. (Brahme and Upadhyaya 1979: 168). However, the declaration of emergency on 25 June 1975, prohibited all forms of democratic mass protests like strikes, *bandhs*, mass rallies, *gheraos*, etc. In many places, leaders were arrested as well as imprisoned and the press was warned not to report agitations (Mies 1976a: 184). Nevertheless, organised strikes took place in Sangamner *talukas* in Ahmednagar district for immediate implementation of the Minimum Wages Act. During this period the farmers' movement demanding remunerative prices, subsidised inputs and remission in loans and

Table 7.2 Agitation during the drought period of 1972–73

Month and year of agitation	Demand for relief work		Demand for declaration of scarcity conditions		Demand for timely wage payment		Demand for increase in wages		Malpractices in relief works		Demand for increase in food supply works		Anti-price rise demonstrations	
	District covered	Talukas covered	District covered	Talukas covered	District covered	Talukas covered	District covered	Talukas covered	District covered	Talukas covered	District covered	Talukas covered	District covered	Talukas covered
September 1972	5	4	1	1	2	1	–	–	2	2	–	–	–	–
October 1972	4	4	2	2	2	2	2	2	–	–	–	–	–	–
November 1972	3	17	–	–	2	2	–	–	2	12	–	–	–	–
December 1972	3	14	1	2	3	2	2	2	2	1	1	1	–	–
January 1973	2	3	1	12	2	12	5	44	1	11	2	2	–	–
February 1973	2	2	–	–	3	3	2	3	4	5	2	2	–	–
March 1973	3	4	–	–	–	–	2	2	1	1	–	–	–	–
April 1973	3	15	–	–	1	12	2	2	4	5	9	29	–	–
May 1973	4	6	–	–	5	3	6	4	5	4	8	8	–	–
June 1973	5	6	–	–	3	3	2	2	6	22	6	2	–	–
July 1973	2	3	–	–	–	–	1	1	3	3	3	1	–	–
August 1973	1	1	–	–	1	1	–	–	1	2	–	–	4	1
September 1973	1	1	–	–	1	1	–	–	1	2	–	–	4	1

Source: Brahme (1983: 124–28).

taxes, extension of credit facilities and other issues gained ground in states like Karnataka, Punjab, and Tamil Nadu. In 1979, *Kisan Sabha* carried out agitations in Ahmednagar on the question of debt relief of peasants.

Following these movements, the state initiated several measures. It introduced the employment guarantee scheme in May 1972 as a drought relief programme and increased the number of relief works from 5,000 in September 1972 to nearly 65,000 by May 1973 (Brahme and Upadhyaya 1979: 72). Maharashtra Debt Relief Ordinance, 1975 was passed to provide relief from indebtedness to marginal farmers and rural labourers. The government appointed a committee in 1976 to enquire into the problem of illicit moneylending and bonded labour, and minimum wage rates were revised in October 1978. It also responded to the concerns of tribal peasants through the introduction of new projects and laws to improve their socio-economic conditions (Calman 1986). The positive impact of the agitations during and following the drought years on peasants has been noted by many authors. (Brahme and Upadhyaya 1979; Joseph 2006; Rao 1988; Omvedt 1975; Mies 1976a, 1976b). It is argued that the 1970–73 drought conditions, despite their severity, did not result in starvation deaths, sharp increase in mortality rates, nutritional deterioration and large-scale migration to other states, as it happened in other places due to organised public action (Dreze and Sen 1989).

Although agricultural conditions improved in the 1980s in terms cropped area and productivity (Table 7.3) with the availability of subsidised credit and other inputs, the peasant movements became more organised and widespread during the 1980s with well-articulated ideology and an organisational structure following the emergence of *Shetkari Sanghatan* under the leadership of Sharad Joshi demanding remunerative prices for agricultural produce.

The logic of remunerative prices was based on the premise that peasants respond rationally to price movements and react to price incentives by increasing acreage and investment by adopting improved technology. It was also based on the idea that peasants should go beyond self-sufficiency to protect themselves against droughts and famines.[47] Joshi's slogan of 'Bharat-India' divide emerged as the new idiom of rural mobilisation. While 'Bharat' represented the villages and refugees from Bharat in the cities (the unorganised urban

Table 7.3 Area, production, and yield in 1980s

Crops	Area (% to gross cropped area)		Production ('000 tonnes)		Yield (kg./hec.)	
	TE 1982–83	TE 1989–90	TE 1982–83	TE 1989–90	TE 1982–83	TE 1989–90
Bajra	8.23	8.88	661.13	1,004.87	409.59	530
Jower	33.40	30.07	4,657.03	5,621.63	710.53	875
Cereals	57.32	52.00	8,903.80	10,168.60	791.64	915
Food grains	71.13	74.76	9,848.20	12,455.87	705.57	780
Cotton*	13.63	12.14	1,447.30	1,684.10	541.07	649
Oilseeds	10.21	12.25	1,097.57	1,617.60	547.61	618
Sugarcane	1.49	1.54	27,910.17	28,163.83	95,321.61	85,405

Source: Agricultural Statistics at a Glance, relevant issues.
Note: Production of cotton ('000 bales of 170 kgs. each).

sector), 'India' depicted the westernised industrial-cum-bureaucratic elites, inheritors of colonial exploitation. To Joshi,

> the British tried to squeeze this country for the benefit of the Empire. The 'Black British' are doing the same thing for the benefit of 'India' at the expense of 'Bharat' . . . our farmer's movement is a movement for a second Independence.[48]

The motto of *Shetkari Sanghatan* as put by Omvedt (1993: 120) was *bhiknako, have gamala dam* (we don't want alms but recompense for our sweat; we are not asking for anything, only that you stop exploiting us). The new agrarianism of *Shetkari Sanghatan* aimed at ensuring for the farmers the highest possible degree of freedom and a life of self-respect on par with that of the non-farming communities. Unlike the earlier peasant movements, which mobilised tenants against landlords, peasants against moneylenders and lower castes against higher castes, the *Shetkari Sanghatan* highlighted the contradictions between Bharat and India, and tried to unite the rural mass across various strata. In the initial years, *Sanghatan* projected itself as an apolitical body and Joshi was critical of all political parties and their leaders.

This ideological position of the *Sanghatan* appealed to the peasants across categories and they responded spontaneously to its call for *rastaroka* in the Chakan-Nashik area in 1980 demanding remunerative prices for onion. To broaden its base, almost six month later, Joshi took up the cause of sugarcane growers and undertook another massive protest in Nashik, leading to a blockade of roads where 30,000 peasants courted arrest. The movement spread from Nashik and parts of Pune to Ahmednagar, Dhule, most of Vidarbha, and parts of Marathwada. Road blockades and demonstrations continued for six weeks at various places in the state. Nine thousand persons were arrested in Nashik alone and 7,000 on a single day in Vidarbha (Lenneberg 1988: 447). Subsequently, the *Shetkari Sanghatan* turned to the *Beedi-Tobacco* growers of Nipani, the bordering area between Maharashtra and Karnataka. This agitation in March 1981 was in the form of blocking Bombay-Bangalore national highway which lasted for as many as 24 days where 10 farmers were killed in police firings, and hundreds were injured. In early 1982, the *Shetkari* Sanghatan held its first official conference in a small *talukas* town in Nashik with 18,000 delegates representing half the districts in Maharashtra. It closed with the warning, "Let the rulers of India be warned: The age of the Peasant has arrived" (Omvedt1993: 113). Gradually, the influence of this movement spread far and wide and its ideological formulations were widely acknowledged in other parts of the country. The major protests organised by it in the 1980s are illustrated in Table 7.4.

It is reported by Dhanagare (2014: 43) that the *Sanghatan* had successfully launched agitations in support of onion, sugarcane, and the tobacco growers of western and southern Maharashtra. The demand for remunerative prices for cotton and *jowar* was taken up by it in1985 and 1986. Farmers, regardless of their internal differences in terms of landholdings and caste identities, were united even though class interest remained quiescent under the impact of the populist ideology (Dhanagare 2016a: 235–236). The movement drew national attention to the farmers' demands and was successful throughout the 1980s, compelling the government to meet the demands of the farmers. It succeeded in getting some concessions from the government (for example, increase of support price paid by the National Marketing Federation). The impact of the movement on farmers was impressive (Dhanagare 1990, 2014; Nadkarni 1987).

Table 7.4 Major activities of Shetkari Sanghatan in 1980s

Year	Date and month	Activities undertaken
1979	9 August	Shetkari Sanghatan was formed at Chakan
1980	1 March	Rastaroka agitation in Chakan demanding remunerative price for onion
	8–16 March	Sharad Joshi undertook hunger strike in Chakan and satyagraha was launched by Sangatan demanding remunerative price for onion
	19 March	Rastaroka agitation in Bombay-Agra Highway demanding remunerative prices for onion where police lathi charged and fired bullets leading to death of two farmers in Pimpalgaon Basawant of Nashik district
	1–6 May	Sharad Joshi started hunger strike protesting NAFED's decision to stop buying onions from farmers
	25 May	Rastaroka agitation in Chakan against NAFED's irregular purchase of onion
	15 August	Meeting of sugarcane cane producing farmers to discuss wholesale price of sugarcane and onion at Niphad Cooperative sugar factory NAFED in Nashik
	27 October	Agitation demanding remunerative price for sugarcane and protest Srigunda Sugar Cooperative Factory's sugarcane procurement procedure where one was killed by police firing
	10 November	Rastaroka in Bombay-Agra Highway and Rail Roko at Baladi accompanied by disruption of electricity and water supply by the activists of *Sanghatan* demanding remunerative prices for sugarcane where two farmers were killed in police firing and 30,000 peasants courted arrest.
	16–24 November	Agitation by the *Sanghatan* and Sharad Joshi started hunger strike in jail opposing states excessive use of police against the agitating farmers.
	10 December	Sanghatan celebrated its victory day in Pimpalgaon, Niphad and Nashik for success of their agitation and decided to extend agitation further.
1981	24 February	Agitation at Nipani of Belgaon district demanding remunerative prices for tobacco growers
	6 April	Agitation in Nipani of Belgaon demanding remunerative prices for tobacco where 12 farmers were killed in police firing
	20 September	Farmers from all parts of Maharashtra attended meetings at Pimpalgaon, Baswant, and Niphad of Nashik district
	10 November	Sugarcane agitation throughout Maharashtra to oppose compulsory reductions leading to death of four farmers in the state

Year	Date and month	Activities undertaken
1982	1–3 January	First Plenipotentiary conference of *Shetkari Sanghatan* held at Nashik
	7 August	Meeting of the main activists from different parts of Maharashtra in Pune
	30–31 October	Representatives of different farmers' unions from other states participated in the conference at Vardha where inter-state coordinating committee was formed
1983	15 Mayw	Agitation throughout Maharashtra demanding remunerative price for Milk producers
1984	17–19 February	Second Plenipotentiary conference of *Shetkari Sanghatan* at Pandharpur
1985	14 January	Sent packets of black *jowar* to Prime minister as sign of protest demanding justice to *jowar* growers
	21–22 January	Third session of *Shetkari Sanghatan* took place in Dhule where Sangatan's role in State Assembly election was discussed (formation of opposition parties united together with *Shetkari Sanghatan*)
	3 February	Organised a token 'rastaroko' at Dhule in support of the demands of the farmers for a remunerative price of agricultural produce all over Maharashtra where all the parties opposing Congress joined.
	22 August	A meeting of women farmers was held at Dabhadi of Nashik to motivate more women participation in agitations
	2 October	Agitation was launched all over Maharashtra against the state government's policy on cotton procurement
	6 October	Agitation for remunerative prices for sugarcane at Rahuri of Ahmednagar to generate awareness among urban people regarding the farmers' problem. The activists of *Sanghatan* offered betel leaf (Pan) and flower along with a pamphlet (in which they mentioned the issues related to farmers' problems) which was popularly known as *pan-phul* protest. They also tried to halt sugar production in sugar factories
	31 October	Meeting at Dhule demanding employment to the farmers of drought affected areas of Maharashtra through EGS

(Continued)

Table 7.4 Continued

Year	Date and month	Activities undertaken
1986	21–22 September	Organised agitation of *jower* producing farmers at Ambejogai of Beed district for remunerative prices
	2 October	Agitation at Akola demanding change in cotton export policy
	9–10 November	Meeting of women farmers at Chandwad of Nashik to discuss issues relating to their empowerment
	23 November	Rastaroka agitation by cotton growing farmers at Chandwad of Nashik district for remunerative prices where female farmers participated in huge numbers
	6–12 December	Rail roko by farmers at Sevagram of Vardha on various issues and militant strategy was adopted. Cotton agitation in Parbhani to oppose the policy of monopoly cotton procurement and protection to synthetic fibre where three farmers lost their lives
1987	15th February	Farmers protested in Mumbai before the chief minister's residence demanding remunerative price for cotton growers
	20 October	Agitation at Latur for remunerative price for jowar
	20 November	*Vidarbha prachar yatra* from Sindkhed to Nagpur for loan waiver to farmers
1988	18 April	Agitation for loan waiver
	26 April	Agitation for loan waiver to farmers
	2 October	Meeting at Kolhapur to intensify agitation against the state indifferent attitude towards farmers

Source: Illustrated History of Shetkari Sanghatan (*Marathi*), accessed from http://sharadjoshi.in/node/147 on 25 March 2016 and www.angelfire.com/in/farmersforfreedom/martyres.html#December 1997 on 25 March 2016.

Neoliberalism and agrarian crisis: farmers' suicides

Since the 1990s, the agrarian conditions in Maharashtra have entered a new economic order under the influence of the neoliberalism that intended to bring about major policy changes in agriculture, particularly in the sphere of production and marketing, following the Agreement on Agriculture (AOA) under the Uruguay Round of negations of GATT (subsequently named WTO). The reforms were aimed to improve internal and external competitiveness through greater private sector participation to accelerate capitalist agriculture leading to the promotion of exports and rapid agricultural growth. As a result, in the post-WTO regime, Maharashtra's agriculture witnessed significant changes.

Public sector gross capital formation in agriculture declined substantially without much increase in private investment. There was deceleration of growth of private investment in agriculture. The estimate made by Mishra (2009: 128) reveals that the growth rate in Maharashtra's agriculture sector declined from 3.5 per cent in the pre-reform period (1980–81 to 1992–93) to 0.8 per cent in the post-reform period (1993–94 to 2004–05). To go by Chand *et al.* (2007: 2531), Maharashtra NSDP in agriculture, which witnessed a growth rate of 7 per cent between 1984 to 1985 and 1995 to 1996, remained almost stagnant during the next 10 years. The share of agriculture and allied activities in the Gross State Domestic Product (GSDP) declined persistently from 16 per cent in 1999–2000 to 12 per cent in 2005–06, accompanied by a decrease in government expenditure on this sector from 10 per cent in 2001–02 to 5 per cent in 2005–06.[49] Regarding the sectoral expenditure of plan expenditure in the state, the share of agriculture and allied services declined from 6 per cent during the Seventh Plan (1985–90) to three per cent during the Tenth Plan (2002–07).[50]

Fertiliser subsidy, directly funded by the union government of India, was reduced considerably during the post-reform period. The subsidy which amounted to 3.2 per cent of the GDP and 6 per cent of the union revenue expenditure in 1991 was reduced to 2.5 per cent and 5.0 per cent, respectively, by 1997–98 (Acharya 2004, as cited in Reddy and Mishra 2009: 22). In addition, the power sector reforms introduced since 1997 at the directive of the World Bank increased power tariffs. The government also brought reforms in the irrigation sector, announced the State Water Policy in 2003 and enacted the Maharashtra Water Resources Regulatory Act, 2005 and the Maharashtra Management of Irrigation System by Farmers Act 2005, which made new provisions for fixing of water charges for agricultural purposes and for the optimal utilisation of surface and groundwater through efficient distribution and delivery mechanism. The distribution of hybrid and improved certified/quality seeds of various crops under the control of the state largely shifted gradually to private sector. Currently, the private seed supply has overtaken seed sourcing from public sources (Mohanty 2013). The production and distribution of quality seeds, which was largely controlled by Maharashtra State Seeds Corporation (MSSC) and National Seeds Corporation (NSC), were dominated by the private seed producers. The Maharashtra Agricultural Produce Marketing (Development and Regulation) Act 1963 was amended in 2005 to include direct marketing, private market, farmer consumers' market, single licenses and contract farming. After the amendment, 165 direct marketing licenses were issued up to January 2014.[51]

Peasants were encouraged to integrate their agricultural activities with floriculture, horticulture, viticulture, and food processing, and service sectors like information technology, banking, and insurance, in tune with economic changes, without properly informing them the associated risks. Since 1990–91 horticultural development has been linked with the Employment Guarantee Scheme (EGS). The area under fruit crop, which was 0.24 million hectares in 1991, gradually increased to 1.8 million hectares in 2013–14 in response to the National Horticulture Mission declared by the union government in 2005–06 to expand the area under horticulture. Besides, the state government established agro-processing and agri-business section to make the raw material available for processing industries throughout the year.

In response to the changing conditions, peasants changed their cropping pattern by shifting to high-valued crops, especially cotton, sugarcane, and oilseeds and horticultural crops (Table 7.5); followed innovative methods of farming; and increased their applications to enhance agricultural productivity.

The use of tractors and share of purchased inputs to the total operational cost of cultivation increased remarkably (Mohanty and Lenka 2016: 173–174). The growing application of high doses of agricultural inputs accompanied by hikes in power tariffs as well as irrigation rate and withdrawal of subsidies led to a rise in the overall cost of cultivation and production (Table 7.6).

The cost of production of major crops was higher than the minimum support price fixed by the Commission for Agricultural Costs and Prices (CACP), which indicates that if purchases were made by the government agencies in the event of a price crash, the peasant was not in a position to recover his cost of production. Further, the state agencies were often slow in entering primary markets and peasants hardly received the necessary support (Mohanty and Shroff 2003). Though minimum support prices increased, it did not benefit the farming community as these prices

Table 7.5 Area, production, and yield during post-reform period

Crops	Area (% to gross cropped area)			Production ('000 tonnes)			Yield (kg./hec.)		
	TE 1994–95	TE 2004–05	TE 2012–13	TE 1994–95	TE 2004–05	TE 2012–13	TE 1994–95	TE 2004–05	TE 2012–13
Bajra	8.53	6.57	3.72	1,367.77	1,056.00	788.67	751	720	949
Jower	27.21	20.91	15.48	5,748.20	3,464.67	2,705.00	989	742	782
Cereals	48.73	40.36	31.71	11,140.57	8,671.90	8,985.53	1,070	963	1,268
Food grains	72.67	55.90	50.70	13,873.37	10,565.93	12,884.80	893	847	1,137
Cotton*	12.04	12.55	18.22	2,311.13	2,871.67	7,783.33	898	1,025	1,911
Oilseeds	12.62	12.86	16.38	1,977.41	2,673.70	4,848.33	733	932	1,324
Sugarcane	1.97	2.00	4.36	34,335.10	29,586.67	76,932.91	81,401	66,239	78,852
Fruit Crops	1.38	1.98	3.02	–	–	–	–	–	–

Source: Agricultural Statistics at a Glance, relevant issues.

Note: Production of cotton ('000 bales of 170 kgs. each).

Table 7.6 The cost of cultivation of major crops in Maharashtra (Rs. per hectare)

Crops	Cost of cultivation	1995–96	2005–06	2012–13
Bajra	Operational Cost	3,634.43	9,280.47	21,481.03
	Total Cost	5,124.1	12,966.62	30,786.45
Jowar	Operational Cost	3,991.11	10,877.97	25,284.16
	Total Cost	5,929.27	14,846.56	35,737.68
Cotton	Operational Cost	7,580.28	15,499.22	53,319.9
	Total Cost	10,374.96	20,793.82	71,079.73
Sugarcane	Operational Cost	26,593.24	63,876.54	11,2826.9
	Total Cost	35,804.34	84,656.83	16,5962.4

Source: *Reports of the Commission for Agricultural Costs and Prices,* Government of India, relevant issues.

were meant to compensate for the rising cost of production in the absence of yield-increasing public investments. As a result of the rising cost of cultivation and production, peasants were increasingly indebted to private moneylenders and traders. As per NSS data, while the proportion of indebted agricultural households increased from 55 to 57 per cent, the average amount of outstanding loans increased from Rs. 16,973 to Rs. 54,700 between 2003 and 2013.[52] Moreover, the magnitude of indebtedness among rural cultivating households was still higher. In 2013, nearly 56 per cent of cultivating households were indebted and the average amount of debt per household with outstanding loans came to as high as Rs. 0.122 million.[53] On the other hand, in the post-reform period, there was a sharp decline in the share of the formal sector in rural credit (Shah *et al.* 2007: 1357). The 'targeted priority lending' or 'directed credit' to agriculture was put on the back burner at the recommendation of the Narasimham Committee (1992) on financial reforms. Moreover, while banks charged just 7 per cent interest rate on loans for purchasing cars and other luxury consumer goods, they charged 9–14 per cent interest on farm loans (Dhanagare 2016b). As a result, the peasants increasingly got into the debt trap of input dealers, moneylenders, and private shopkeepers (Mohanty 2001a, 2005; Dhanagare 2016b).

Though area under crops, particularity commercial crops, increased consistently, the production and average yield of many crops followed an erratic pattern (Table 7.5).

A cursory glance at the year-wise data on area under crops and average yields indicate fluctuating trend more clearly (Figure 7.1 and 7.2). Agriculture was largely dependent on vagaries of the monsoon and the area under irrigation did not improve substantially,[54] and in many districts, particularly in Vidarbha and Marathwada, it was negligible. Capital-intensive crops like cotton were cultivated without irrigation. Due to irregular rainfall, crop failures were frequent in many parts of the state. The adverse climatic conditions caused heavy crop losses in several years (Mohanty 2001a; Mohanty and Shroff 2004) as a result of which the peasants remained in a perpetual deficit. In addition, the consequence of economic liberalisation lowered the prices of many agricultural products like cotton due to pressure in the international market. For example, the subsidies offered by the United States government to its cotton growers slashed the price of Indian cotton in the international market. With the opening up of the economy in 1994, cotton lint exports were placed under an Open General License (OGL) – that is, they were freely importable – though since 1970, imports had been canalised through the Cotton Corporation of India. Further, from July 2001, raw cotton exports were also under the OGL which made India's cotton economy susceptible to price shocks from the world market

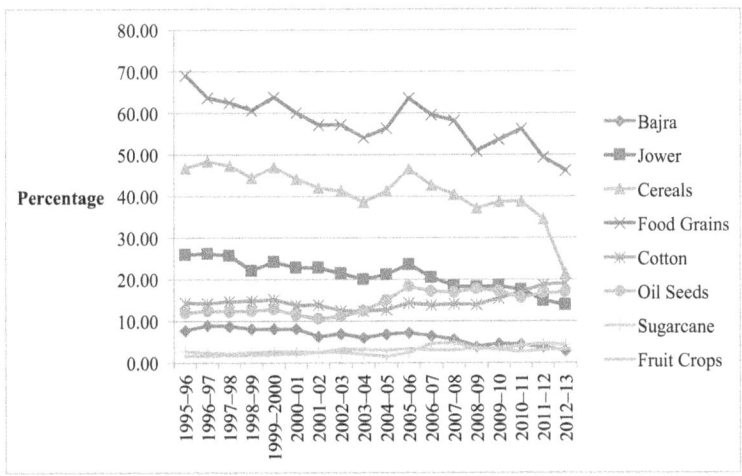

Figure 7.1 Area under crops (percentage to gross cropped area)

Source: *Agricultural Statistics at a Glance,* relevant issues.

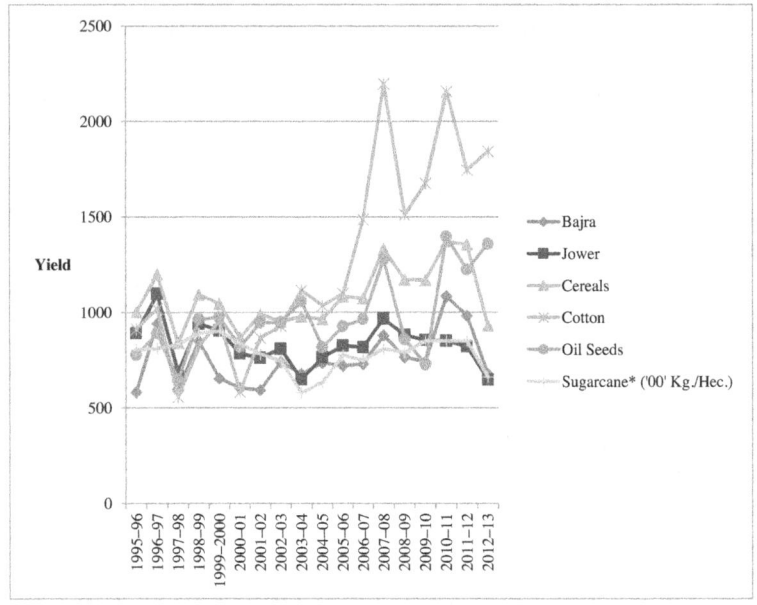

Figure 7.2 Average yield of major crops (kg./hec.)

Source: *Agricultural Statistics at a Glance*, relevant issues.

Note: Production of cotton ('000 bales of 170 kgs. each).

(Mitra and Shroff 2007). In short, in the post-liberalisation period the peasants faced not only yield risk but also price risk. Gupta (2005: 752) aptly observed,

> Indian agriculture has always lurched from crisis to crisis. If the monsoons are good then there are floods, if they are bad there are droughts, if the production of mangoes is excellent then there is a glut and prices fall, if the onion crops fail then that too brings tears. The artisanal nature of agriculture has always kept farmers on tenterhooks, not knowing quite how to manage their economy, except to play it by (y)ear.

As a consequence, agriculture as an occupation witnessed gradual decline. As per the census report, the number of cultivators which was 46 per cent in 1991, reduced to 42 per cent of the main workers in 2011 in rural Maharashtra. Moreover, a recent analysis of

NSS data made by Rawal (2013) indicates that the rural households which did not cultivate any land in Maharashtra increased from 43 per cent in 1993–94 to 50 per cent in 2011–12. To put it succinctly, agrarian conditions in the post-reform period multiplied the risk and uncertainty in agriculture, leading ultimately to a prolonged crisis that pushed the peasants to perpetual economic hardship.

Although distressful conditions of the peasants continued for several years, hardly any major political party of the state, including the communist parties (both CPI and CPI-M), who were instrumental in mobilising peasants during the drought crisis of the 1970s, effectively organised the peasants to undertake any political action. The distressed peasants were unable to organise themselves due to rising contradictions among them on the lines of caste, class, kinship, and region. A class of rich farmers, which emerged in the countryside in the 1970s and 1980s appropriating benefits of the green revolution and development measures launched through the cooperatives and Panchayati Raj institutions reaped the benefits of expanded production by orienting their agriculture to the demands of economic liberalisation, thereby leading to a greater polarisation in rural society. Apart from this, the rising assertiveness of the members of the lower castes, in view of their wider mobilisation and organised activities, created a kind of hostility between their members and of the higher castes (Mohanty 2001a). The continuous prejudice of the upper castes against the former untouchables increased the isolation of the low-ranking new agriculturalists. The growing regional disparity in agriculture backed by caste and kinship network (Mohanty 2009) also acted as a hindrance for the development of statewide solidarity among the peasants. Instead, agriculturalists competed among themselves to enhance productivity or grow new crops that could fetch better market prices. Moreover, the rural population has gradually entered the phase of consumerism en masse having far-reaching implications on the way they organise their lives and relationships. The traditional family, neighbourhood, and community linkages are being replaced by corporate structures by the new spirit of consumerism. With the unprecedented growth in the consumption of goods and services a new kind of village society has emerged under the impact of neoliberalism. In a recent study on rural Maharashtra, Johnson (2005: 43) observed,

> The process of consumerism, has now permeated much of rural Maharashtra. Throughout the 1990s, as villages moved into

more cash-based systems of production, consumer goods have become more accessible. Goods such as televisions, electric fans and irons, motorcycles, telephones, VCD, and VHS players are becoming more common in village homes.

As a result, the collective identity of the village was fragmented leading to unprecedented growth of individualism and competitiveness among the peasants.

Peasants who enthusiastically joined *Shetkari Sanghatan* in the 1980s under the leadership of Sharad Joshi started withdrawing from it due to its changing ideology. The *Sanghatan's* inconsistency towards electoral politics, particularly its frequency in making alliances, confused the farmers (Dhanagare 2014: 49). The distressed peasants realised that it is biased in favour of the rich farmers and corporate interests. Omvedt (1980: 2042) rightly noted,

> It is safe to say that not only the movement being led by rich peasant but (is) also a movement basically of the rural rich which is in contradiction to the interests of the majority of rural poor families . . . even middle peasant, who buy on the markets as much as they sell have little to gain.

In the post-reform period, the class character of the *Shetkari Sanghatan* movement was explicitly reflected in its activities (Dhanagare 1994; Lindberg 1994; Omvedt 1994b). The *Sanghatan* openly defended the *Dunkel* proposal and favoured an uncontrolled market in agricultural produce and international free trade in both inputs and outputs in agriculture despite wide protests from other famer organisations in the country. It endorsed the WTO rules of multilateral trade, disinvestment in non-viable public sector units and generalise entry for foreign direct investment and also foreign institutional investment. Joshi stated, "We fully support the *Dunkel* proposals and a totally free economy. We shall seek an alliance with other forces which stand for a free economy".[55] While responding to a question whether patents would not make peasants slaves of multinationals, Joshi replied that it would be better than current slavery to the Indian state ("if we have to be somebody's slaves, we prefer that they be more competent people").[56] In addition, the *Sanghatan* also strongly supported the formation of SEZs despite wide criticisms that the establishment of these zones led to the displacement of peasants. It spread its wings to support

a number of agriculture-related industries like pesticides, micro-
and macro-sprinkler irrigation systems, plantations, and crop
processing industries.

In response to growing peasant indebtedness, when the government agreed for the loan waiver but decided to extend it to only
those farmers having a loan amount up to Rs. 50,000 and having
less than 2.5 hectare of landholdings, the *Sanghatan* did not accept
these conditions. While commenting on debt relief and loan waiver
scheme, Joshi (2010: 97) states,

> The smallness of the holding has very little economic signifi-
> cance. If agriculture is a losing proposition, the small holder
> should logically be a smaller loser than the larger holder. . . . It
> should not be forgotten that it is not the small and the marginal
> farmer who made the country self-sufficient in food-grains.

Apart from this, Joshi, the leader of the movement who at the
initial period claimed *Sanghatan* was an apolitical organisation and
was critical of other political parties and their leaders,[57] later joined
electoral politics by making alliances with political parties in the
1990s.[58] In 1990, he decided to enter electoral politics and a new
party the *Swatantra Bharat Pakshya* (SBP) was founded in 1994. It
is widely believed that he used the *Shetkari Sanghatan* movement as
the means to achieve political power. To quote Dhanagare (2016a:
249), "Joshi consciously tried to get to the centre stage of political
power . . . apparently did so by sacrificing farmers' movement at
the altar of his ambition for political power and position". Moreover, the election manifesto of SBP which was prepared for 1999,
2004, 2009, and 2011 elections was hardly relevant to the farmers'
existential demands. It overlooked the central issues concerning the
peasants and there was no reiteration of its earlier demand either
for MSP for farm produce or loan waivers for farmers. Instead,
the peasants were advised to take the advantage of emerging market opportunities, set up their own agro industries and marketing
through their own agencies to export their farm produce ranging
from *papads*, sauces, and *chautneys* to neatly packed grains, *dals*,
dried fruits, and spices (GO 1993: 2708).

As a result, by the 1990s the farmers' agitations organised by the
Sanghatan declined. Joshi's mass base gradually eroded and SBP,
which had five assembly seats in 1990, came down to only one
seat in the 2004 elections and the total votes polled also declined

(Dhanagare 2016a). The great protest marches that were organised by *the Sanghatan* did not take place on the same scale and spirit during the 1990s. By the mid-1990s, the activities of the *Sanghatan* declined remarkably and the focus shifted from the core issues afflicting the peasants (Table 7.7).

Table 7.7 Major activities of Shetkari Sanghatan in the 1990s

Year	Day and month	Activities undertaken
1990	2–27 January	Phule-Ambekar rally was organised from Satara to Nagpur where the issues related to loan waiver, remunerative price for agricultural produce and women problems were raised. The rally was inaugurated by V P Singh
1991	4 January	Rastaroka in many places of Maharashtra demanding remunerative price for sugarcane growers
	15 May	*Shetkari Sanghatan* Joined with BKU in Chandigarh and participated in Jail Bharo Andolan demanding remunerative price for wheat
	10 November	Establishment of farmers' advisory committee under the leadership of Sharad Joshi and demanded new a NAP
1992	8 February	Statewide *Rastaroka* agitation in support of neoliberal policy and free economy
	10–11 October	Executive council meeting of *Sanghatan* at Aurangabad to oppose state's closed economic policy. Gheraoed Nehru port in Mumbai and *Rastoroko* agitation throughout the state
	14 November	Protest t against the state's Nehruniti and for agriculture development
1993	30 January	Meeting of *Sanghatan* workers at Bapu Kutir in Sevagram of Vardha where large number of farmers participated and farmers were advised to be fearless and follow Gandhiji's *mahamantra*
	31 March	Protest march to Delhi in support of economic liberalisation and free economy
	29–31 October	Fifth session of *Shetkari Sanghatan* at Aurangabad where decision was taken to support Dunkel Draft in support of economic liberalisation
1994	4 July	Agitation against corruption and remission of loan
	23 November	Agitation for cotton export policy
	10–12 November	Agitation against state and in support of free economy and liberalisation policy

Year	Day and month	Activities undertaken
1995	12 December	Agitation against government cotton export policy and restrictions.
	14 December	Bus roko andolan for various agricultural issues
1996	24 March	Agitation in Mumbai by *Shetkari Sanghatan Mahila Aghadi* opposing the Beijing World Women's Conference
1997	28–29 April	*Sanghatan* joined the rail roko and rasta roko agitation organised by Punjab and Haryana farmers against the wheat import policy
	1 May	*Sanghatan* joined the agitation organised by Punjab farmers in Chandigarh demanding remunerative price for wheat
	9 August	Organised agitation in all parts of state against the corrupt bureaucrats
	13–14 December	Cotton agitation in Amravati to oppose the policy of monopoly cotton procurement leading to the death of three farmers.
1998	10–12 December	Agitation by people's assembly demanding self-respect and freedom of the farmers
1999	9 August	Seventh session of *Shetkari Sanghatan* at Nanded to discuss the demand for free economy, *Sanghatan's* role in next Lok Sabha election
2000	7 February	Boycott of tax payment and electric bill payment by the famers until their demand for loan waiver is fulfilled
	11–12 November	Eighth session of *Sanghatan* at Sangli to protest against state's *Nehruniti* and to make the farmers aware of ill effects of *Nehruniti*
2001	17 January	Protest at Nagpur demanding remunerative price for Paddy
	10 November	Rail roko and rasta roko at Vardha demanding remunerative price for cotton
2003	9–11 November	Ninth session of *Sanghatan* at Chandrapur and agitation for a separate Vidarbha state
	25 December	Agitation at Chandrapur for a separate Vidarbha state

Source: Illustrated History of Shetkari Sanghatan (*Marathi*), accessed from http://sharadjoshi.in/node/147 on 25 March 2016 and www.angelfire.com/in/farmersforfreedom/martyres.html#December 1997 on 25 March 2016.

Lamenting on the decline of the farmers' movement in Maharashtra, Vijay Jawandhia, the prominent leader of *Shetkari Sanghatan*, said,

> The leaders of the farmers' movement misguided and mislead the movement; and it lost its sense of direction . . . the net outcome of the movement is that those who feed the country have now started committing "suicide as form of protest and movement". What could be a greater misfortune for the country?
>
> *(Jawandhia 2008, as cited in*
> *Dhanagare 2016a: 255)*

Though attempts were also made by peasants to explore the possibility of joining all India farmers' organisation like BKU, later it was found that this organisation also did not represent their true interests. Rather it represented the interests of the rich farmers. To quote a farmer young activist who had gone to Tikait's 1988 Meerut protest,

> We went ticketless, with only 24 rupees between us. . . . You should have seen the milk and *shirra* (a sweet dish made of wheat) they were getting! And they treat their agricultural labourers so cruelly. They really are rich peasants. Not only that, they have to talk of Ram and Krishna in every sentence they utter.
>
> *(Omvedt 1994b: 138)*

Having no other alternative, the distressed peasants felt helpless and many of them committed suicide. The number of suicides increased year by year indicating an alarming trend (Table 7.8).

Response to agrarian crisis: a comparative analysis

A comparative analysis of the conditions of the agrarian crisis during three different periods (Figure 7.3) provides insight to understand peasant behaviour in varied contexts. During the 1860s and 1870s, excessive land revenue and the cotton boom following the American Civil War motivated peasants to go for commercial crops and make risky investments by borrowing from moneylenders, and not to make profit but to secure their subsistence. But when the conditions of peasants deteriorated threatening their bare subsistence,

Table 7.8 Incidences of farmers' suicides in Maharashtra

Year	No	Year	No
1995	1,083	2005	3,926
1996	1,981	2006	4,453
1997	1,917	2007	4,238
1998	2,409	2008	3,802
1999	2,423	2009	2,872
2000	3,022	2010	3,141
2001	3,536	2011	3,337
2002	3,695	2012	3,786
2003	3,836	2013	3,146
2004	4,147	2014	4,004

Source: Accidental Deaths and Suicides in India, National Crime records Bureau, Ministry of Home Affairs, Government of India, relevant issues.

they expected the colonial state and moneylenders to be lenient. Contrary to their expectations, neither of them offered any remission. Instead, while the state enhanced the revenue, the moneylenders became more avaricious and used the crisis as an opportunity to take possession of movable and immovable properties of the peasants.

Given the cohesiveness among peasantry (who were drawn largely from the *Kunbis*) as well as the support of the rural elites and Poona *Sarvajanik Sabha*, the peasants agitated against the moneylenders. However, before undertaking violent resistance, they tried other alternatives like boycotting the moneylenders, putting carcasses of animals in moneylenders' houses, etc. Finally, when these efforts went in vain and the moneylenders did not soften their stand, peasants rioted.

On the other hand, during the 1970s, peasants expanded the area under cultivation and adopted new technology, taking the advantage of the available opportunities with a hope to improve their subsistence living. However, when they experienced devastating crop loss for consecutive years leading to a perpetual crisis, they started looking for opportunities to sensitise the state for undertaking ameliorative measures. In fact, for the first two years of the crisis (1970–71 and 1971–72), they tried to cope with the situation and revive, but

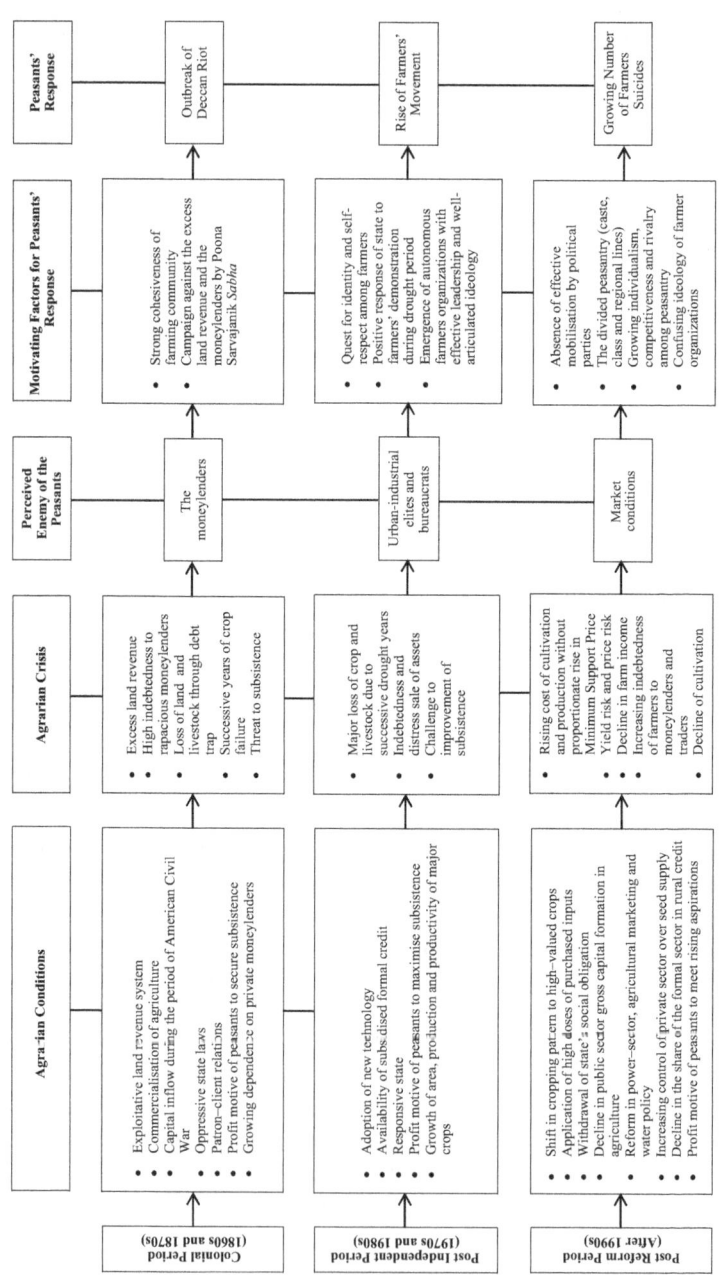

Figure 7.3 Agrarian conditions, crisis, and peasants' response during colonial, post-independence and post-reform period.

when the crisis continued for the third year, their condition worsened. They joined political mobilisation, which was undertaken by many political and non-political bodies against the state, demanding first relief measures and then raising wider issues of agricultural development. Interestingly, even though their conditions improved after the crisis, they joined movements in the 1980s organised by *Shetkari Sanghatan* when they found the ideology and leadership of the organisation had the agenda to offer them a new identity and better farm income by challenging the dominance of urban-industrial elites and bureaucrats. They demanded higher remunerative prices and minimum support prices to improve their subsistence.

Their conditions in the post-reform period were different. In order to meet the rising aspirations, they shifted to high-valued crops and followed innovative farming methods based on high doses of agricultural inputs, taking loans from moneylenders and traders. The issue of subsistence living was not in question as the demands of democratic electoral politics made it difficult for the state to deny them minimum.[59] Peasants were interested in raising farm income to meet their growing quest for social prestige and self-esteem in the context of growing competitiveness, individualism, and rivalry across social groups. However, the rising costs of cultivation and production coupled with uncertainty and risk associated with production and marketing pushed the peasants to a prolonged crisis. The state started withdrawing from agriculture and attributed the crisis to market and climatic conditions. Resistance from within the peasants became a distant dream due to growing division in the rural society on the lines of caste, class, and region. Farmers' organisation also betrayed the peasants by changing its ideology and joining electoral politics to secure power and position for the leaders and favoured large holders, traders, and corporate groups. Given these inherent socio-political weaknesses, the peasants found it difficult to organise themselves against the market conditions. The other alternatives like migration to other places, involvement in organised crimes (which was found earlier in history), etc., were not feasible as they were against their self-respect and dignity. The cultivating communities like *Kunbis* were conscious of their self-respect and always tried to improve their position through hard work and innovative farming. SC and OBC farmers were also motivated to challenge the hegemony of the upper castes in the rural areas through innovative farming with the support of new technology (Mohanty 2001a).

Conclusion

The analysis of three types of peasants' responses to agrarian crisis does not neatly confirm the arguments based on moral economy made by Scott or made by Popkin in the line of political economy approach. The peasants' behaviour goes beyond the Scott-Popkin debate which is based on economics' narrow and self-imposed rejection of alternatives and its incapacity to see more than a single dimension of most phenomena. The stereotypical binary between profit and subsistence, rational and moral, risk and safety, and individual and collective welfare overlooks many discursive social dimensions of peasants' response to agrarian change. True, peasants follow safety first principle, they go by subsistence ethics and moral values. On the contrary, they also make risky investments and do not necessarily resist change. In the colonial period they went for commercial crops, particularly during the American Civil War, during the post-colonial period they adopted new technology, and during the post-reform period they changed their agricultural practices in response to market conditions. When they experienced the adverse effects of these changes, they responded differently considering the existing socio-political conditions and their immediate enemies. Peasants made cost-benefit calculations but not in pure economic sense, as argued by Popkin. They were motivated to make profit to secure subsistence, to maximise subsistence or to meet the rising aspirations in the changing socio-economic context.

To sum up, peasants did protest in response to the crisis, but crisis was not sufficient for protests. The motivation for protest was determined by a variety of socio-political preconditions and types of enemies. Scott (1976:4) also acknowledged this when he said,

> If anger born of exploitation were sufficient to spark a rebellion most of the Third World (and not only the Third World) would be in flames. Whether peasants who perceive themselves to be exploited actually rebel depends on a host of intervening factors – such as alliances with other classes, the repressive capacity of dominant elites, and the social organization of the peasantry itself-which are not treated except in passing here.

Peasants neither protested to restore the traditional system nor to check the market and bureaucrats. Rather they protested to improve their individual as well as collective well-being. They rebelled during the colonial period when they realised that there was a remote possibility of reviving subsistence (either through the intervention of the state or leniency of the moneylenders or through extra labour), and there existed a kind of cohesiveness among fellow peasants (who were drawn from similar socio-economic background) and external support. They joined movements during the post-green revolution period, keeping in view the positive response of the state to their earlier agitations during the drought crisis, emergence of effective organisation and leadership with appropriate ideology and a kind of unity in the rural society seeking self-respect and identity in the context of rising rural-urban divide. During the post-reform period they did not rebel or join movements because of the absence of unity among them due to growing individualism, competitiveness, and rivalry accompanied by ambiguous ideologies of farmers organisations and ineffective leadership. Therefore, it would be appropriate to argue that peasants were always rational while responding to agrarian crisis whether they rebelled, joined movements, or committed suicide.

Notes

1 This is a revised version of the paper presented as "Professor Rama-krishna Mukherjee Commemorative Special Lecture" held in March 2016 at Indian Statistical Institute, Kolkata.
2 It is estimated that nearly a quarter of a million farmers committed suicide in the countryside between 1995 and 2007 (see Nagaraj 2008; Sainath 2009).
3 Peasants also rebelled in many states at the time of independence and thereafter for many years against the exploitative landlords and moneylenders (Gough 1968; Mukherji 1979; Dhanagare 1983).
4 Though crisis is an over-used word implying varied meanings covering a spectrum from, at one end, expression of irritation with petty difficulties to, at the other end, meltdown and reconfiguration meaning different things to different people. For the purpose of present analysis agrarian crisis is viewed in terms of following major indicators which are by and large commonly accepted: (a) decline in the growth rate of agricultural production and productivity, accompanied by a steep rise in the cost of farming; (b) increasing preponderance of marginal holdings with poor returns from cultivation; (c) growing impoverishment and indebtedness, particularly to informal agencies; and (d) rising rural unemployment leading to distress migration.
5 Though conceptually peasants and farmers are not the same, as the former are oriented towards subsistence and latter associated with capitalist economy, they often overlap in Indian context in a very complex

manner. The agrarian economy of India continues to have landlord-tenant, and peasant and capitalist elements, though a trend towards market-oriented economy, have emerged in the post-green revolution period (Mencher 1978; Harriss 1982). The suicide victims are mostly small-scale producers and are closer to the characteristics of peasants, though they possess some elements of farmers.

6 This debate has its roots in the classics of political economy (in the writings of Henry Maine, Karl Marx, Max Weber, and V. I. Lenin).

7 Scott uses the term 'subsistence ethic' which defines the minimal needs that must be met for members of the community which is ultimately rooted in peasants' existential needs. The 'ethics of subsistence' provides the framework for the social and economic arrangements to minimise risks and ensures the right of a peasant to a bare livelihood.

8 To Scott (1976: 167), "reciprocity serves as a central moral formula for impersonal conduct. The right to subsistence, in effect, defines the minimal needs that must be met for members of the community within the context of reciprocity".

9 However, Scott also indicated that high taxes are accepted so long as they are structured to allow peasants to survive in bad agricultural years. He believes that the test for the peasant is more likely to be "What is left?" than "How much is taken?"

10 Scott responded to Popkin's criticism and clarified his position in his subsequent two essays which he had written two years after *The Moral Economy of the Peasant* and after the publication of Samuel Popkin's volume *The Rational Peasant*. See Scott (2012: 134n2).Besides this, in an interview with Richard Snyder in Durham in 2001 on Popkin's criticism, Scott made the following remark to strengthen his argument: "When Popkin wrote his book, he sent me the manuscript, and I sent him back fifteen or sixteen pages of criticism. He revised the manuscript to some extent in light of what I had to say and then sent back a second version. I started to read it, got four pages into writing my criticism . . . and sent Popkin a letter saying", "Sam, you're on your own. I'm not going to help you sharpen your sword to cut me up". For details, see Munck and Snyder (2007: 351–391).

11 Lieberson (1981) has comprehensively summarised this debate. Also see Colburn (1982);Moise (1982), Keyes (1983), and Peletz (1983).

12 Though Greenough (1983) examines the relevance of moral and political economy debate in the context of peasant behaviour during the crisis in Bengal caused by famines during the colonial period, he did not relate it the peasant resistance movements. A recent analysis made by Mitra (1992) on Gujarat and Orissa tangentially touched upon this debate.

13 Though Khandesh is usually included in Bombay Deccan, its geographical features are different from the rest of Bombay Deccan, and the agriculturalists of this region have cultural affinity with Gujarat (see Choksey 1955: 1–5).

14 For details, see A. T. Etheridge, Report on past famines in the Bombay Presidency (Bombay 1868), p. 76n and H. H. Mann (1967: 131–134).

15 See T. Marshall, Statistical Reports, pp. 133–34. As cited in Guha (1985: 12).

16 *Baluta* was the payment made in terms of cereal or money or a little land to people against rendering certain necessary services and those who received it were called *balutedars*. They were classified into three categories. While the first category included those whose work closely related to production and maintenance of agricultural implements (for example *maharas*, *chaugulas*), the second category refers to service providers like potter, rope maker and barber. Artisans rendering other services, religious or personal like washer man, water carrier, were included in the third category. There were three broad kinds of remuneration for baluta-servants. One was the remuneration in kind or in cash; the second was additional minor remunerations in kind or in cash variously called *hak* (rights), *Iavajima* (perquisites), or *man-pan* (privileges), and the third was revenue-free inam land which was given to the more important *balutedars*, and such land is called *vatan* and the holder was the *vatandar*. Not only *vatan*-holding servants but also strangers were entitled to the baluta-remuneration so long as they offered specific service to the village. See Fukazawa (1972: 27–28).

17 T. Coates account of the state of villages in 1820 reveals that as many as 79 peasant households out of 84 in Lony had taken small loans ranging from Rs. 50 to 200 from six *vanis*. See Kumar (1965: 614n)

18 For more details, See Kumar (1965: 614).

19 Between 1772 and 1800, years of administration of Nana Fadnavis, management of the Peshwa's land revenue was perhaps more efficient than at any other time where *mamlatdars* were in charge of the main divisions of district organisations and chosen from families of character and respectability. Due to their influential position, they were able to get thorough knowledge of the people and were careful in collecting revenue.

20 As reported by the Superintendent of the Ahmednagar Survey, Colonel Anderson. See RDRC (1878: 10).

21 The expenditure in the Public Works Department and Irrigation Department in Poona District, which, as has been shown, was 3.1 million in 1868–69, but in 1873, it had fallen to about 1.1 million. See RDRC (1878: 26).

22 The effect of revision in the *talukas* in Poona District was as follows:

Talukas	Former demand	Revised demand	Increase	Percentage
	Rs.	Rs.	Rs.	
Indapur	81,184	125,845	44,661	55.01
Bhimthari	81,475	133,131	51,656	63.40
Haveli	80,475	133,174	52,699	65.48
Pábal	92,350	139,315	46,956	50.84
Súpa	59,926	78,788	18,862	31.47

Source: *Gazetteer of Bombay Presidency*, Vol. XVIII, Part II (Poona), p. 118n.

23 Registered land transactions in disturbed villages (Bhimthari *Talukas*) for the years 1865 to 1875 are given next:

Year	Mortgages		Sale		Total	
	No.	Values (Rs.)	No.	Values (Rs.)	No.	Values (Rs.)
1865	14	7,037	4	2,700	18	9,737
1866	18	6,624	9	2,104	27	8,728
1867	28	9,594	6	1,089	34	10,683
1868	37	16,826	7	2,423	44	19,249
1869	66	18,779	24	3,546	90	22,325
1870	86	22,868	21	3,990	107	26,858
1871	73	16,992	22	5,710	95	22,702
1872	63	14,556	36	6,888	99	21,444
1873	97	28,744	44	6,271	141	35,015
1874	64	20,297	18	5,827	82	26,124
1875	13	2,508	7	936	20	3,444
Total	559	164,825	198	41,484	757	206,309

Source: RDRC (1878: 28).

24 See *Gazetteer of Bombay Presidency*, Vol. XVIII, Part II (Poona), pp. 118–119.

25 District Deputy Collector, Khandesh, Para 11, no. 108 dated 12th July 1874. See RDRC (1878: 168).

26 Talegaon, sub-judge, dated 14 June 1875. See Deccan Commission Report (1878: 169).

27 Sholapur, sub-judges letter, No. 9 dated 27 March 1875. See RDRC (1878: 169).

28 While justifying the necessity of attaching the crop, Giberne informed the Parliamentary Committee of 1847–48, "If you allow the *ryots* to carry it away you would seldom see any revenue at all" (see Guha 1985: 26).

29 As described by (Keatinge 1912: 82*n*), the person (mostly a Brahmin) sitting on *dharna* insistently demanded payment, abused the debtor, refused to eat or drink till the debt was paid and sometimes he also placed a stone on his head or tying his hair to a peg in the debtors door. The debtor was made responsible for all these self-inflicted tortures which were considered a great sin.

30 It was observed by the Deccan Riots Commission that the law of India appeared to be the only modern law which allows such unlimited resort, and under it, the debtor and his family are liable in person and property to an extent which was practically unlimited. To quote the report (RDRC 1878: 41), "When we compare the law of India with that of other countries we find that not one is so oppressive as the Civil Procedure Code in this respect, not even the oldest law in the world, the law of Moses, which allowed the debtor a discharge after serving seven years".

31 For example, in 1871, the number of warrants of arrest printed was 172,600, while the number of arrests was only 7,135. So it would seem probable that somewhere about 150,000 warrants had been used as threats only. See RDRC (1878: 47).

32 Details of number of peasants arrested and imprisoned in jail due to indebtedness in the districts of Bombay Presidency during the year 1872 and 1871 are given next:

Districts	Persons arrested	Prisoners committed to jail	Average no. of days confined in jail
Ahmadabad	1,283	295	26
Surat	428	220	34
Tanna	1,389	204	37
Ratnagiri	264	40	22
Dhulia	1,017	181	12
Ahmednagar	317	80	29
Poona	428	427	25
Sholapur	359	106	34
Satara	461	226	12
Belgaum	284	41	38
Kaladgi	–	21	31
Dharwar	88	20	37
Karwar	817	16	28
Total	7,135	1877	30
Results in 1871	–	1958	37

Source: RDRC (1876) Appendix B, p. 124.

33 The following narration by a peasant in Supa, as documented in the RDRC is given next:

> I owe Rs. 150 to Jagjivan Bhaichand on a registered bond. My house was mortgaged. No decrees have been given against me. Rs. 150 were not received by me in cash. The debt is an old debt from my father's time. I got nothing myself, not even when the bond was passed. I wrote the bond because he was threatening to file a suit against me and get me sent to jail. The amount of the bond was Rs. 150 principal. I don't know the rate of interest. He told me verbally that the interest was ½ an anna in the rupee. There were previously 2 bonds bossed about 6 years ago. There was no mortgage in either. One of them, for Rs. 30, was paid off 3 years ago in presence of us both, and was torn up. We only paid the principal, Rs. 30. The banker let us off the interest. I don't know why my father borrowed the money. I know that he did not get cash. It was for an old debt. He got nothing in cash. No old bond was shown. No accounts were shown. He probably got some money to buy donkeys. The Rs. 30

bond was probably at ½ an anna an interest too. The bond for Rs. 25 was passed on the same day. No money was got for that either. The reason for passing 2 bonds was that the term was different. The Rs. 30 one was for 4 months, and the Rs. 25 one was for 6 months. My father paid for the stamps of both. One of the papers was for 2 annas and one for 4 annas. The Rs. 25 bond was liquidated about 4 years ago by the bond for Rs. 150. First of all the Rs. 25 bond became one for Rs. 40. This was signed by both my father and myself. I paid Rs. 20 of this bond at the time the Rs. 40 was payable, namely 7 or 8 months after it was executed. I further passed a bond for Rs. 60. On this the bond of Rs. 40 was torn up. My father may have got some turban or something from the *Marwari* as well. I myself have received nothing of any sort whatever. My father is alive. About 2½ or years ago the banker said his bond was becoming time-expired, and he threatened to sue us. On that one bond was passed for Rs. 50 on mortgage of one of our houses, and another was passed, I believe, for Rs. 100 on the other. Both were registered. Jagjivan keeps accounts. About 1½ year or years ago he asked for payment of the Rs. 50 on a threat of suing us. I don't know if my father passed a new bond then or not, as I went to another village, Pandeshwar, to live with my father-in-law. If he had not threatened us with a civil suit and the jail afterward, we would not have passed the bonds. We paid nothing except the Rs. 50. We gave him 500 to 1,000 bundles of karbe and Rs. 10 or Rs. 12 worth of grain out of our balote. We did not pay this every year. One year it might be Rs. 10 and another Rs. 20. Once we paid him 1500 bundles of karbe on the occasion of his marriage. He said he would give us a receipt after the wedding was over, but he did not do so. On another occasion, when we gave him 500 bundles, he said that we owed him a great deal and should give him that amount for nothing. He gave us no receipts for the annual amounts, as he said we owed him Rs. 6 for rent. Our property consists of 2 houses and 1 donkey. We have no land, nor ornaments of any sort, nor cooking pots except about Rs. 5 worth. I can't say what the houses are worth exactly. One of them consists of 10 khans and the other of 6 khans (Rs. 100 + 60?). . . . I see that unless some house is built or something turns up for me to do, I must lose my house. I will then beg.

(RDRC, Appendix B: 1–3)

34 For details, see *Gazetteer of Bombay Presidency*, Vol. XVIII, Part II (Poona), pp. 119–120.

35 For details, see *Gazetteer of Bombay Presidency*, Vol. XVIII, Part II (Poona), pp. 121–122.

36 For details of land reform legislations in Maharashtra, See Rajasekaran (1996) and Deshpande (1998).

37 The principal objectives of this Act were (a) to bring about equity in bargaining power among the agriculturists and traders, (b) to promote mutual confidence between the two parties, (c) to prevent malpractices

in trading, and (d) to give a fair deal to the peasants. See Mohanty and Shroff (2004: 5600).

38 The number of primary agricultural credit societies in Maharashtra increased from 18,998 in 1960–61 to 20,091 in 1969–70 accompanied by rise in the membership from nearly two million to three million during this period. See Brahme and Upadhyaya (1979: 102).

39 Moreover, 70 per cent of water resources were devoted for cultivation of sugarcane and horticultural crops, 10 per cent for other cash crops and 20 per cent for food grains. In 1971 the sugarcane crop absorbed nearly 60 per cent of the water resources covering only about 1.5 per cent of the gross cropped area (See Brahme and Upadhyaya 1979: 122).

40 The number of districts, villages, and people affected is given next:

Year	Affected districts	No. of scarcity declared villages	Population affected (million)	Loss of crops %
1970–71	21	23,061	18.8	60
1971–72	20	14,687	11.2	70
1972–73	25	30,780	30.0	60

Source: *Review of Scarcity in Maharashtra up to 21 December 1973*, Lokrajaya, 1 October 1974, p. 24.

41 Index of the Value of Output (1960/61–1961/62 = 100)

Year	Jower	Sugar	Cotton	Groundnut
1968–69	170	345	136	206
1969–70	160	213	122	231
1970–71	103	272	60	231
1971–72	128	351	135	205
1972–73	128	487	169	128
1973–74	347	646	313	459

Source: Oughton (1982: 191).

42 For more details, see Brahme and Upadhyaya (1979: 104).

43 Index of Real Rates of Payment (1960/61 = 100)

Year	Agricultural		Village artisans3	
	Labourers	Carpenters	Blacksmith	Cobbler
1969–70	119	117	134	201
1970–71	110	112	111	168
1971–72	103	99	103	149
1972–73	–	76	73	74

Source: Oughton (1982: 184).

44 For details, see Joseph (2006: 5150–5151).

45 The struggle began with a united left conference of 8,000 agricultural labourers and poor peasants in Ahmednagar district in February 1970, followed by a march in Bombay of 30,000 tribals and poor peasants on 4 March. Between March and December 1970, there were as many as 34 demonstrations, marches, strikes, and meetings where 53,000 people participated and again between January and October 1971 over 0.15 million rural poor participated in marches, *gheraos*, road closings along with simultaneous demonstrations of 50,000 poor peasants before 119 *talukas* government offices on May 1. There were 1,529 marches, 171 processions, 97 *bandhs* and 52 *gheraos* in the first eight months of 1973 and 27,000 persons were arrested for breaching the law. The most widespread and united action in the history of Maharashtra took place on 15 and 16 May, 1973 with the active participation of rural mass. For more details, see Omvedt (1975: 47).

46 For details of the historical background of this movement, its organisation, leadership, and achievements, see Mies (1976a, 1976b).

47 For *Shetkari Sanghatan*, "self-sufficiency is the virtue of less cerebral species" and "trade and exchange are beneficial for attaining higher levels of production and higher standards of living". For more details, see *Shetkari Sanghatan*, www.sharadjoshi.in/.

48 Bentall's interview with Joshi, 26 March 1992, as cited in Bentall and Corbridge (1996: 30–31).

49 See *Economic Survey of Maharashtra 2006–07*. Mumbai: Directorate of Economics and Statistics, Planning Department, p. 199.

50 See *Annual Plan*. Planning Department. Government of Maharashtra, 2009.

51 See *Economic Survey of Maharashtra 2013–14*. Mumbai: Directorate of Economics and Statistics, Planning Department, p. 94.

52 For details, see (1) *Situation assessment survey of farmers: Indebtedness of Farmer Households (2005)*, NSS, 59th Round (January–December 2003); and (2) *Key Indicators of Situation of Agricultural Households in India* (2014), NSS 70th round (January– December 2013).

53 To go by the NSSO 70th Round report, 56 per cent of cultivating households are indebted with an average debt of Rs. 68532. It also reveals the average amount of debt per household with outstanding loan is Rs. 122818 with debt asset ratio of 2.72 per cent. For details, see Key Indicators of Debt and Investment in India (NSS KI (70/18.2) January–December 2013, P. Appendix-A (A-2).

54 The irrigation intensity which was 14.40 in TE 1994–95 increased to 18.70 in TE 2011–12. See *Agricultural Statistics at a Glance*, relevant issues, Govt. of India.

55 See *Hindustan Times*, 17 February 1993, as cited in Gupta (2006: 316).

56 See Roy Chaudhuri and Shankar (1993), as cited in Omvedt (1994b: 155).

57 Joshi expected the followers to "remove their 'political shoes' outside before entering the temple of the farmers' movement". He also went to the extent of saying, "If individually I ever approach you (voters) to

seek your votes for myself then you should beat me with *chappals* and shoes" (Narade 1994: 7, as cited in Dhanagare 2014: 43).

58 Dhanagare (2014) gives a comprehensive account of *Shetkari Sanghatan*'s involvement in electoral politics and its varying strategies Joshi adopted to make alliance with various political parties ranging from Left wing parties like the CPI-M to the right-wing BJP and Shiv Sena.

59 It is argued that the demands of electoral politics compel the state to undertake minimum measures for the welfare of the peasant mass to counter the adverse effects of neoliberal reforms. For example, see Chatterjee (2008).

Chapter 8

"We are like the living dead"

Farmer suicides in Maharashtra[1]

There can be no question that the current spate of farmer suicides in a number of Indian states is an accurate indicator of problems afflicting the rural economy and society. While the respective state governments attribute these self-inflicted deaths mainly to crop failure, the media highlights factors such as the rising cost of cultivation, indebtedness, and bottlenecks in agricultural marketing. Studies undertaken in Andhra Pradesh, Karnataka, and Punjab also emphasise these same linkages (Parthasarathy and Shameem 1998; Shiva and Jafri 1998; Prasad 1999; Vasavi 1999; Bose 2000; Iyer and Manick 2000; Deshpande 2002; Rao and Gopalappa 2004; Sarma 2004)[2] Some of these studies have hinted that what might be termed social issues are features that contribute to these deaths. Broadly speaking, however, the available literature subscribes to the view that farmer suicides in India are an effect of economic hardship caused by crop loss and indebtedness.

Although in essence correct, it is necessary to qualify this view somewhat. In his analysis of suicide, Durkheim gave equal weight to social causes.[3] According to him, the act of suicide is an individual phenomenon the causes of which are nevertheless located in the wider socio-economic context. As he (Durkheim 1952: 299) himself put it, "The victim's acts which at first seem to express only his personal temperament are really the supplement and prolongation of a social condition which they express externally".[4] In this connection, it is important to emphasise that his methodological quarrel is not with economists per se, but rather with psychologists, who reduce suicide to an act that cannot be understood by reference to any phenomena beyond the individual self. Thus for psychologists the act of suicide has no wider socio-economic meaning.[5]

The presentation which follows utilises this Durkheimian approach to examine the meaning and causes of farmer suicide in the state of Maharashtra, in western India. It focuses on the districts of Amravati and Yavatmal, where the incidence of suicides was the highest.[6] The first section looks at the theoretical debate about suicide, with particular reference to the work of Durkheim on this subject, and its connection with his views about the transition from pre-industrial to industrial societies. The second part outlines the historical background of Amravati and Yavatmal districts in Maharashtra, while the third focuses on the case studies of farmer suicides. A final section briefly relates the findings from Maharashtra to instances of farmer suicide elsewhere.

Durkheim, suicide, and agrarian transition

The concept 'suicide' is one of those deceptively common-sense terms that everyone knows but the meaning and causes of which are strongly disputed. In the social sciences it is one of the few concepts that is debated across disciplines, engaging as it does the attention of psychologists, philosophers, and sociologists. The reason for this is not difficult to discern. It is because its meaning – and therefore its cause – extends from the self outwards to society, and there is consequently little agreement as to where precisely the element of determination for such agency lies. Occasionally it is difficult to distinguish suicide (= a voluntary act) from murder (= an involuntary act): what has the outward appearance of taking one's own life may in fact be the result of irresistible pressures exerted by those in the wider context. The latter may extend from family, kinship or ethnic group, via the village community and neighbourhood, to what is termed 'public opinion' circulating within a given society. This in turn is compounded by the very different meanings attached by societies to suicide. In some cultures, therefore, it is regarded as an act which confers honour, while in others the opposite holds true: it is stigmatised and brings shame, both on the person carrying out such an act and on those linked to them.

The conceptual complexity of this term, and consequently the need for caution when analysing it, can be illustrated by reference to the different meanings attached to suicide. Thus, in some contexts, suicide might in certain circumstances be seen as a virtuous and ennobling act.[7] Both the state and the nation recognise such acts of conscious self-sacrifice that result in deaths of many enemy combatants during wartime as heroic.[8] Equally, they may receive the same kind of recognition from religious teachings and/ or authorities, which classify them as instances of martyrdom that merit rewards in heaven. Suicide is in these cases viewed positively, as an unselfish and sacred act that elicits reciprocity either in temporal or spiritual terms. In European countries where historically Roman Catholicism has been a very conservative ideological force, however, religious teaching and authority strongly condemns suicide.[9] Here, by contrast, it is an act that is viewed negatively, not least because the church categorises it as a mortal sin (= the gravest and most unforgivable of offences against God), and as such one that consigns its subject to eternal damnation in Hell.

In India such debate as there has been about the meaning and causes of suicide has focused on two specific phenomena, each of which is gendered. One is sati, or the act of self-immolation by a widow. The other is dowry death, an act which is officially categorised as suicide, the victim of which is also female.[10] Whereas the latter is mainly an urban phenomenon, the former by contrast is not merely a rural one but an act that is regarded as part of a longstanding cultural tradition. It is this latter aspect which informs the analysis presented here, an approach which is in keeping with that of the main sociological theorist of suicide, Emile Durkheim. As he observed (Durkheim 1933: 246), in pre-industrial societies, suicide "is not an act of despair but of abnegation. If with the ancient Danes, or Celts, or Thracians, the old man at an advanced age put an end to his life, it was because it was his duty to free his companions from a useless burden. If the widow of an Indian did not survive her husband, nor the Gaul the chief of his clan, if the Buddhist has himself torn on wheels of the carriage carrying his idol, it is because moral or religious prescriptions demand it. These voluntary deaths are therefore no more suicides, in the common sense of the word, than the death of a soldier or doctor exposing himself knowingly because of duty".[11]

For Durkheim, therefore, suicide in pre-industrial society is a culturally virtuous act, an effect of – and, indeed, designed to reproduce – its socio-economically integrated nature. His ideas about suicide, its meaning and causes, however, have to be located epistemologically and politically within the framework of the negative views he held about the wider process of socio-economic transformation.[12] These relate to the breakdown of both the division of labour and the 'organic solidarity' based on this, both attributed by him to pre-industrial societies, and the corresponding rise of individualisation connected with the spread of industrialisation. According to him such a development necessarily leads to social disintegration. In non-industrial (or 'primitive') societies, Durkheim notes, suicide is the effect of social cohesion expressed in terms of community disapproval, whereas in more industrialised contexts it is a result of an absence of community, as manifested in individualisation coupled with alienation. In rural contexts, therefore, suicide is altruistic, an effect of the rootedness of the subject in their community.[13] This form is termed by him 'egoistic' suicide, in that it prevents the assertion of individuality, thereby reaffirming the importance of community. For Durkheim, 'egoistic' suicides occur when the ties binding the individual to others are weakened owing to the absence of adequate social integration. The greater the social isolation, the lesser the individual participates as a social being. As a result, life lacks purpose and meaning, the subject experiences a loss of direction, a sense of apathy and finally an absence of attachment to life itself.

Individualisation – or 'egoism' – refers to the loosening or dilution of traditional social ties binding the members of a group to one another, and results in isolation (Giddens 1966: 278). Durkheim stresses further that the degree of individualisation varies in relation to the domestic environment. The larger the size of the family, the greater the degree of protection against suicide, because this institutional form represents a powerful kind of social cohesion due to stronger sentiments and memories (Morrison 1995: 174). The duties and obligations as well as the demands and expectations in the family generate an attachment to life. A disposition to suicide is, therefore, greater among those either belonging to small families (particularly when they face widowhood, separation, and childlessness) or who are unmarried (Durkheim 1952: 180–216).

By contrast, in industrialised societies where such rootedness has been eroded by individuality, suicide is paradoxically a consequence of this lack of rootedness. Termed anomic suicide, it occurs as a result of the estrangement of the self from what Durkheim takes to be a 'natural' community, and it is in this transformation that the clue to his politics lie. His belief is that in such contexts normative boundaries control the upper and lower limits of aspirations of its members, and an anomic situation arises when these boundaries go haywire (Marks 1975: 333).[14] This situation generates feelings of failure and disappointment which lead to the growth of what is termed a 'suicidogenic impulse'. Most significantly, Durkheim views anomie as the chronic state of a modern socio-economic system – that is, capitalism – and so he considers such suicides as the characteristic feature of industrial society.[15]

Like so many others writing at the end of the nineteenth century, therefore, Durkheim deplored the negative impact of industrialisation on what he saw as the 'organic solidarity' of rural society, a perception that emerges clearly from his work on the division of labour.[16] Hence the association by him of increasing specialisation at work with a decline of happiness, a position then linked to the opinion that rural inhabitants of pre-capitalist societies (hunters, gatherers, peasants) are more content than their urban counterparts, so-called civilised humanity (Durkheim 1933: 233ff.).[17] This in turn structures and is responsible for the increased incidence of suicide in industrial societies.[18] The relevance of this analysis for explanations of farmer suicides in Maharashtra stems from this dynamic. That is, the Durkheimian view that the incidence, the meaning and the cause of anomic suicide are all linked to the 'new' conditions faced by rural producers as a result of the rapid economic growth in general, and in particular the spread of neoliberalism.[19]

Broadly speaking, it could be argued that pre-industrial agrarian communities – the 'before' of the Durkheimian approach to suicide – were indeed present in recent Indian history. Traditional rural society was hierarchical in terms of caste relations, and each caste was an occupational group. As part of this caste-based occupational structure, members of lower castes did not own the land they cultivated. As peasant tenants, they depended on limited sources of income, mostly from labour services provided as rent to landowners. The ascribed socio-economic duties limited their hopes and aspirations. They witnessed the prosperity and experienced the domination of upper caste landholders who controlled agriculture, the major source of income and employment.

The social movements that began to challenge the upper caste dominance and the subsequent protective and ameliorative measures introduced in the post-independence phase made this deprived class aspire to a better socio-economic position. The acquisition of land through reform measures and state intervention appeared to them a means of fulfilling their long-cherished desires. The prospect of a modernised agriculture, based on the high yielding varieties introduced at the time of the 1960s green revolution, held out further promises. By contrast, the new social order challenged the historical dominance exercised by the higher castes, who lost some portion of their land to the lower castes due to the promulgation of land ceiling legislation. To retain their social position, these higher castes began to look beyond agriculture to trade, finance, and politics (Baviskar 1980; Dhanagare 1994; Rutten 1995; Punalekar 1998; Mohanty 1999, 2001a).

This, then, forms the theoretical and historical background to the issue of farmer suicides in Maharashtra during the late 1990s. Before turning to the details of the latter, however, it is necessary to examine the agrarian history of the two districts in which fieldwork was conducted.

Amravati and Yavatmal districts, Maharashtra

The districts of Amravati and Yavatmal (see Map 8.1) form a part of Vidarbha region of Maharashtra and were until 1803 a part of the Maratha kingdom, being transferred subsequently to the Nizam of Hyderabad and coming under the direct rule of the British in 1853.[20] Based historically on a form of labour-service (the *balutedari* system), the agrarian structure of the two districts witnessed changes during the British Raj, when the colonial state initially reduced the land revenue demand, but then enhanced it heavily through periodic settlements. An elaborate bureaucratic network was created to ensure prompt revenue collection, a task delegated by the colonial authorities to the upper castes.[21] Adopting the *ryotwari* system extant in Bombay, the new land tenure system vested property rights in the cultivators, who were henceforth legally entitled to occupy the land permanently and dispose of it freely.[22]

Not surprisingly, a result of this was that the better-off castes (Brahmins, Rajputs) began acquiring ownership rights to land, though it was the *Kunbis* who gained control over the largest amount of agricultural holdings.[23] Those belonging to lower castes became tenants, while those at the bottom of the rural hierarchy – untouchables/'outcastes' such as the *Mahar*, *Chamar*, and *Mang* communities – who together constituted around 20 per cent and 15 per cent of the population of Amravati and Yavatmal, respectively, were for the most part landless labourers.[24] Although some of the latter had access to *vaten* land, this was generally of a poor quality (Brahme and Upadhyaya 1979).

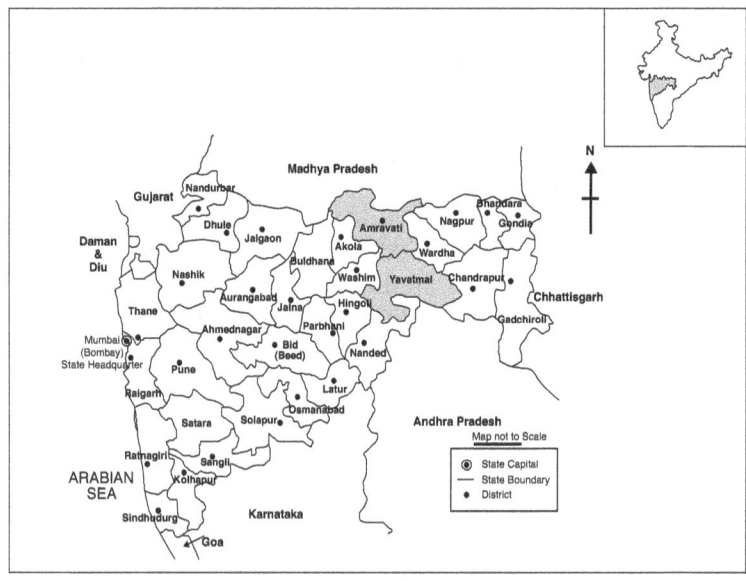

Map 8.1 Map of the districts of Maharashtra

Source: Mohanty (2005)

Cotton, colonialism, and crisis

The area of cultivated agricultural land increased under the British, who pressed for the extensive farming of commercial crops, especially cotton.[25] This was particularly true of the *jari* variety grown in the Berar region, much in demand in the Bombay market for export to England, the requirement for which increased yet further when prices for American cotton were higher (Borpujari 1973: 71). During the American Civil War, the area in Maharashtra planted with cotton increased rapidly, as American exports of this crop ceased and Lancashire faced a cotton famine, the result being a sudden rise in the British demand for Indian raw cotton (Guha 1972; Benjamin 1973; Borpujari 1973). Even after the Civil War ended, the sown area went on increasing, as demand for this commodity in both foreign and Indian markets expanded, and the opening of the Nagpur branch of the Great Indian Peninsula Railway in 1866 facilitated its transportation. Producers expanded the cultivation of this crop, and Alfred Lyall – writing in 1870 – described the agrarian situation of Berar in the following words:[26]

The land revenue increased and multiplied with marvelous rapidity under the combined stimuli of good Government, railways and the Manchester cotton famine. Cultivation spread over the land like a flood tide.

The area under cotton, which was only 38 per cent in Amravati and 29 per cent in Yavatmal during 1891–92, increased gradually to 57 and 45 per cent, respectively, in the years 1925–26.[27] Since output was dependent on the monsoon, however, agricultural production was subject to crisis, both districts experiencing a near total crop failure in 1896–97 and 1899–1900, due largely to low rainfall. Early in the twentieth century, for example, there were four years (1922–23 to 1925–26) of continuous crop failures in Amravati district, a result of which was high priced food grains combined with lower wage rates.[28] In such periods of crisis many low caste farmers and landless labourers had to borrow from large landholders and/or grain/cotton dealers belonging to the *Brahmin*, *Komti*, *Marwari*, and *Kunbi* communities, to whom they then owed debts carrying a high rate of interest.[29] One consequence of agrarian crisis coupled with high indebtedness among rural labourers was that many low caste peasants and workers migrated to Bombay, Nagpur, and other cities in search of employment in the cotton mills, docks, and railways.[30] Not only did these developments generate a process of socioeconomic differentiation, but they also provoked a specific kind of political response among the lower castes in the area. Many of them became religious converts, the Christian population in Amravati increasing more than threefold and in Yavatmal by more than eightfold during the period from 1891 to 1931.[31] As significant was the fact that members of lower castes also joined the *Satyasodhak* Movement organised by Jyotiba Phule, which gained a substantial following in the districts of Amravati and Yavatmal (Omvedt 1976; O'Hanlon 1985). The target of the *Satyasodhak Samaj* was the socio-economic power exercised by the higher castes, and in particular the way it took the form of an unequal distribution of land and the practice of usury. This from-below rural challenge to caste power continued with the emergence of Ambedkar and his Dalit Panther movement, many lower caste adherents of which converted to Buddhism.[32] Among other things, religious conversion on the part of poor peasants and landless labourers underlined forcefully their rejection both of the low status ascribed to them by the caste hierarchy and of the structural immutability of this kind of subordination.

Independence and the 'revolution of rising expectations'

Independence not only marked a formal political break with colonial rule, and thus the realisation of national sovereignty, but – perhaps more importantly – was also seen by many of the rural poor as the start of a reform process that would transform their lives for the better. Nehruvian economic planning linked to state intervention embodied this optimism, promising as it did to redress the balance of power in the Indian countryside by a combination of land distribution, resource provision, employment policies, new technology, and infrastructural improvement. This development strategy was premised on the expropriation of the landlord class and the transformation of tenant usufruct rights into ownership, objectives enshrined in many legislative ordinances promulgated during the late 1950s and early 1960s.[33] Notwithstanding the many loopholes, these laws had a positive impact on the redistribution of land (Nanekar 1968; Rao 1972; Rajasekaran 1998; Deshpande 1998; Mohanty 2001b).

In the case of Maharashtra, the Vidarbha region generally recorded a positive outcome in terms of the land reform process, as did the districts of Amravati and Yavatmal within this context (Rajasekaran 1998; Mohanty 2001c).[34] Furthermore, many lower caste peasants and agricultural labourers who migrated to Nagpur and Bombay succeeded in purchasing small amounts of land with their new sources of income (Omvedt 1994a). Consequently, the number and area of holdings operated by the SCs increased noticeably in both districts. Between 1985 to 1986 and 1990 to 1991, therefore, the number and area of SC landholdings in Amravati increased from 9 to 12 per cent and from 7 to 9 per cent, respectively. In Yavatmal, the corresponding increases were from 8 to 9 per cent in terms of holdings, and from 6 to 7 per cent in terms of area (Mohanty 2001c). In other words, there has been an increase in the number of peasant cultivators, especially among the SCs.[35]

The extent to which agriculture in the two districts has been transformed during the post-1947 era can be illustrated by changing cropping patterns and in particular the increasing importance of oilseeds and sugarcane in relation to cotton (see Table 8.1).

Table 8.1 Amravati and Yavatmal districts: area under principal crops 1970–2010–11 (% gross cropped area)

Years	Amravati				Yavatmal			
	Food grains	Oilseeds	Cotton	Sugarcane	Food grains	Oilseeds	Cotton	Sugarcane
1970–71	39	7	51	1<	49	6	44	1<
1975–76	45	6	46	1<	51	6	41	1<
1980–81	44	6	47	1<	51	5	43	1<
1985–86	46	5	45	1<	43	4	51	1<
1990–91	46	8	40	1<	47	5	47	1
1995–96	39	13	41	1<	43	5	49	1
2000–01	39	20	30	1<	42	9	46	1<
2005–06	43	25	32	1<	35	26	37	1<
2010–11	40	30	21	1<	30	21	49	1<

Source: *Season and Crop Report* (Government of Maharashtra, Director, Government Printing, Stationery, and Publications), various issues.

Note: *The data for this year are from the Agriculture Department, Government of Maharashtra, Pune.

The expansion in the sown area of oilseeds is due directly to policies of the Indian state promoting the adoption of high-yielding varieties, a result of which has been a corresponding rise both in the use of chemical fertilisers and pesticides, and in production costs.[36] To meet the demand for capital, and also to curtail usury, agricultural cooperative credit societies have been established and expanded their membership base and operations in each district.[37] Due to these and government procurement policies, therefore, the latter half of the twentieth century has witnessed a steady rise in crop yields and farm commodity prices, particularly for cotton.[38]

This success story has a downside, however, in the shape of the continuing dependence of agriculture on the monsoon rains as a source of irrigation.[39] For this reason, crop losses are still a major problem facing most farmers in this region. Between 1967 to 1968 and 1997 to 1998, therefore, smallholding peasants in Amravati and Yavatmal experienced crop losses on no less than 19 occasions.[40] This is not to say that such economic setbacks automatically correlate with the incidence of farmer suicides, particularly since in rural areas these acts are a relatively common occurrence, and their rate continues to increase steadily (Table 8.2).

Table 8.2 Rural suicides in Amravati and Yavatmal districts 1981–2015 (per 100,000 inhabitants)

Year	Amravati	Yavatmal
1981	1.77	1.60
1982	2.77	4.00
1983	2.14	3.87
1984	n.a.	n.a.
1985	3.43	3.19
1986	3.07	4.44
1987	3.86	4.06
1988	4.67	2.88
1989	n.a.	n.a.
1990	6.27	4.41
1991	6.53	4.82
1992	6.87	3.44
1993	5.73	5.22
1994	6.07	4.72
1995	7.06	6.56
1996	5.63	5.00
1997	5.50	7.16
1998	7.06	7.05
1999	6.31	7.63

Source: *Annual Vital Statistics Report* (Government of Maharashtra, Directorate of Health Services, and Chief Registrar of Births and Deaths), various issues.

In other words, the incidence of farmer suicide does not appear to vary simply according to whether or not agricultural production is in crisis (Table 8.3), which suggests in turn that social causes are as important as economic ones (Table 8.4). In order to understand the nature of this link better, it is necessary to examine both incidence and cause of farmer suicide in more detail, at the grassroots level.

The causes of farmer suicides

The microlevel analysis which follows is based on 66 reported cases[41] of farmer suicide occurring in the districts of Amravati and Yavatmal during 1998.[42] Research itself was conducted between

Table 8.3 Agricultural situation and suicides in Amravati and Yavatmal districts, 1980–98

Year	Amravati Agricultural situation	Suicide rate	Yavatmal Agricultural situation	Suicide rate
1980–81	Bad	Lower	Normal	Lower
1981–82	Normal	Lower	Normal	Lower
1982–83	Normal	Lower	Good	Higher
1983–84	Bad	Lower	Bad	Higher
1984–85	Normal	n.a.	Normal	n.a.
1985–86	Normal	Moderate	Good	Moderate
1986–87	Bad	Moderate	Bad	Higher
1987–88	Normal	Moderate	Normal	Higher
1988–89	Normal	Higher	Bad	Moderate
1989–90	Good	Higher	Good	Higher
1990–91	Good	Higher	Normal	Higher
1991–92	Normal	Higher	Normal	Higher
1992–93	Good	Higher	Good	Moderate
1993–94	Good	Higher	Good	Higher
1994–95	Good	Higher	Normal	Higher
1995–96	Good	Higher	Good	Higher
1996–97	Good	Higher	Good	Higher
1997–98	Bad	Higher	Bad	Higher

n.a. = Not available
Source: Mohanty (2001a: 162).

Note: The annual agricultural situation of the two districts has been assessed on the basis of the per acre average yield of cotton and jowar, the two crops which most accurately reflect the well-being (or otherwise) in each context. The failure of either crop matches the failure pattern for other rainfed crops. The normal yield of these two crops, and divergences either side of this, has been estimated on the basis of the average yield for normal rainfall years. The latter are calculated in terms of the minimum value of co-efficient of variation of rainfall across the months (June to December) and variation in terms of per cent of actual to average rainfall. Similarly, the suicide rate is treated as moderate when it is closer to the average annual rate than to levels above/below this.

January and April in 1999.[43] In order to evaluate the impact of agricultural distress, farmers who committed suicide were differentiated into three categories – small, medium, and large cultivators – on the basis of the amount of land owned.[44] By far the largest

Table 8.4 Amravati and Yavatmal districts: suicide by causes, 1998–2000 (%)

Causes*	Amravati			Yavatmal		
	1998	1999	2000	1998	1999	2000
Economic causes	8	7	4	7	11	8
Social causes	86	57	63	75	56	69
Other causes	1	28	4	12	23	10
Cause unknown	6	8	29	5	10	12
Total	100	100	100	100	100	100

*The various causes cited in accidental deaths and suicides in India have been divided into three categories.

Source: Mohanty and Shroff (2004: 5603).

Note: District level data prior to 1998 are not available.

number of suicides were small farmers (64 per cent), medium and large producers accounted for much lower figures (23 and 14 per cent, respectively). Unsurprisingly, most small farmers (86 per cent) derived the major portion of their income from a combination of subsistence cultivation and the sale of their labour-power to others, whereas the income of 78 per cent of large proprietors and over 60 per cent of medium ones was obtained from trade (Table 8.5).

Those farmers who died in this fashion were mostly cotton growers, with around 60 per cent of their cropped area devoted to this commodity, with the figure being even higher in the case of small producers (Table 8.6).

Since large and medium cultivators possessed relatively more irrigated land, their diversified cropping pattern included other high-value commodities (sugarcane, wheat, and vegetables). What all the farmer suicides had in common agriculturally, however, was an extensive area under HYVs, and correspondingly higher production costs incurred as a result of expenditure on technical inputs. Prolonged and unseasonal rains from October to December 1998 damaged many of the crops, but especially cotton. Although producers in each category suffered crop losses as a result, the impact was less severe in the case of large and medium cultivators who had a smaller area under cotton.

Estimated crop losses indicate that, whereas most small farmers (93 per cent) were unable to recover their production costs (Table 8.6), a majority among the medium and large producers

Table 8.5 Amravati and Yavatmal districts: socio-economic characteristics of farmers who committed suicide

Socio-economic characteristics	Small	Medium	Large	All
	N = 42	N = 15	N = 9	N = 66
Farmers who committed suicide (%)	64	23	14	100
Average family size of the deceased farmers	5.19	6.80	6.89	5.79
Average number working family members	3.07	3.13	2.88	3.06
Farmers above 60 years of age (%)	14	27	56	23
Unmarried farmers (%)	5	33	0	11
Farmers widowed or separated (%)	2	20	33	11
Farmers belonging to higher castes (%)	2	80	89	32
Farmers belonging to medium castes (%)	10	0	0	6
Farmers belonging to lower castes (%)	88	20	11	62
Average landholding (acres)	4.11	13.72	28.02	9.56
Tenants or tenants-cum-owner cultivators (%)	38	13	0	27
Farmers with non-agricultural incomes (%)	14	60	78	33
Non-literate farmers (%)	40	27	0	32
Farmers having television/radio sets (%)	10	53	67	27
Average value of major assets (Rs.)	75,839	23,5748	438,299	161,608

Source: Field survey.

managed to generate a profit when all economic activity is taken into consideration. In terms of cultivation only, the per acre loss of expected income was Rs. 5,690 for small farmers, Rs. 4,173 for medium cultivators, and Rs. 3,948 for the large cultivators. Although in no category did expectation match agricultural income, this of itself is not generally problematic, especially in the semi-arid tropics where shortfalls in output are the norm. The element of crisis stems from the different levels of economic exposure: whereas small producers were almost totally dependent on agriculture, and therefore on climatic vagaries governing output, large and medium proprietors by contrast were cushioned against this by income generated from non-agricultural sources. For this reason, crop failure

Table 8.6 Amravati and Yavatmal districts: agricultural economy of farmers who committed suicide

Particulars	Small	Medium	Large	All
	N = 42	N = 15	N = 9	N = 66
Cropping intensity	120	109	103	108
Area under cotton (%)	65	57	59	61
Area under jowar (%)	21	21	20	21
Area under HYVs (%)	70	82	63	71
Per acre use of fertilisers (kg)	84	81	71	78
Per acre use of pesticides (litres)	1.11	1.35	1.09	1.18
Per acre labour input (days)	66	58	51	58
Family labour employed (%)	73	53	35	55
Per acre cost of cultivation (Rs.)	3,282	3,349	3,245	3,290
Farmers unable to recover cost of cultivation (%)	93	40	44	74
Per acre loss by farmers unable to cover cost of cultivation (Rs.)	1,333	2,409	2,569	1,886
Average loss by farmers unable to cover cost of cultivation (Rs.)	5,911	28,911	54,590	12,701
Per cent of farmers recovering cost of cultivation	7	60	56	26
Average profit of the farmers who recovered cost of cultivation (Rs.)	1,940	3,854	5,461	3,989
Per acre profit by farmers whore covered cost of cultivation (Rs.)	469	360	265	320
Farmers failing to realise expected income (%)	100	100	100	100
Per acre loss of expected income (Rs.)	5,680	4,173	3,948	4,610
Average loss of expected income (Rs.)	25,075	47,067	82,475	37,900
Farmers gaining more than half of expected agricultural income (%)	0	53	56	21

Source: Field survey.

and the economic crisis it generated had a differential impact on farmers in Amravati and Yavatmal districts, large, and medium producers escaping its worst effects.

Death, debt, and agrarian distress

This cushioning process was itself linked to access by large, medium, and small producers to distinct forms of agricultural credit, and, consequently, to the different levels and outcomes of indebtedness. In keeping with the pattern found in many parts of rural India (Sahu *et al.* 2004), large cultivators borrow from formal lending agencies (cooperatives, the state) while smallholders depend mainly on informal sources for credit (moneylenders, traders, better-off proprietors).[45] As Table 8.7 confirms, among those who committed suicide, it was the category of small farmers (83 per cent) that was most in debt to non-institutional sources.

Whereas large and medium proprietors who also borrowed from informal sources did so for non-agricultural purposes and without having to provide collateral, small peasants by contrast were compelled to mortgage their land and/or jewellery as sureties for such loans, destined for agricultural production.[46] Moreover, the debt ratio was low in the case of large and medium farmers, the borrowing of whom from formal and informal sources amounted to no more than 22 and 11 per cent of their assets, respectively. The same ratio, however, was high in the case of small peasants, whose level of debt represented no less than 69 per cent of their total assets. It is difficult to avoid the conclusion that loss of collateral because of non-repayment of loans was an important factor in the suicide of such cultivators.[47]

These findings raise in turn three issues, all of them methodological: the extent of the connection between the timing of suicide and harvest failure, why some farmers affected by agrarian distress but not others resort to this act, and how close family members interpret the reasons for suicide. On the first point, since the harvest for the major crops extends from October till February, the assumption is that suicides occasioned simply by economic distress arising from crop failure would be most likely to fall within this period. Although it is obviously true that farmers experience hardship as a result of income loss throughout the year, the fact of income loss itself is felt most acutely at harvest time, when the full extent of economic crisis faced by a peasant household registers itself.

Table 8.7 Amravati and Yavatmal districts: farmers debt and assets sold

Particulars	Small	Medium	Large	All
	N = 42	N = 15	N = 9	N = 66
Farmers borrowing from formal agencies (%)	57	87	89	68
Average loan by formal agencies to indebted farmers (Rs.)	12,940	30,500	25,305	20,211
Per acre debt of farmers to formal agencies (Rs.)	2,857	2,848	1,289	2,246
Farmers borrowing from informal sources (%)	83	67	67	77
Average debt owed by farmers to informal sources (Rs.)	14,279	27,350	26,000	18,221
Per acre debt owed by farmers to informal sources (Rs.)	3,233	2,346	1,393	2,425
Borrowing from formal agencies for agricultural purposes (%)	97	100	100	100
Borrowing from informal sources for agricultural purposes (%)	55	6	12	33
Ratio of total debt to total assets (%)	69	22	11	21
Farmers selling assets (%)	74	40	44	62
Average value of assets sold (Rs.)	20,457	47,750	36,343	26,001
Assets sold in relation to total major assets (%)	31	19	7	19

Source: Field survey.

That almost all suicides by small farmers (93 per cent) coincided with the harvest period underlines the economic importance of crop failure for those in this category. Against this, however, it is necessary to note that suicides by farmers in the other two categories generally occurred at a different point in the agricultural cycle: between March and June, when preparatory operations (harrowing, sowing) for agricultural production were being undertaken. This would seem to suggest that in the case of these kinds of farmer, explanations of suicide causation must seek to combine the element of economic distress with other factors.

The second issue concerns a rather obvious point: why is it that not all farmers adversely affected by a similar process of economic distress – that is to say, crop failure, debt, and actual/potential loss of assets – end up committing suicide? To answer this, recent field-work conducted by Mohanty and Shroff (2003) in the districts of Amravati, Yavatmal, and Wardha compared farmers who commit-ted suicide with a control group composed of those who – although subject to a similar pattern of economic distress – did not.[48] The conclusion was that, although it is undeniably a significant con-tributory factor (especially with regard to smallholding peasants), economic distress of itself does not result in suicide.

This leads to the third issue, which concerns the way in which those immediately and directly affected by the suicide perceived causality. Thus information provided by family members of the suicide victims tended to confirm the view that – together with indebtedness – crop failure and attendant economic distress was the single most impor-tant reason why small farmers took their own lives. It is necessary, however, to invoke a methodological caveat, since in such instances there were other contributory factors, and in their answers family members of suicide victims understandably highlighted the more general and less stigmatised causes. In the case of large and medium farmers, by contrast, crop failure as a cause of suicide was scarcely mentioned by close relatives of the victim. Among the reasons given by them in reply to questions about cause were old age, illness, fam-ily problems and business failure (Table 8.8).

These methodological issues merely serve to underline the com-plexity of attributing cause to the act of suicide. It is a truism that, especially when economic causes are identified, reasons given by family members of the victim for its occurrence tend to focus nar-rowly on processes operating at the grassroots in the immediate vicinity – such as crop failure, plus the fact and/or level of debt – an explanation that is valid as far as it goes. The problem is that even apparently straightforward economic causes are themselves the effect of determination that is not always self-evident. Thus, for example, policies pursued by government may accentuate an otherwise survivable agrarian crisis, by deregulating agricultural prices and wages, the impact of which will be particularly severe on those whose livelihoods depend on either or both of these forms of remuneration. In Maharashtra, and more generally India, it is precisely small farmers who are most at risk in this regard. It is this systemic element of causality, a structural origin which frequently

Table 8.8 Amravati and Yavatmal districts: reasons given for suicide (%)

Reasons	Farmer category			
	Small	Medium	Large	All
Crop loss and indebtedness	88	40	11	67
Family disputes and problems	2	33	22	12
Old age and illness	0	7	33	6
Business losses	0	13	33	8
Miscellaneous	10	7	0	8
All	100	100	100	100

Source: Field survey.

goes unmentioned by researcher and interviewee alike, that must be taken into account when explaining farmer suicide as the outcome of agrarian distress.

Suicide, agrarian struggle, and declining expectations

In attempting to situate the rising incidence of suicide within this multiple and complex pattern of causality, therefore, it is necessary to ask the following questions. Why were small farmers so vulnerable to crop failure, and why did this have such a deleterious impact on them? Answers focus on the issue of technical knowledge/skill required for cultivation, plus the continuing efficacy of caste ideology, and thus an upper caste discourse about low caste religious conversion as betrayal.

Most of those who committed suicide were smallholders belonging either to medium castes (*Telis*, *Beldars*, and *Banjaras*) or lower castes (from the *Mahar*, *Nav-Bouddh*, *Matang*, *Chamar*, and *Dhangar* communities).[49] The large and medium cultivators were *Marathas*, *Kunbis*, and *Rajputs*. Although engaged in a wide variety of agricultural activities, lower castes were not in the main cultivators.[50] For this reason, many lower-caste smallholders did not themselves initially possess the skills and knowledge necessary for the kind of more advanced agricultural operations they wished to conduct, a problem where intensive cultivation of HYVs is concerned.

Of the lower-caste farmers who committed suicide, therefore, more than 36 per cent were either tenants or proprietors who – because they owned very small plots – had leased in additional land on a fixed rental. A third of them had received extra holdings as a result of government redistribution schemes involving either wastelands or surplus properties acquired through land ceiling legislation. Most significantly, over 85 per cent of these lower-caste farmers had less than ten years of experience in cultivation (Table 8.9).

In this regard, the problem facing small farmers from the lower castes is simply stated. The cultivation of cotton requires extensive knowledge that was virtually new to such producers, not least because more than 58 per cent of them had been engaged in this highly competitive commercial economic activity for less than five years. Elsewhere in Maharashtra, the same kind of difficulty has surfaced, in the shape of lower-caste farmers being driven to suicide due to crop losses resulting from inadequate technical knowledge about the growing of commercial crops (Omvedt 1999). Significantly, it was a *Mahar* farmer in Amravati district who drew attention to this situation, about which he made the following comment:

> They [Maratha/Kunbi large farmers] are the real agriculturists, we are so by name only. They know well which type of soil suits which crop, the appropriate seed varieties, the timing of application of manure, fertilizers, pesticides and their quality and quantity. Even with the same crop on the same quality of land they manage to produce more than we do. In adverse climatic conditions also they manage to obtain some margin of profit.

Lest it be thought that such a view amounts to blaming-the-victim, it must be emphasised that lower-caste farmers were not slow in their attempts to acquire the necessary knowledge, and thus to remedy the situation of which they themselves were fully aware. Their attempts at self-improvement, however, encountered what in the circumstances were a fairly predictable series of socio-economic obstacles.

To begin with, government sponsored agricultural extension services focused on rich farmers belonging to the upper castes, at the expense of lower caste producers. Consequently, the latter were compelled to rely on local shopkeepers for crucial information about the recommended doses of chemical inputs and the appropriate methods of cultivation. Owning neither television sets nor radios, moreover, lower-caste farmers were similarly beyond the

Table 8.9 Amravati and Yavatmal districts: deceased farmers according to caste group

Particulars	Caste			
	Higher	Medium	Lower	All
	N = 21	N = 4	N = 41	N = 66
Average landholding (acres)	17.53	6.91	5.72	9.56
Irrigated area (%)	23	5	9	17
Cropping intensity	104	104	115	108
Area under cash crops (%)	61	63	65	63
Area under HYVs (%)	59	73	71	65
Per acre cost of cultivation (Rs.)	3,080	3,695	3,548	3,290
Family labour employed (%)	34	73	65	49
Per acre debt owed to formal agencies (Rs.)	1,725	1,437	1,631	1,676
Per acre borrowed from informal sources (Rs.)	886	2,695	2,545	1,613
Per acre gross income from agriculture (Rs.)	3,255	843	999	2,268
Per acre net loss/gain (Rs.)	174	−2,852	− 2,550	− 1,022
Per acre loss of expected agricultural income (Rs.)	3,045	6,100	5,752	4,236
Farmers dependent on agriculture and working for others (%)	38	100	78	67
Average income received from non-agricultural sources (Rs.)	6,838	1,225	1,701	3,307
Average income expected from non-agricultural sources (Rs.)	26,595	1,475	1,543	9,511
Farmers with more than ten years of experience in agriculture (%)	86	50	17	41
Farmers taking agricultural decisions independently of family (%)	19	75	83	62
Debt repaid to formal agencies (%)	8	11	11	10
Debt repaid to informal sources (%)	40	28	19	27

Source: Field survey.

reach of vital information about new methods of production disseminated through the mass media. More important than either of these factors, however, was the element of antagonism informing social relations between cultivators from high and low castes in Amravati and Yavatmal districts.

For rather obvious reasons, high caste farmers who were used historically to regarding and treating members of the low castes as innately subordinate menials were unwilling to share their agricultural knowledge and/or skill with those in the latter category. Not the least important considerations in this regard were the views held by higher castes both that lower-caste farmers had already benefitted sufficiently from government policies and that any shared knowledge about agricultural production would serve merely to close the traditional gap between the top and bottom of the caste hierarchy. In other words, high caste farmers saw those from lower castes as potential economic competitors who had already gained advantages as a result of state intervention.

This antagonism was itself aggravated by the perception on the part of upper caste cultivators that, as a result of social reform movements led by Phule and Ambedkar, lower castes were becoming 'uppity'. Such hostility was evident in the comments of a village Brahmin – a landlord – in Amravati district, who observed,

> Ambedkar only spoiled them [= the Mahars]. Earlier they were all like our brothers. They were taking all kinds of help from us. There are several instances where we have rescued them. For example, my father had leased out three acres of land to a Mahar before which he was not getting food for survival. When he was on the point of death, I took him to hospital. My elder brother also waived the borrowed amount of Rs. 300/- keeping his bad economic condition in mind. On another occasion, my father intervened and resolved the crisis in which they were fighting with one another because of a family dispute. But finally he [the Mahar] claimed our land under tenancy reform laws. Now who would help these betrayers? They also betrayed their own Hindu religion and became Buddhists.

This symptomatic utterance, an almost classic statement of upper caste ideology, combines a number of beliefs. First, a fear of further threats from below to existing property relations. Second, resentment about what better-off producers have in the past depicted as patronage extended by them to the less well-off, which is framed in the discourse about the betrayal of reciprocity (along the lines of "we helped them when they needed help, and now they repay us by trying to take our land"). And third, the theme of betrayal is reinforced by virtue of being extended from the economic sphere to the domain of the sacred. Hence the conversion of lower-caste farmers to Buddhism was perceived as the final proof of rejection and ingratitude.

In the absence of any assistance from higher castes, small farmers were compelled to rely on local dealers and private agencies for information about new methods of agricultural production. Most of the lower- and medium-caste farmers who committed suicide had followed the advice of local shopkeepers regarding doses, quality, and timing of chemical inputs. Not surprisingly, these commercial enterprises recommended high quantities of the most expensive inputs that were not of the best quality. The inevitable result was that small producers experienced a twofold disadvantage: since purchases of these inputs were made on credit, their indebtedness to shopkeepers and/or traders spiralled, while their agriculture suffered from the poor quality of the inputs purchased.

That many low caste farmers strived hard in order to improve their livelihoods by cultivating cotton is beyond dispute. The following instance, encountered in the course of fieldwork in Maharashtra, provides an example of this:

> N, 32, was a Mahar small farmer. After graduation, he tried for a government job but failed. N returned to the village and helped his father and younger brother in agricultural activities. From 1995 he took over the entire responsibility for them, and managed the finances independently. He extended the area under cotton and invested a higher amount, borrowing from cooperatives and moneylenders. But he could not repay the loans. Gradually his indebtedness increased to rupees twenty two thousand. In 1996 and 1997 he was unable to recover the amount he had invested. His family faced severe economic crisis. N blamed himself for this failure and committed suicide.

Other cases of a similar nature abound:

> D, a Mahar, worked as an agricultural labourer. Though many of his friends left for Nagpur in search of better paying employment, he preferred to stay on in the village. His aim was to

improve the standard of living for his family through cotton cultivation. In 1999 he acquired from the government four acres of
land, of poor quality and without irrigation. D applied for and
obtained a loan of Rs. 48,000/- to dig a well, but failed to find
a water source. He then received money from the government
under the Indira Awas Yojana, but used it for cotton cultivation.
When the crop failed, he fell into despair and committed suicide.

What these case studies demonstrate is the gap between expectation
and realisation. Possessing neither sufficient experience as cultivators nor the requisite technical knowledge to make a success of new
methods of production, small farmers nevertheless subscribed to the
view that HYVs would lead to higher productivity and hence greater
profitability. After all, this appeared to be consistent with the pattern
of economic progress and social development that was central to so
many government programmes in the period following the end of
colonialism. For this reason, among others, lower-caste farmers over-
extended themselves financially, by borrowing in order to invest,
their production costs and levels of indebtedness increasing as a
result. Crop failure and spiralling debt negated in a dramatic fashion
the optimism generated since independence, a situation compounded
by the adoption at the national level of neoliberal policies for agriculture. Together with the antagonism expressed towards them by
upper castes, this all contributed to a harsher social and economic
situation faced by small farmers in the Indian countryside.

In keeping with Durkheimian theory about suicide, there were
additional factors contributing to a process of social alienation. The
first was that the new economic circumstances in which small farmers found themselves – commercial agriculture, increased financial
costs, higher levels of debt – fuelled the trend towards individualisation, which Durkheim linked to family and/or community disintegration. Lower- and medium-caste farmers belonged to small families,
and crop failure served to underline their perception of themselves as
being personally to blame for the loss, and its impact on the well-being
of the family. This was compounded by the fact that the intensity of
small farmer involvement in agricultural operations meant that family members remained unaware of borrowing and debt. Hence the
lament by the father of one small farmer who had committed suicide:

He was rarely interested in matters other than agricultural ones.
He did not tell us the amount he spent on buying fertilisers and

pesticides, and the sources of his borrowings. It was only after his death that we learned he had borrowed Rs. 25,000/- from a moneylender of the neighbouring village. Had he told us about this, we would have sold our land to repay the loan.

As significant, perhaps, was the second factor contributing to a sense of social alienation: religious conversion. Although members of the lower castes embraced Buddhism to escape the stigma and restrictions inherent in the rigid caste hierarchy, this displacement from Hindu society nevertheless gave rise to a perception of 'rootlessness'. Having broken with a longstanding and ideologically powerful sense of 'belonging' to a rural community from which they drew concepts of selfhood and value – however oppressive these might have been – lower- and medium-caste farmers who converted to Buddhism experienced an unavoidable element of social isolation. An instance of this is as follows:

> K was a small farmer. He was originally a Mahar and became a Nav-Buddhist. His father and elder brother opposed his conversion, after which his brother moved out of their home. Since then K's wife and children have faced ostracism and criticism. Many people started addressing K himself as "Lord Buddha". The Brahmin landlord who had leased out nine acres of land to K transferred them to his – K's – elder brother. When K's father died suddenly and inexplicably, K's elder brother and others in the village blamed K for the death of his father, a punishment for his having adopted Buddhism. The following year K's younger son also died, and K's financial condition gradually deteriorated. Then his wife also fell sick. Finally, when he faced crop losses for two years in succession – 1996 and 1997 – he committed suicide.

That Nav-Buddhists formed 36 per cent of small farmers who committed suicide underlines the extent to which estrangement of selfhood was an effect of religious conversion.

Suicide and the loss of esteem

Suicides by large and medium farmers raise issues of a different kind, but ones which nevertheless feed ultimately into the same epistemological roots identified by Durkheim. As the following two case studies indicate, farmers from this end of the socio-economic spectrum took their own lives because of reverses experienced in the trading and political arenas:

> Q, a Rajput large farmer had 22 acres of land, which he culti-vated with hired labour. The major portion of his income came from cotton trading. He collected this crop from local farm-ers to whom he had supplied credit and sold the commodity in Andhra Pradesh. His business was flourishing, and he had advanced more than two hundred thousand rupees to the cot-ton growers. Many of the farmers who owed him money went bust as a result of crop failures. Q lost all hopes of getting back his investment and committed suicide.
>
> Z was a Kunbi large farmer who owned 22 acres of land along with a tractor. In the locality he was considered as one of the richest persons. He was Sarpanch [head man] of the Gram Panchayat for a term, but in the following election he contested he was defeated. He contested the election for a third time, again with high hopes of success, spending around one hun-dred thousand rupees in campaigning. He lost the election, and despairing of ever again being elected, took his own life.

Like small farmers, medium, and large ones also commit suicide as a result of economic reverses. In the first case, ironically, this out-come was because the economic fate of a better-off producer was itself linked to – and, indeed, determined by – that of smallhold-ers who were themselves victims of agrarian distress. This merely serves to underline the interconnected nature of the modern agrar-ian economy in India. Just as the first case involved the death of a better-off farmer who lost his investment in agriculture, so the sec-ond case entails a similar loss by the same kind of farmer, but in the political sphere. Given the close connection between politics and agriculture at the village level, it is possible to classify this particular case as possessing roots in the sphere of economic production.

Another cause of farmer suicide in the case of large and medium cultivators was old age and illness, factors which were themselves linked to a loss of esteem (and a perception of 'worthlessness') on the part of elders within rural communities. The latter, too, stemmed from the process of economic modernisation, since his-torically the old within rural communities were the repositories not just of wisdom but – more importantly – of knowledge about the intricacies of agricultural technique and production. It was to these age-specific attributes that the esteem accorded to elders was linked, and in time of economic crisis, it was they to whom younger and more inexperienced cultivators turned for advice. In a very real sense, however, such a role has been rendered obsolete by much of the new technology and its accompanying inputs. It is perhaps difficult to think of a more apt illustration of the effects of anomic suicide, as conceptualised by Durkheim with regard to the displacement of traditional values/knowledge in the course of the spread of agrarian capitalism.

The way in which these processes combine – the displacement of traditional knowledge about agriculture, and how this age-specific status diminution culminates in a sense of 'worthlessness' – to reproduce isolation within the context of rural community and reinforce thereby the loss of esteem is demonstrated by the following example:

> B, 78, was a Kunbi widower who belonged to a large landholding group. When he was active, he supervised its agricultural activities. When he fell sick, this involved expenditure of Rs. 300/- on his medical treatment every month. Gradually B's condition deteriorated. His son became indifferent to his situation, and his daughter-in-law ill-treated him. When finally they also deprived him of the company of his grandchildren, B killed himself.

The view expressed by an aged farmer in Yavatmal district, about both the fact and the impact of the loss of esteem on people like him is particularly apposite in this connection:

> We are like the living dead. We are considered as old-fashioned. Now-a-days there is no concern for the elders. See L's case. What has he not done for his son? But when he suffered from illness, his son did not even take him to the hospital. As long as a person is physically well at this age it is all right, but the moment he falls sick it is better to end life.

The theme of kinship also informs disputes about property, as yet other case studies confirm:

> G, 62, a Kunbi large farmer had two sons. His elder son worked on the farm and looked after agriculture and household

related expenses. His younger son B, a graduate, did not like to work in the fields, spending his time with friends and playing cards. This caused conflicts in the family. B asked his father for Rs. 50,000/- to start a business. G was reluctant. Then B asked for the sale of his share of land. To avoid the division of the family property, G agreed to provide his younger son with Rs. 30,000/-. Then his elder son also demanded the same amount, whereupon G committed suicide.

V, a Kunbi medium farmer, was a drunkard. He had an affair with a woman who operated the liquor shop. V gave her money and other valuables secretly. When his wife learnt this, she went to her father's house with her children and the matter was made public. V felt humiliated, and ended his life.

In a real sense, this kind of expression of regret by large and medium farmers mirrors the idioms that structure their discourse in relation to lower-caste farmers who reject the values of the Hindu caste system by converting to Buddhism. Common to each is the view of reciprocity negated, and consequently of respect – earned (as a farmer) or innate (as a member of a higher caste) – denied. Just as lower-caste farmers are said to repay better-off farmers for the generosity the latter extend to the former 'by trying to take our land', so the sons, daughters, wives and/or daughters-in-law are perceived to be negligent for non-fulfilment of obligations to respect/honour the head of the household. In short, both constitute a betrayal of traditional values/norms, one in the realm of caste, the other in the domain of kinship.

Concluding comment: Durkheim in Maharashtra

The findings presented here are consistent with the argument put forward by Durkheim linking suicide to what happens when rural communities become more urban and industrial. His objection to economic interpretation concerned explanations not of suicide but of the benefits attributed to the process of industrialisation and in particular the advantages this offered when compared to what he (and others) saw as the benign aspects (e.g., cooperation) of pre-industrial societies. The act of suicide, which in pre-industrial contexts prevented the emergence of what he regarded as a malign individualisation, was in industrial contexts an effect of this. For

Durkheim, therefore, the explanation of suicide is not a question of social as distinct from economic causation. Rather, the clue to explanations of both incidence and meaning lies in a process, one that involves change from one set of benign socio-economic characteristics (pre-industrial, rural, community) to another and less benign set (industrial, urban, individual).

This analytical framework is useful in explaining the current incidence of farmer suicides in the state of Maharashtra. Low- and medium-caste farmers in the two districts (Amravati and Yavatmal) where fieldwork was conducted were mostly smallholders. Such producers aspired to improve their socio-economic position as a result of the land reform programmes and the planned agriculture that characterised the strategy of successive Indian governments throughout the post-1947 era. The majority of those committing suicide were small farmers who had been adversely affected by the introduction of neoliberal economic policies during the early 1990s. Some of these had over-extended themselves financially. Not only had their agricultural income declined as a result of crop losses but also their costs of cultivation and indebtedness to informal agencies had increased. In short, what these smallholders experienced was an increasing trend towards individualisation, an identity conferred on them by the market. Suicides among large cultivators were due to a combination of business failures, family quarrels (over land), plus diminished political influence and social esteem. In their case, suicide was a consequence of estrangement linked to the breakdown in traditional values and norms.

Similar processes are evident elsewhere. In the United Kingdom, for example, the rising incidence of suicide among farmers and agricultural workers has been linked to the process of economic liberalisation that began in the 1980s.[51] Among other things, it involved the withdrawal of state subsidies, in the form of price controls on farming inputs combined with price supports for farm-produced commodities. These in turn entailed cutbacks in farming incomes generally, as small producers had to pay more for purchased inputs (energy costs, fertiliser, pesticides, etc.) on the one hand, and on the other received less for their output (crops). The latter was compounded by adverse global market conditions, a result of overproduction of farm produce. Stress leading to suicide among farmers in the United Kingdom was linked to loss of financial resources and status within what were already isolated rural communities.

More recently, China has witnessed an analogous phenomenon, but with a different victim.[52] As in the case of small farmers in the United Kingdom, poor peasants in China have been adversely affected by economic liberalisation, which in the Chinese context involved the privatisation of land in rural collectives and the decline of the latter as a secure form of employment and income.[53] According to the Beijing Suicide Research and Prevention Centre, incidences of suicide in China during 2004 were not only higher than the global average, but the rate for the countryside was three times that for urban centres. Most significantly, those who took their own lives were rural women who remained behind in villages as males in the family migrated to towns and cities in search of work.

What these instances from other national contexts all suggest is that, where the connection between farmer suicide and rapid economic transformation in rural areas is concerned, India may not be unique.

Notes

1 This is the revised version of the article "'We Are Like the Living Dead': Farmer Suicides in Maharashtra, Western India" published earlier in *The Journal of Peasant Studies*, volume 32, no. 2, 2005, pp. 243–276.

2 Two surveys carried out in Maharashtra (one jointly by the Rambhau Mhalgi Prabodhini, Mumbai, and Indian Agro-Economic Research Centre, New Delhi, and the other by Dash *et al.* (1998), also highlight crop loss and indebtedness as the main causal factors in farmer suicides.

3 Hence the view (Durkheim 1952: 245) that "economic distress does not have the aggravating influence often attributed to it, . . . it tends rather to produce the opposite effect. There is very little suicide in Ireland, where the peasantry leads to wretched a life. Poverty-stricken Calabria has almost no suicides; Spain has a tenth as many as France. Poverty may even be considered a protection".

4 This view has been very influential on all subsequent studies (Miley 1972).

5 This of course is the subject of much debate. For a discussion of Durkheim's objections to psychologistic reductionism where explanations of suicide are concerned, see MacIntyre (1971: 222ff.).

6 Farmer suicides are regularly reported in these districts. In 1999, they accounted for over 50 per cent of the total farmer suicides in the state, and by 2002 the figure had risen to 57 per cent. For details, see Mohanty (2001a) and Mohanty and Shroff (2003).

7 An example is the concept of the 'suicide bomber', as deployed by the Japanese during the 1939–45 war, and by the Palestinians (and others) currently in the Middle East.

8 It is a truism that large-scale conflict between nations has a profound impact on the meaning not only of death but also of suicide. The latter is not used to classify soldiers in battle obeying military commands that they know will result in their death, as happened in the case of soldiers 'going over the top' in the trenches during the 1914–18 war. Even where this possibility is acknowledged, however, the act of suicide is invariably recast as one of self-sacrifice, in the process shifting the meaning from a negative to a positive one. This is true of most warfare.

9 Examples would include Ireland and Spain. It is important, however, to note a methodological caveat in this connection. Because in such contexts suicide is stigmatised, its true incidence remains hidden. Hence the observation by Brody (1973: 100) in his classic study of rural Ireland: "Suicides, though carefully concealed in a Roman Catholic society wherever possible, are not rare".

10 On dowry death in India, see among others Kumari (1989).

11 When considering the incidence of suicide in relation to women, Durkheim (1933: 247) manages to invoke all the usual prelapsarian gender stereotypes. Females are less prone to suicide because, in his opinion, "(w)oman has had less part than man in the movement of civilization . . . She recalls, moreover, certain characteristics of primitive natures".

12 As Lockwood (1982) has pointed out, Durkheim's conservative vision of society is underwritten by a concept of 'fatalism', a term which ascribes to its subject an inability to change his/her surroundings. This, as Lockwood also notes, is particularly true of his conceptualisation of 'suicide', with its inference about wholesale resignation to the status quo, either in pre-industrial or indeed industrial contexts. This is ironic, since it is also possible to interpret self-inflicted death as the opposite kind of phenomenon: that is to say, as an act of resistance, or agency designed to change the status quo. The latter is a view which structures analyses of the slaves in plantation societies.

13 An illustration of this kind of interpretation is the case-study outlined by Malinowski (1926: 77–79) of a suicide in the Trobriands as a result of breaking the tribal rules of exogamy. Malinowski (1926: 80) qualified the Durkheimian approach by pointing out that it was only when scandal occurred that suicide resulted. In terms of causality, therefore, it was not so much the breaking of clan totemic exogamy as it becoming a topic of gossip that led to ostracism and eventually suicide.

14 Where social wants exceed the possible means for attaining them, the subject is confronted with a perpetual gap between aspirations and achievements. If the ends of action are contradictory, inaccessible or insignificant, a condition of anomie arises (Powell 1958).

15 Durkheim (1952: 259) also discussed the crises of widowhood and divorce as also their debilitating effects on the subject. These are among the domestic reasons giving rise to anomic suicides.

16 The theoretical connection between Durkheim's views about suicide and his work on the social division of labour have been noted by many previous writers, including Giddens (1971). None, however, have

placed his analysis of suicide in the wider context of agrarian transition, and linked it to his concerns about the impact of economic development in general – and that structured by a laissez-faire philosophy in particular – on rural communities.

17 As is clear from Besnard (1983), in France the leading Durkheimians – such as Mauss, Le'vy- Bruhl and Bougle' – were anthropologists, or scholars whose theoretical concerns were precisely the 'otherness' of pre-industrial societies.

18 This key theoretical linkage is found in Book Two, Chapter 1, Pt. 2 of the Division of Labour, (European) country, one observes the same relation. Everywhere suicide rages more fiercely (entitled "The Progress of the Division of Labour and of Happiness"). There, for example, one encounters the following explicit claim (Durkheim 1933: 247): "In the interior of each (European) country, one observes the same relation. Everywhere suicide raises more fiercely in the cities than in the country. Civilization is concentrated in the great cities, suicide likewise . . . The classes of population furnish suicide a quota proportionate to their degree of civilization".

19 Significantly, a target of the critique contained in the Division of Labour was Spencer, an enthusiastic advocate of laissez-faire economic policy (Durkheim 1933: 200ff.). It is Important to note that this attack came not from the political left but rather from the political right. What is often forgotten, therefore, is that there is a strong anti-capitalist tradition among conservatives, who decry the destruction of 'natural' rural communities. Like many conservatives, Durkheim objected to the erosion of 'primitive' society – the repository of virtue – by untrammelled market forces.

20 Rural society was based on a rigid caste structure, the main cultivators being the *Kunbis*, *Malis*, and *Baris* agricultural production was mostly for local needs, the area under cultivation being adjusted to meet the requirements of a changing local population. Since peasant cultivators were not only undifferentiated economically but also prohibited from accumulating wealth, the major form of surplus extraction they faced was the land revenue demand. See (1) *Central Provinces District Gazetteers – Amraoti District*, Volume A. (1911) Descriptive (edited by S.V. Fitzgerald and A.E. Nelson), Bombay: Caxton Works, pp. 180–81. (2) *Central Provinces District Gazetteers – Akola District*, Volume A. (1910), Descriptive (by C. Brown), Calcutta: Baptist Mission Press, p. 192.

21 The occupational distribution of selected castes is contained in the 1931 Census of India. See *Census of India* (1933), Volume XII: Central Provinces and Berar, Part I – Report by W.H. Shoobert, Nagpur: Superintendent of Census Operations, pp. 265–269.

22 Significantly, improvements in productivity did not attract extra revenue demands.

23 Nagpur Settlement Record, as cited in Central Provinces District Gazetteers – Amraoti District Volume A, *op.cit.*, p. 131. About *Kunbi* landownership, a *Marathi* proverb (Russell 1916: 49) observes, "Wherever it thunders, there the *Kunbi* is a landholder and tens of millions are dependent on the *Kunbi* but the *Kunbi* depends on no man".

24 The percentage distribution of different Castes in the population of Amravati and Yavatmal districts according to the 1911 census is set out next:

Castes	Amravati	Yavatmal
Higher castes	7.1	4.8
Higher cultivating castes	36	27.6
Higher artisan or trading castes	2.0	2.4
Serving castes	2.5	1.3
Lower cultivating, artisan, trading, and miscellaneous castes	9.1	16.5
Untouchables	19.0	15.0
Others	24.3	32.4
All castes	100	100

Source: Government of British India (1927a: 8–11, 1927b: 7–11).

25 As long-stapled American cotton was preferred to short-stapled Indian cotton for use in Britain (Benjamin 1973; Borpujari 1973; Desai *et al.* 1978), attempts were made to introduce the former variety.

26 Quoted in *Central Provinces District Gazetteers – Amraoti District*, Volume A, *op.cit.*, p. 301.

27 For details on area under crops in these districts from 1891 to 1992 through 1925 to 1926, see Mohanty (2001a: 153).

28 The retail prices of *jowar*, wheat, paddy, cotton, and gram from 1891 to 1892 through 1925 to 1926 broadly indicate a rising trend. The monthly wages of agricultural labourers were almost constant from 1891 to 1892 through 1902 to 1903, but lower during the famine years. For details, see (1) *Central Provinces District Gazetteers – Amraoti District* (Statistical Tables 1891–1926), 1927, Nagpur: Government Press. (2) *Central Provinces District Gazetteers – Yeotmal District* (Statistical Tables 1891–1926), 1927, Nagpur: Government Press.

29 To encourage the cultivation of cotton and other cash crops, the colonial state advanced loans to substantial proprietors under the Land Improvement Act and the Agriculturists Loans Act, even in periods when harvests were poor. See (1) *Central Provinces District Gazetteers – Amraoti District* (Statistical Tables 1891–1926), *op.cit.* (2) *Central Provinces District Gazetteers – Yeotmal District* (Statistical Tables 1891–1926), *op.cit.* Such credit facilities were of course not available to landless labourers.

30 For evidence of labour migration from rural areas to Bombay, Nagpur, and other cities, see Government of British India (1911: 109) and Omvedt (1994a: 8).

31 See *Census of India*, Volume XII: Central Provinces and Berar, Part I – Report, 1933, by W.H. Shoobert, Nagpur: Superintendent of Census Operations. p. 341.

32 For Ambedkar's decision to embrace Buddhism, and how this embodied the aspirations of lower castes to achieve social equality, see Benjamin and Mohanty (2001).

33 Thus, for example, the Madhya Pradesh Abolition of Proprietary Rights Act (1950) abolished all the proprietary rights in estates, mahals, and alienated lands, while the 1961 Maharashtra Agricultural Lands (Ceiling on Holdings) Act prescribed an upper limit to the amount of land owned. The Bombay Tenancy and Agricultural Lands Vidarbha Region and Kutch Area Act (1958) provided for the transfer of leased land into the ownership of the tenant. Other measures included the consolidation of holdings and the distribution of land to scheduled castes and tribes.

34 Up to 1995, ceiling surplus land amounting to 10,751 holdings consisting of 15,963 hectares in Amravati and 4,635 holdings of 8,178 hectares in Yavatmal were distributed to landless families of various categories (Rajasekaran 1996). A total of 4,018 holdings covering an area of 5,728 hectares were distributed to the scheduled castes in Amaravati, the corresponding figures for Yavatmal being 1,302 holdings with an area of 2,337 hectares.

35 Census data indicate that scheduled caste cultivators in relation to workers in rural Amaravati rose from 11 per cent in 1971 to 14 in 1981. Although the 1991 census suggests there was an apparent decline in the overall proportion of scheduled caste cultivators (Government of India 1993: 442), this was due to the inclusion of Nav-Buddhists in this group. The proportion of scheduled caste cultivators in Yavatmal increased from 8 per cent in 1971 to 14 per cent in 1981 and 17 per cent in 1991 (Government of India 1975: 92,1983a: 160,1993: 458). See (1) Census of India 1971: Series- ii Maharashtra Part II-C(i), Social and Cultural Tables, Mumbai: Directorate of Census Operations, p. 92. (2) Census of India 1981: Series-12 Maharashtra Scheduled Castes and Scheduled Tribes, Primary Census Abstract, Mumbai: Directorate of Census Operations, pp. 158 and 160. (3) Government of India 1993. Census of India 1991: Series-14 Maharashtra Part II-B(ii) Primary Census Abstract, Scheduled Castes and Scheduled Tribes, Mumbai: Directorate of Census Operations, p. 458.

36 Data from the Department of Agriculture in the Government of Maharashtra (various issues) covering the period 1980–81 to 1997–98 show that in both districts the area under HYVs increased over the years, to more than 90 per cent in the case of crops like wheat, bajra, jowar, and cotton. The per hectare cost of cotton cultivation in Maharashtra increased from Rs. 2,1444 in 1981–82 to Rs. 6,341 in 1995–96 in real terms (Government of India 1991: 127,1998: 279). Similarly, the cost of cultivation of *jowar* rose from just Rs. 716 in 1981–82 to Rs. 2,119 in 1995–96. See (1) *Reports of the Commission for Agricultural Costs and Prices on Price Policy for the Crops Sown in 1985–86 Season, New Delhi: Department of Agriculture and Cooperation*, p. 70. (2) *Reports of the Commission for Agricultural Costs and Prices for the Crops Sown during 1998–99 Season, New Delhi: Department of Agriculture and Cooperation*, p. 282.

37 There are also licensed moneylenders who advance loans under the provision of the Bombay Moneylenders Act (1946). Though their number has decreased owing to the expansion of formal credit network, the amount of loans advanced by them has increased over the years (Mohanty 2001a). The expanding importance of agricultural cooperative societies is evident from the information presented next:

Particulars	Amravati				Yavatmal			
	1980– 81	1990– 91	1993– 94	2009– 10	1980– 81	1990– 91	1993– 94	2009– 10
Number of societies	667	682	682	679	538	597	597	595
Membership (000s)	215	241	278	382	215	257	300	399
Outstanding average loan (Rs.)	727	852	933	2,040	955	1,258	1,151	8,717

Source: Government of Maharashtra, Directorate of Economics and Statistics, various issues.

38 Annual crop reports issued by the government of Maharashtra (various issues) indicated that from 1970 to 1971 through 1998 to1999 harvest prices for all the major crops grown in the two districts have increased sharply throughout this period. The monopoly procurement scheme for cotton was introduced in Maharashtra during 1972, since when the price of cotton has gradually increased.

39 Estimates for the agricultural season 2000–2001 indicate that in Amravati district the irrigated area was only eight per cent of the sown area, the corresponding figure for Yavatmal district being six per cent. In the entire period from 1970 to 1971 through 2000 to 2001 most of the major crops were cultivated almost without recourse to irrigation.

40 In 1997–98, for example, financial losses due to crop failure amounted to Rs. 324 million in Amravati and Rs. 16 million in Yavatmal (Mohanty 2001a; Mohanty and Shroff 2003).

41 All the suicides were reportedly committed by male farmers. As the households in the rural areas were headed by males and land was mostly owned by them, suicides committed by female members, if at all any, were possibly not reported as 'farmer suicides' by the government or media reports. Many other studies on Maharashtra (Mohanty and Shroff 2003; Mishra 2006a) also did not come across any incidences of suicides by female farmers.

42 A list containing 72 cases of farmer suicides was obtained from the Agriculture Department, Government of Maharashtra. Village level fieldwork revealed that four of these cases involved landless labourers, and a further two cases could not be investigated because of insufficient details.

43 Research methods employed in the course of fieldwork were as follows. Background information was gathered on villages where suicides were reported, covering a variety of socio-economic issues (patterns of agricultural production, other economic activities). This included discussions with the officials of the agricultural extension services, cooperative societies and other formal lending agencies, as well as moneylenders, local traders and businessmen. Detailed information was then collected from the adult family members of the suicide victims, a few cases being selected for more qualitative information. Friends, relatives, and neighbouring households of farmer suicides were also interviewed. Information about production related to the agricultural year of 1997–98.

44 Suicide victims were categorised into three groups according to the amount of land operated. Those cultivating less than ten acres were classified as small farmers, those with holdings between ten and 20 acres as medium farmers, while those with land in excess of 20 acres were categorised as large farmers. Interviews conducted in villages where suicides were reported indicated that on average a small farmer household of five to six members, with three to four adults and two to three children, required an additional source of income for survival in a normal harvesting year. In bad harvesting years such smallholdings were unable to survive without recourse to other sources of income. Medium households, by contrast, were able to reproduce themselves economically without access to additional sources of income. Even in cases of a partial crop loss, these units were able to manage without much difficulty. Households with excess of 20 acres not only generated surpluses but were easily able to survive crop failures and – more generally – ride out most kinds of financial exigencies.

45 This different borrowing pattern stems from the fact that better-off producers generally have a privileged access to formal lending institutions. Loans from the latter are not regarded by producers as onerous, since both interest and repayment rates are less burdensome when compared with informal sources of credit. Significantly, many large farmers borrow from institutional sources without actually needing to do this, using the low-cost credit for non- agricultural income-generating ventures. Thus, for example, four farmers (two each from the medium and large categories) who borrowed from formal agencies were themselves moneylenders. Cooperative officials pointed out that rich farmers frequently default on loans, a problem that is encountered not just in Maharashtra but also in other parts of India (Dhanagare 1975; Sarap 1991; Mohanty 1999, 2000b).

46 Around half of the large and medium producers who obtained loans from non-institutional sources in fact borrowed from their friends and relatives at either a nominal rate of interest or interest-free. Of the 24 medium and large farmers, 21 – or 88 per cent – borrowed money from informal sources without any collateral.

47 Three quarters of indebted small farmers were required to sell live-stock, agricultural implements, land, and household items amounting on average to Rs. 20,458, whereas only 40 and 44 per cent of medium and large farmers were forced into similar distress sales to liquidate

debts of Rs. 47,750 and Rs. 36,342, respectively. Similarly, the value of assets sold by small farmers amounted to 31 per cent of their total, compared with the figures of 19 and 7 per cent for medium and large farmers.

48 This study was based on a sample of 30 cases of farmer suicide, 10 from each district. An equal number of control cases possessing a similar range of socio-economic characteristics (landholding, cropping pattern, etc.) was chosen from the villages to which suicide victims belonged. Like those who had committed suicide, the farmers in the control group were categorised into three groups (small, medium, and large) on the basis of their landholding size.

49 A large number of suicides by lower caste farmers, who were mostly small and marginal landholders, has also been reported in the neighbouring state of Andhra Pradesh (Nirmala 2003).

50 *Mahars* were traditionally the village watchmen, who also undertook tasks such as cutting wood and fencing land, while *Chamars* were leather workers and *Dhangars* were the hereditary herdsmen. For details, see Russell (1916).

51 Details about this began to emerge at the end of that decade. See, for example, "Down and Out on the Farm", the *Guardian* (London), 12 March 1987.

52 See "China Finally Faces up to Suicide Crisis", the *Observer* (London), 21 November 2004.

53 On this point see in particular the recent collection edited by Ho, Eyferth, and Vermeer (2003).

Concluding reflections

Having reached the end of our journey through the trajectory of agricultural transformation, its economic gains and diverse social costs, it is now time to briefly recapitulate our findings and draw together various threads and themes of the discussion carried out in the preceding chapters for making broad generalisations. The central focus of the book is to explain how and why the process of agrarian transformation intended to bring economic gains turned counter-productive evident in a variety of undesirable social consequences. As has been discussed, all three phases of agrarian transformation, colonial, post-colonial, and neoliberal, were pursued under different institutional frameworks, though aimed at expansion of agricultural output involved social costs of varying degrees.

Such agricultural modernisation, as was introduced by the British Raj, to enhance agricultural growth was based on exploitative land, labour, and credit relations, and was promoted with an eye for colonial interests. It encouraged capitalist production in agriculture, which enhanced productivity and developed international markets for commercial crops. As a consequence, while the privileged landowning groups prospered, peasants particularly belonging to the lower segments of the rural society were impoverished and led a miserable life. Social inequalities became widespread, leading to tension and violent conflicts.

Consequently, at the time of independence, India inherited an agrarian economy that was characterised by glaring social inequality and stagnant agriculture, particularly in food grain production. Although arguably the post-independent Indian state continued agrarian transformation on capitalist lines, it was a kind of 'welfare capitalism' as it intended to maximise agricultural growth on one hand and ensure social equality on the other. A comprehensive package of institutional measures like land reforms, community development programmes, Panchayati Raj, and cooperatives were introduced to combat the inequalities and barriers to productivity of the colonial class structure. However, most of them met with much less success. Subsequently new technology based on water-seed-fertiliser strategy backed by land- and crop-based subsidised input and formal credit facilities was launched to enhance agricultural growth. As the new technology was introduced without the earlier inequalities being removed, social equality across regions and social groups remained elusive.

Protests cutting across social categories were organised time and again, demanding state intervention to protect the interests of farming populations. In response to these agitations and movements measures like the Small Farmers' Development Agency, provisions for minimum support price, Tribal Sub-Plan, Special Component Plan, and many poverty alleviation schemes[1] were launched. Notwithstanding loopholes of these measures, the state intervention was partly successful in improving the lives of socially and economically deprived sections of the society. Rapid rise in social inequality was to some extent controlled. It is rightly remarked that we did not have an agrarian upper class anything like England's landed aristocracy, Russia's *Junkers*, or the Latin American *Latifundistas* (Ahmed 2017: 34). Yet, the performance of Indian economy was also incomparably better in the post-colonial period[2]

The neoliberal reforms, introduced in 1991, brought about major changes in the macroeconomic policy framework of the planned economy that existed for four long decades (from 1950 to 1951 through 1990 to 1991). The new institutional framework characterised by strong private property rights, free markets and free trade with minimum state intervention significantly altered the dynamics of agrarian production and made fundamental changes in the previous state policy to achieve high 'growth at all costs'. The search for a return to profitability for global capitalism through the integration of agriculture into the global financial networks became the immediate goal. The new policy reversed the earlier distributive component of land reform; rolled back input subsidies, extension, and support services in agriculture; dismantled the support mechanisms to protect farmers from the uncertainties of the market; reoriented agricultural credit towards large agri-business companies; privatised several infrastructure and services; and opened domestic markets to international trade.

The analysis in the foregoing chapters showed that the costs of this transformation fell heavily on those least able to cope. The so-called dazzling growth came with a rapidly growing inequality in terms of caste, class, and gender. A good number of studies on rural India also show significant rise in inequality across social categories in terms of access to agricultural land, ownership of assets, and consumption expenditure of rural households during the period of economic reforms (Deaton and Dreze 2002; Dev and Ravi 2007; Jayadev et al. 2007; Walker 2008; Banerjee 2012). In addition, regional inequality became more prominent, and the dominant agrarian communities emerged as a unified political class using their caste and kinship network to further their regional interests using the state apparatus. Many scholars held the view that regional inequality in India, which remained largely unchanged during the 1980s, rose dramatically after the 1991 reforms (Chandrasekhar and Ghosh 2003; Kar and Saktivel 2007). It is rightly argued that neoliberalism actually thrives on the very fact of inequality and the increased inequality was construed as necessary in order to drive growth by encouraging entrepreneurial risk, innovation, and competitiveness (Canterbury 2007; Harvey 2007a).

Though the combined effects of post-independent land reform measures and sustained movements against the economic and social dominance of the privileged communities enabled the members of SCs and STs to improve their landholding positions, they found it difficult to utilise these lands to improve their situations under the conditions of neoliberal economic order. Due to poor land-based asset and resource ownership positions, they were unable to cope with modern farming, which is heavily dependent on purchased inputs. As a matter of fact, their agriculture remained at the subsistence level and they struggled to meet a bare minimum of necessities. The neoliberal policy instead of expanding further the little benefit which the members of SCs and STs received earlier, in fact, made it redundant. Without the required capital to succeed in a free market economy, the SCs and STs were marginalised further. Similarly, women were marginalised all over again. They lagged far behind their male counterparts in resource ownership, employment, distribution of productive activities, wages, and earnings and were pushed from the secured self-employed to casual labour category. Though the movement of male workers out of agriculture necessitated the women to stay in agriculture, they hardly remained as independent farmers and were rarely allowed to participate in major farm-related decisions. The findings of many studies from different parts of India also subscribe to this view. As discussed in Chapter 6, there has been a structural shift in the composition of the rural labour force. Over the years, the size of the rural proletariat increased rapidly and a class of labour emerged in the countryside. The expansion of labour force is accompanied by a marked gender division as there has been feminisation of agriculture. Worse still, the shift to wage labour has gone hand in hand with the growth of temporary or seasonal wage labour. While permanent employment opportunities tend to be the preserve of men, women are pushed to low-paid temporary and even less-recognised jobs. In a nutshell, the neoliberal reforms imposed greater degree of insecurity of labour, particularly for women, and widened gender disparity in employment, wages, and earnings. The rest of India is not very different from Maharashtra (see, e.g., Jackson and Rao 2009; Garikipati and Pfaffenzeller 2012). Women in Maharashtra, however, are relatively better off in terms of access to land,[3] mainly due to the effects of their movements against gender inequality in land rights.

That said, the policy based on strategy of 'growth at all costs' in the process of restructuring the production, exchange, and market relations impoverished millions of peasants. As we have seen in Maharashtra, a large section of rural population is in a state of perpetual poverty. In the country as a whole, the incidences of rural poverty are still on the higher side (discussed in Chapter 6). Though the recent official estimates indicate a declining trend, independent estimates made by economists like Utsa Patnaik cast serious doubt on official measurement. She argues that both comparisons over time of the all India and state-level estimates of poverty, based on a method that renders irrelevant the question of a nutrition norm, are unacceptable. A comparison of the consumption expenditure and associated nutritional intake data shows worsening poverty in terms of the percentage of people unable to reach the minimum required calories energy intake (Patnaik 2007,2013).[4] Others have also argued that the number of poor in the country increased and there has been a slowdown in the rate of poverty reduction after 1993 (Sen and Himanshu 2005). The rising trend of distress migration, starvation deaths and malnutrition does not show any sign of decline in rural poverty. After all, contrary to expectations, the agricultural sector in India neither experienced any substantial growth nor did it derive the expected benefits from trade liberalisation in the aftermath of economic reforms (Bhalla and Singh 2009; Reddy and Mishra 2009; Tripathy 2014).

The growing size of rural proletariat has not accelerated the rural – urban migration as in the stylised scenario of economic development. Agriculture and rural non-firm sectors still absorb a large section of labour force. Migration is poverty-driven and it takes place within the agricultural sector from backward to developed pockets and within the different sectors of rural economy. A large body of literature on labour migration reveals this trend (see Chapter 6). The observation made by Li (2014: 3) in this context is worth quoting here:

> For billions of rural people, the promise that modernization would provide a pathway from country to city, and from farm to factory, has proven to be a mirage. Lacking an exit path, they stay where they are, but all too often the old set of relations that enabled them to live and work in the countryside has disappeared, and the new ones – increasingly capitalist in form – do not provide a viable livelihood.

The much-acclaimed employment guarantee scheme MGNREGA considerably failed to arrest the distressed migration. Evaluations have pointed out that in most states implementation of the scheme is fraught with problems such as delays in the payment of wages, a lack of worksite facilities (particularly for women) and corruption in the execution of work and maintenance of records (Dreze and Khera 2009).[5]

Not surprisingly, neoliberalism has successfully demobilised the forces of peasant resistance. As was discussed (Chapter 7), the peasant farmers, instead of mobilising themselves against the neoliberal policies that are unleashing inequality, poverty, and distress migration, and pushing them to all kinds of hardships, as well as risks, committed suicides. Ironically, it was the same peasant farmers who earlier had shown rebellious potentialities during the British Raj in response to its exploitative agrarian policies and organised mass movements during the post-independence period demanding state interventions to protect their interests. The conditions of neoliberalism have prevented popular mobilisation of peasants by dividing them based on caste, class, gender as well as region and infused the values of competitiveness, rivalry, and consumerism. The peasant organisations and radical political outfits which were instrumental in mobilising peasants earlier failed to do so now. It is not only the case with Maharashtra but many other the states like Punjab, Karnataka, Andhra Pradesh, and Tamil Nadu, which have experienced farmer suicides also have a longstanding background of peasant movements during the colonial and post-colonial periods.[6]

To recap from Chapter 8, these farmer suicides were caused by a process of socio-economic 'estrangement' from agrarian communities experienced by rural producers in the context of rapid economic growth stimulated by neoliberalism. The small holders, particularly belonging to the SCs, that had acquired land through land reforms and tried to modernise their agricultural practices found themselves trapped between enhanced aspirations and the reality of neoliberalism (rising debt, declining income). It is disheartening to note that the members of lower castes who were *depeasantised* during the colonial period and *repeasantised* during the post-colonial period finally ended up committing *suicides* in the neoliberal era. Farmer suicides are no more confined to states like Maharashtra, Andhra Pradesh, Punjab, and Karnataka; the syndrome has spread far and wide. Even advanced pockets of many backward states like Odisha and Chhattisgarh reported such incidences. It is commented that, "the areas of capitalist intensive agriculture are today better known as suicide belts" (Walker 2008: 572).

In sum, the agrarian transformation which began on capitalist axis during the colonial period and was promoted during the post-colonial phase though 'welfare capitalism' reached the current phase of global capitalist development under neoliberal reforms. The scale and speed of this transformation led to a kind of unprecedented crisis in the countryside having far-reaching social implications. The weaker sections of the society, particularly the members of SCs and STs, women, the small peasants and the rural poor, who did not cause this crisis were the ultimate sufferers. The state compelled them to pay for the crisis by socialising the cost through its apparatus. Touraine (2001: 24) rightly notes, "The triumph of capitalism has been so costly and intolerable that everyone, on all sides, is trying to find a way out of the 'neoliberal transition'".

Like India, the distressing social consequences of agrarian transformation during and after the implementation of neoliberal reforms was experienced by many other countries of the Global South. A large body of literature outlined the process of social differentiation that has set into the countryside of African countries destroying millions of livelihood in the rural sectors (Raikes 2000; Barrett *et al.* 2001; Bryceson 2009; Oya 2010). While narrating Africa's experience, Oya (2010) notes, ". . . structural adjustment and liberalization altered the rules of the game and weakened states to an extent that agriculture has increasingly become a space of 'social Darwinism' where farmers' differentiation has become more visible and ruthless". A growing temporary and seasonal wage labour, feminisation of wage labour, replacement of tenants by wage labour and transforming settled peasants into seasonal rural proletariat are reported in Latin American countries like Chile and Mexico (Kay 2004; Kurtz 2004). Similar observations were also made in many East Asian countries (for example, Li 2009, 2014). A study of 23 countries in Asia, Africa, and Latin America revealed that in almost every case, the neoliberal programmes undermined food production, increased rural poverty, suppressed wages, and diminished employment.[7] In a nutshell, almost all the developing countries experienced the deleterious effect of agrarian change under neoliberalism. To put it in the words of Harvey (2007b: 39), "The wave of creative destruction neoliberalisation has visited across the globe is unparalleled in the history of capitalism".

Our analysis thus contributes to at least three major ongoing debates on social costs of agrarian transformation. The first concerns the trend of inequality and differentiation. Some argue that agrarian change in the wake of neoliberal reforms has created new opportunities for all sections of peasants, and the emerging growth trends have made a dent on rural poverty and inequality (Dollar and Kraay 2002; Sala-i-Martin 2002; World Bank 2002; Bhagwati 2004). Others challenged this view and held that the ongoing transformation is polarising the rural society where the usual winners are the few capitalists and rich farmers living closer to urban areas while the losers are the poorer peasant farmers who, with little competitive potential, struggle to subsist, and those dependent on rural wages whose working conditions have become more miserable (Gibbon *et al.* 1993; Bryceson 1999; Kay 2002). The thesis that runs through several chapters of this book is the persistence of tremendous inequality in terms of caste, class, and gender, and it underlines the complexities grounded both in colonialism as well as in the state-directed development planning in post-independent India. The response of peasantry and rural population to agricultural transformation varies; it is not only region-specific but also specific to social categories and is conditioned by historical as well as contemporary features. As we have seen in Maharashtra, while the advanced western region changed its agriculture in tune with the demands of neoliberal policies, the other backward regions struggled to catch up with the changing situation. Women experienced more discrimination in advanced pockets than the backward ones in many respects. Similarly, though the SCs and STs in general are disadvantaged and unable to cope with modern farming, SCs are relatively in a better position. Moreover, the degree of social inequality varies with the presence and absence of social movements.

The second debate relates to the enduring controversy on persistence and disappearance of peasantry. Though the classical Marxist and non-Marxist debates on the relationship between capitalism and peasantry died down due to evidences on the persistence of peasantry from different parts of the world, it generated a renewed debate recently in the context of neoliberal global capitalist development. While one group of scholars argues that the emergence of conditions of new relations of production under the neoliberal regime leads to the death of peasantry (Hobsbawm 1994; Bernstein 1994; Friedmann 2006; Huang 2015), the other group emphasises on the persistence of peasantry (Johnson 2004; Van der Ploeg 2010; Vanhaute 2012). Our analysis does not indicate any sign of immediate disappearance of peasantry. Rather, the number of cultivators belonging to the SCs and STs increased over the years. Though the effects of neoliberal reforms has pushed small farmers to the edge by declining their farm income and making them vulnerable to price and yield risks, they continue to struggle to cope with the emerging situation by increasing their agricultural applications. Even though they went to the extent of killing themselves they still do not show any sign of leaving agriculture. Rural-urban migration did not take place as it happened in case of the developed countries of the West.

The third debate is on the impact of neoliberalism on farmers' resistance. Quite a few scholars view that neoliberalism has spawned a swath of resistance movements, many of which are radically different from earlier movements (Harvey 2007b; Li 2014; Klassen 2015). On the contrary, it is argued by some that neoliberalism has demobilised peasants and diluted their resistances (Petras and Veltmeyer 2011; Chang 2015). In the case of India, as we found, farmers' movements miserably declined. Farmers who had participated in movements with great enthusiasm during 1970s and 1980s withdrew themselves when they found the leaders instead of representing their genuine problems were supporting capitalist and corporate interests. Far more seriously, the discontentment of farmers was manifested by a mounting number of suicides, which distinguishes most glaringly Indian experience from that of other developing countries.

Thus, the social costs of agrarian transformation have become far more complex and nuanced in India than is commonly assumed. Though the adverse impact of agrarian transformation can be traced out in the dynamics of productive relations all over the world in different spatial and historical contexts, what is perhaps distinctive about Indian agrarian transformation in the current context of neoliberal globalisation is its severe impact on socially weaker sections. In India, as many people as in whole of sub-Saharan Africa, largely drawn from the socially disadvantageous groups, live in hunger, and their day to day survival is intrinsically linked to agriculture. Therefore, for decades, the national planning was prioritising the development of agriculture and it was under state protection. The recent attempt to facilitate the end of agricultural 'exceptionalism' has affected masses of socially weaker sections who recently became increasingly dependent on agriculture with the support of state-led welfare measures. Moreover, lack of required skill and education handicaps them further to take up other opportunities available in urban and industrial pockets in the wake of new economic reforms. In sum, the Indian trajectory of agrarian transformation and its social costs, though similar to many other countries in Asia, Africa, and Latin America in some respects, has certain features which make its experience unique at the global level. Therefore, the existing debates on the social costs of agrarian transformation need to be drawn into broader national as well as global space so these social costs and their implications could be further explained. It is aptly commented that when 'one-size-fits-all' neoliberal agenda was promoted regardless of the marked diversity across countries, crops, and classes of farmers, and without consideration of external opportunities and constraints (Oya 2010: 99), it was bound to generate differential experiences.

The analysis in the book suggests that all characteristics of the current agrarian crisis in India could be attributed to the effects of the ongoing global capitalist transition, which is being facilitated by the neoliberal policies. Although the policy in the current neoliberal phase rehashes the objectives of the nationalist and colonial phases, there are significant differences. While the earlier attempts to bring about economic change were concerned with rural underdevelopment and relied heavily on state intervention, the current neoliberal policy is oriented to bring change by adopting a laissez-faire approach which makes the market supreme to decide what is necessary. In this situation, the state is not a mere interventionist per se, rather it tries to ensure and enforce the will of the market. Such kinds of ever-increasing disembeddedness of economy, to use Karl Polanyi's term, may entail a series of social costs and their social fallout in the rocky decades ahead.

True, as the classic experience in England and elsewhere reminds us that successful transformation is painful for agricultural sector in all societies, the magnitude of its social costs in a country like India proved to be unmanageable for the growth levels achieved that fail to attain the prosperity and privileges which the developed countries in Europe and America did. The political hegemony of Europe and America in global economy successfully reduced, if not eliminated competition and created lucrative captive markets, colonial import of agricultural produce and the like. Moreover, many of the developed countries of the West overcame the problem of rising unemployment inherent in their early capitalist growth, mostly by exporting their unemployment abroad, which is not feasible in any way for the present large labour-surplus economies like India. It is well-known that the social crisis in Western Europe and America following agrarian transformation was controlled and regulated by strong state intervention in the form of protective support to farmers as opposed to domestic consumers. The United States has chosen to open its agriculture more fully to market forces in pursuit of perceived competitive advantage in the world trading system. The European Union countries have taken a more differentiated approach, assisting some farm operators to participate in world markets, while shifting structural policies to pursue social and environmental objectives. Interestingly, powerful developed countries were largely able to retain their protectionist agricultural policies while preaching the pro-free-market rhetoric to the rest of the world.

Needless to say, the intensity of adverse social consequences of capitalism is relative to the extent to which capital is regulated and disciplined through state intervention. Consider the case of China. The state has been an actively interventionist and growth-promoting. Capitalism is growing slowly and has been pursued with due caution (Byres 2005: 90).[8] Similarly, in Argentina, steps were taken to deal with the negative consequences of privatisation of social policies, for example, through the nationalisation of the pension system (Arza 2012). Likewise, the Caribbean countries, which are highly dependent on external markets, have undertaken a series of measures to respond to social impacts of the crisis (Downes 2012). However, in the case of India, the agrarian crisis remains because the state has no desire to resolve in future. The state withdrew itself and the peasants were exposed to the fury of global capitalism without any protection. In his landmark book *Making Globalisation Work*, Joseph E Stiglitz (2006: 50) observed, "Development is about transforming the lives of people, not just transforming economies".

The people who are left stranded by capitalist processes need protective measures, but the state under present conditions can seldom initiate such measures on its own unless forced to. Consider the achievements of earlier movements during the colonial or post-colonial period; all of them invariably followed some welfare measures. The recent few protest movements also successfully challenged many neoliberal projects.[9] Therefore, the alternative to the dismal failure of the neoliberal state to improve the conditions of the farming population and the rural poor is the main argument for reviving the option of revolutionary change, however difficult it may appear in the current conjuncture. Ultimately, if the policies do not change, then politics need to change (Griffin 1974: 201). Suicide of farmers, whether it is a mark of protest or helplessness, has not yielded any effective response from the state. The quick-fix approach to address the problem through measures like *Agricultural Debt Waiver and Debt Relief Scheme* implemented in 2008 hardly served any purpose.

In short, it is not through suicides, rather only through strong resistance movements that the peasant mass may overcome this crisis. Many farmer groups from different states[10] have already expressed their solidarity while protesting for the violations of their rights brought about by unjust neoliberal policies. The recent week-long march by thousands of farmers in Maharashtra[11] demanding higher prices for farm produce and loan waiver echoes the growing discontentment in the countryside. Possibly an honest leadership is awaited to lead the way.

Notes

1 Several target-group-oriented schemes were started which were finally culminated in the Integrated Rural Development Programme and National Rural Employment Programme. These programmes emphasised the expansion of employment opportunities and distribution of renewable assets among the rural poor. Attempts were also made for the transfer of income to the poor in indirect ways through food subsidies and 'dual pricing' of essential commodities.

2 It is reported that from the Independence till 1991 the average annual growth rate was around 3.5 per cent as against zero to negative growth rate during the colonial period (Dreze and Sen 2013, as cited in Hensman (2015: 207).

3 The state is ahead of the national average in terms of area as well as number of holdings operated by female members. The area owned by women in Maharashtra is 13 per cent as against the 10 per cent at national level in 2010–11. The same could be said about the number of holdings also. For details, see *Agricultural Census of India*, 2010–11.

4 The proportion of rural population unable to access 24,00 calories a day climbed from 75 per cent in 1993–94 to as high as 87 per cent in 2004–05 and 91 per cent in 2009–10. It was further noted that the percentage of rural population that did not have access to 2200 calories increased from 59 per cent in 1993–94 to 76 per cent in 2009–10. For details, see Patnaik (2013: 43–58).

5 Also see "Implementation of the National Rural Employment Guarantee Scheme: Submission to the Parliamentary Standing Committee on Rural Development", New Delhi: Indian School of Women's Studies and Development.

6 Punjab witnessed strong peasant resistance during the colonial period, like the uprising of 1890–1900 and the massive agrarian unrest of 1906–07 in Lahore, Amritsar, and Rawalpindi. Further, in 1924, Punjab peasants successfully agitated against the water rate. In the 1930s the Kisan Sabha movement also mobilised the peasantry on issues of water and land revenue. Between 1928 and 1946, Andhra Pradesh experienced a series of peasant movements against the exploitative colonial land revenue policies. Many peasant Associations were formed during this period. Among them was the Coastal Andhra Rytu Sangham of 1928, which spearheaded the movement against the resettlement rates fixed by the British government in several parts of the state. A union of agricultural labour in the Nellore district also joined the anti-Zamindari struggle during 1942. Subsequently, in the year 1946, the state experienced massive peasant resistance (popularly known as the Telengana movement) against the feudal practices of extra-economic coercion (Vetti system) and highly iniquitous agrarian relations. Similarly, Karnataka peasantry also revolted in several occasions during this period (e.g. the Nagar Peasant Revolt, 1830–31). A chain of agrarian movements took place during the 1930s and the 1940s in various parts of Tamil Nadu (e.g. in Thanjavur, Tiruchirappalli, Madurai, Coimbatore, and Salem). During the post-independence period, farmers' movements were mobilised on a wider scale in these states. Under the leadership of Narayanaswamy Naidu, the Tamil Nadu Agriculturalist's Association launched an agitation in 1973 demanding reduction in electricity tariff for the farmers, remission of loans taken from cooperatives and government sources, adequate supply of fertilisers, and other agricultural inputs at subsidised rates (Nadkarni 1987: 60–65; Dhanagare 2016a: 62). Similar kinds of movements also took place in Punjab and Haryana under the banner of Bharat Kisan Union (BKU). In Punjab, farmers' agitation started with the boycotting of mandis and refusal to

sell off wheat at the low procurement price offered by the government authorities. The farmers' movement in Karnataka also started during the early 1980s, which was led by Karnataka Rajya Raitha Sangha (KRRS), demanding remunerative farm prices. Now in all these states farmer mobilisations have become weak and there is a rising trend of farmer suicides.

7 See "The All-Too-Visible Hand: A Five-Country Look at the Long and Destructive Reach of the IMF", 1999, Washington, DC, Development Group of Alternative Policies.

8 In fact, China accepted neoliberalism to the extent it serves national interests, and it pursued inward rather than expansionist development, which was not dominated by the capitalists but insulated from them. For details, see Chang (2015: 186–187).

9 For example, in the case of Nadigram in West Bengal in 2007, the Tata project was finally closed down following strong opposition by peasants, and similarly in Odisha, the Vedanta and Posco had to withdraw due to strong protests by the local peasants.

10 Indian framers' groups led by the Indian Coordination Committee on Farmers Movements and endorsed by Bharatiya Kisan Union (BKU), Karnataka Rajya Raitha Sangha (KRRS), Tamil Nadu Farmers Association (TVS), Kerala Coconut Farmers Association, and South Indian Coordination Committee of Farmers Movements (SICCFM) submitted an open letter against WTO citing numerous instances to its director general, Roberto Azevedo, who was in Delhi, on 8–9 February 2017. See "Bharatiya Kisan Union backs protesting Tamil Nadu farmers in Delhi, calling for their demands to be met immediately" by Abhilash Babu, Via Campesina, 27 March 2017. Accessed from http://lvcsouthasia.blogspot.in/2017/03/bharatiya-kisan-union-backs-protesting_27.html on 8 February 2017.

11 More than 35,000 of farmers from different parts of the state undertook a five days long march and reached Mumbai on 12 March 2018 to gherao the state assembly demanding enhancement of minimum support price for the farm produce, waiver of loan and electricity bill and pension for the farmers. See "Read the Distress Signals", *The Hindu*, 22 March 2018. Accessed from www.thehindu.com/opinion/lead/read-the-distress-signals/article23314876.ece on 27 April 2018 and "Why Maharashtra's Farmers Are Protesting and Why Mumbaikars Are Supporting Them: 10 Points", *Times of India*, 12 March 2018. Accessed from https://timesofindia.indiatimes.com/india/why-maharashtras-farmer-are-protesting-and-why-mumbaikars-are-supporting-them-10-points/articleshow/63263394.cms on 27 April 2018.

Bibliography

Acharya, S. and Panwalkar, V. G. 1988. *Labour Force Participation in Rural Maharashtra: A Temporal, Regional and Gender Analysis*, Studies on Women Workers (2), Asian Employment Programme Working Papers (ARTEP), New Delhi: ILO.

Agarwal, B. 1983. *Mechanisation in Indian Agriculture: An Analytical Study Based on Punjab*, New Delhi: Allied Publishers Private Ltd.

Agarwal, B. 1984. Rural Women and the High Yielding Variety Rice Technology. *Economic and Political Weekly*, 19(13): A39–A52.

Agarwal, B. 1994. *A Field of One's Own: Gender and Land Rights in South Asia*, Cambridge: Cambridge University Press.

Agarwal, B. 2003. Gender and Land Rights Revisited: Exploring New Prospects Via the State, Family and Market. *Journal of Agrarian Change*, 3(1–2): 184–224.

Agarwal, B. 2016. *Gender Challenges, Selected Essays by Bina Agarwal*, 3 vols, New Delhi: Oxford University Press.

Ahluwalia, M. S. 1978. Rural Poverty and Agricultural Performance in India. *Journal of Development Studies*, 14(3): 298–323.

Ahluwalia, M. S. 1996. New Economic Policy and Agriculture: Some Reflections. *Indian Journal of Agricultural Economics*, 51(3): 412–426.

Ahluwalia, M. S. 2002. Economic Reforms in India Since 1991: Has Gradualism Worked? *The Journal of Economic Perspectives*, 16(3): 67–88.

Ahmed, W. 2017. From Mixed Economy to Neoliberalism: Class and Caste in India's Policy Transition. In W. Ahmed (ed.) *India's New Economic Policy: A Critical Analysis*, New Delhi: Rawat Publications, pp. 33–56.

Akram-Lodhi, A. H. 2007. Land Markets and Neoliberal Enclosure: An Agrarian Political Economy Perspective. *Third World Quarterly*, 28(8): 1437–1456.

Alexander, K. C. 1978. *Agricultural Labour Unions: A Study in Three South Indian States*, Hyderabad: National Institute of Community Development.

Ali, I. 1987. Malign Growth? Agricultural Colonialism and the Roots of Backwardness in Punjab. *Past and Present*, 114(1): 110–132.

Allen, R. C. 1992. *Enclosure and the Yeoman: The Agricultural Development of the South Midlands 1450–1850*, Oxford: Clarendon Press.

Apthorpe, R. 1977. Technology and Peasant Production: A Comment. *Development and Change*, 8: 370–373.

Araghi, F. 2009. The Invisible Hand and the Visible Foot: Peasants, Dispossession and Globalization. In H. Akram-Lodhi and C. Kay (eds.) *Peasants and Globalisation: Political Economy, Rural Transformation and the Agrarian Question*, London: Routledge, pp. 111–147.

Arnold, D. 2000. *Science, Technology and Medicine in Colonial India*, Cambridge: Cambridge University Press.

Arun, S. 2012. We Are Farmers Too: Agrarian Change and Gendered Livelihoods in Kerala, South India. *Journal of Gender Studies*, 21(3): 271–284.

Arza, C. 2012. Policy Change in Turbulent Times: The Nationalization of Private Pensions in Argentina. In P. Utting (ed.) *The Global Crisis and Transformative Social Change*, New York: Palgrave Macmillan, pp. 123–140.

Attwood, D. W. 1979. Why Some of the Poor Get Richer: Economic Change and Mobility in Rural Western India. *Current Anthropology*, 20(3): 495–516.

Attwood, D. W. 1987. Irrigation and Imperialism: The Causes and Consequences of a Shift from Subsistence to Cash Cropping. *Journal of Development Studies*, 23(3): 341–366.

Attwood, D. W. 1992. *Raising Cane: The Political Economy of Sugar in Western India*, Boulder: Westview Press.

Badiani, R. and Safir, A. 2009. Circular Migration and Labour Supply: Responses to Climatic Shocks. In P. Deshingkar and J. Farrington (eds.) *Circular Migration and Multilocational Livelihood Strategies in Rural India*, New Delhi: Oxford University Press, pp. 37–57.

Balisacan, A. 1993. Agricultural Growth, Landlessness, Off-Farm Employment, and Rural Poverty in the Philippines. *Economic Development and Cultural Change*, 41(3): 533–562.

Banerjee, A. 2012. From Agrarian Crisis to Global Economic Crisis: Neoliberalism and the Indian Peasantry. In P. Utting *et al.* (eds.) *The Global Crisis and Transformative Social Change*, New York: Palgrave Macmillan, pp. 177–198.

Banerjee, A. and Iyer, L. 2005. History, Institutions, and Economic Performance: The Legacy of Colonial Land Tenure Systems in India. *The American Economic Review*, 95(4): 1190–1213.

Bansode, P. K. 2011. *Seasonal Rural Migration: Quality of Life at Destination and Source: A Study of Sugarcane Cutter Migrants*, Pune: Gokhale Institute of Politics and Economics.

Bardhan, P. K. 1973. Variations in Agricultural Wages: A Note. *Economic and Political Weekly*, 7(21): 947–950.

Bardhan, P. K. 1984. *The Political Economy of Development in India*, New Delhi: Oxford University Press.

Barraclough, S. 1974. Politics First. *Ceres*, 41: 24–28.

Barrett, C. B. *et al.* 2001. *Heterogeneous Constraints, Incentives and Income Diversification Strategies in Rural Africa*, Department of Applied Economics and Management Working Paper, WP 2001–25, Ithaca: Cornell University.

Batliwala, S. and Dhanraj, D. 2004. Myths That Instrumentalise Women: Stories from the Indian Frontline. *IDS Bulletin*, 35(5): 11–18.

Baviskar, B.S. 1980. *The Politics of Development: Sugar Cooperatives in Rural Maharashtra*, New Delhi: Oxford University Press.

Bendix, R. 1978. *Max Weber: An Intellectual Portrait*, London: Heinemann.

Beneria, L. 2003. *Gender, Development, and Globalization: Economics as If All People Mattered*, New York: Routledge.

Benjamin, N. 1973. Raw Cotton of Western India: A Comment. *The Indian Economic and Social History Review*, 10(1): 64–82.

Benjamin, N. and Mohanty, B. B. 2001. Ambedkar's Quest for the Right of Social Equality: An Interpretation. *Social Action*, 51(2): 132–150.

Bentall, J. and Corbridge, S. E. 1996. Urban-Rural Relations, Demand Politics and the 'New Agrarianism'. *Transactions of the Institute of British Geographers*, 21(1): 27–48.

Bernstein, H. 1994. Agrarian Classes in Capitalist Development. In L. Sklair (ed.) *Capitalism and Development*, London: Routledge.

Bernstein, H. 2009. V.I. Lenin and A.V. Chayanov: Looking Back, Looking Forward. *The Journal of Peasant Studies*, 36(1): 55–81.

Besnard, P. (ed.) 1983. *The Sociological Domain: The Durkheimians and the Founding of French Sociology*, Cambridge: Cambridge University Press.

Beteille, A. 1965. *Caste, Class, and Power: Changing Patterns of Stratification in a Tanjore Village*, Berkeley, CA: University of California Press.

Bhagat, R. B. 2016. Nature of Migration and Its Contribution to India's Urbanisation. In D. K. Mishra (ed.) *Internal Migration in Contemporary India*, New Delhi: Sage Publications, pp. 26–46.

Bhagat-Ganguly, V. 2016. Introduction. In V. Bhagat-Ganguly (ed.) *Land Right in India Policies, Movements and Challenges*, New Delhi: Routledge, pp. 1–12.

Bhagwati, J. 2004. *In Defence of Globalisation*, New Delhi: Oxford University Press.

Bhalla, G. S. and Chadha, G. K. 1983. *Green Revolution and the Small Peasant: A Study of Income Distribution among Punjab Cultivators*, New Delhi: Concept Publishing Company.

Bhalla, G. S. and Singh, G. 2009. Economic Liberalisation and Indian Agriculture: A Statewise Analysis. *Economic and Political Weekly*, 44(52): 34–44.

Bhalla, G. S. and Singh, G. 2012. *Economic Liberalisation and Indian Agriculture: A District-Level Study*, New Delhi: Sage Publications.

Bharadwaj, K. 1989. *On the Formation of the Labour Market in Rural Asia*, World Employment Programme Research Working Paper 98, Geneva: International Labour Organisation.

Blyn, G. 1966. *Agricultural Trends in India, 1891–1947: Output, Availability and Productivity*, Philadelphia: University of Pennsylvania.

Blyn, G. 1983. The Green Revolution Revisited. *Economic Development and Cultural Change*, 32(3): 705–725.

Borpujari, J.G. 1973. Indian Cottons and the Cotton Famine 1860–65. *The Indian Economic and Social History Review*, 10(1): 37–49.

Borras, S., Jr. 2003. Questioning Market-Led Agrarian Reform: Experiences from Brazil, Colombia and South Africa. *Journal of Agrarian Change*, 3(3): 367–394.

Bose, A. 2000. From Population to Pests in Punjab: American Boll Worm and Suicides in Cotton Belt. *Economic and Political Weekly*, 35(38): 3375–3378.

Boserup, E. 1970. *Women's Role in Economic Development*, London: George Allen and Unwin Ltd.

Bradnock, R. W. 1984. Agricultural Development in Tamil Nadu: Two Decades of Land Use Change at the Village Level. In T. Bayliss-Smith *et al.* (eds.) *Understanding Green Revolution: Agrarian Change and Development Planning in South Asia*, Cambridge: Cambridge University Press.

Brahme, S. 1973. Drought in Maharashtra. *Social Scientist*, 1(12) (July): 47–54.

Brahme, S. 1983. *Drought in Maharashtra-1972: A Case for Irrigation Planning*, Pune: Orient Longman and Gokhale Institute of Politics and Economics.

Brahme, S. and Upadhyaya, A. 1979. *A Critical Analysis of the Social Formation and Peasant Resistance in Maharashtra Vol. I, II and III (Mimeo)*, Pune: Shankar Brahme Samaj Vidnyan Grathalaya.

Brara, R. 2014. Shaping Land Rights: Tenurial Class, Lineage, and Gender in Malerkotla, India. *Asian Journal of Women's Studies*, 20(4): 7–38.

Breman, J. 1974. *Patronage and Exploitation: Changing Agrarian Relations in South Gujarat*, Berkeley, CA: University of California Press.

Breman, J. 1985. *Of Peasants, Migrants and Paupers: Rural Labour Circulation and Capitalist Production in West India*, New Delhi: Oxford University Press.

Breman, J. 1996. *Footloose Labour: Working in India's Informal Economy*, Cambridge: Cambridge University Press.

Breman, J. 2007. *The Poverty Regime in Village India: Half a Century of Work and Life at the Bottom of the Rural Economy in South Gujarat*, New Delhi: Oxford University Press.

Breman, J. 2010. *Outcast Labour in Asia: Circulation and Informalisation of the Workforce at the Bottom of the Economy*, New Delhi: Oxford University Press.

Breman, J. *et al.* (eds.) 1997. *The Village in Asia Revisited*, New Delhi: Oxford University Press.

Breman, J. *et al.* 2009. *India's Unfree Workforce: Of Bondage Old and New*, New Delhi: Oxford University Press.

Brocheux, P. 1983. Moral Economy or Political Economy? The Peasants are Always Rational. *Journal of Asian Studies*, 42(4): 791–803.

Brody, H. 1973. *Inishkillane: Change and Decline in the West of Ireland*, London: Allen Lane, the Penguin Press.

Brown, W. 2006. American Nightmare, Neoliberalism, Neoconservatism, and De-Democratisation. *Political Theory*, 34(6): 690–734.

Bryceson, D. F. 1999. *Sub-Saharan Africa Betwixt and between: Rural Livelihood Practices and Policies*, ASC Working Paper 43/1999, Leiden: African Studies Centre.

Bryceson, D. F. 2009. Sub-Saharan Africa's Vanishing Peasantries and the Specter of a Global Food Crisis. *Monthly Review*, 61(3): 48–62.

Byres, T. J. 1972. The Dialectic of India's Green Revolution. *South Asian Review*, 5(2): 99–116.

Byres, T. J. 1981. The New Technology, Class Formation and Class Action in the Indian Countryside. *Journal of Peasant Studies*, 8(4): 405–454.

Byres, T. J. 2005. Neoliberalism and Primitive Accumulation in Less Developed Countries. In A. Saad-Filho and D. Johnston (eds.) *Neoliberalism: A Critical Reader*, London: Pluto Press, pp. 83–90.

Calman, L. J. 1986. *Protest in Democratic India: Authority's Response to Challenge*, Boulder: Westview Press.

Canterbury, D. 2007. Caribbean Agriculture Under Three Regimes: Colonialism, Nationalism and Neoliberalism in Guyana. *The Journal of Peasant Studies*, 34(1): 1–28.

Cardoso, J. L. *et al.* 2014. *Economic Development and Global Crisis: The Latin-American Economy in Historical Perspective*, London: Routledge.

Carter, A. 1975. *Elite Politics in Rural India: Political Stratification and Political Alliances in Western Maharashtra*, New Delhi: Cambridge University Press.

Catanach, I. J. 1966. Agrarian Disturbances in Nineteenth Century India. *Indian Economic and Social History Review*, 3(1): 65–84.

Catanach, I. J. 1970. *Rural Credit in Western India: 1875–1930 Rural Credit and the Cooperative Movement in the Bombay Presidency*, Berkeley, CA: University of California Press.

Chakrabarti, A. *et al.* 2016. *The Indian Economy in Transition Globalization, Capitalism and Development*, New Delhi: Cambridge University Press.

Chakrabarty, G. 1998. Scheduled Castes and Tribes in Rural India Their Income, Education and Health Status. *Margin*, 30(4): 100–130.

Chakravarti, A. 2001. *Social Power and Everyday Class Relations: Agrarian Transformation in North Bihar*, New Delhi: Sage Publications.

Chakravarti, A. 2016. If We Had Land, We Would Be Human: The Implications of Landlessness in a Bihar Village. In V. Bhagat-Ganguly (ed.) *Land Right in India Policies, Movements and Challenges*, New Delhi: Routledge, pp. 15–30.

Chakravorty, S. 1975. Farm Women Labour: Waste and Exploitation. *Social Change*, 5(1/2): 9–16.

Chand, R. *et al.* 2007. Growth Crisis in Agriculture: Severity and Options at National and State Levels. *Economic and Political Weekly*, 42(26): 2528–2533.

Chandrasekhar, C. P. and Ghosh, J. 2003. Per Capita Income Growth in the States. *The Hindu Business Line*, August 12, 2003. Accessed from (www.thehindubusinessline.com/2003/08/12/stories/2003081200110900.htm) on 31. 03. 2017.

Chang, D. 2015. The Rise of East Asia: A Slippery Floor for the Left. In L. Pradella and T. Marois (eds.) *Polarizing Development: Alternatives to Neoliberalism and the Crisis*, London: Pluto Press, pp. 180–191.

Chari, A. 2006. Guaranteed Employment and Gender Construction: Women's Mobilization in Maharashtra. *Economic and Political Weekly*, 41(50): 5141–5148.

Charlesworth, N. 1972. The Myth of Deccan Riots of 1875. *Modern Asian Studies*, 6(4): 401–421.

Charlesworth, N. 1982. *British Rule and the Indian Economy, 1800–1914*, London: Palgrave Macmillan.

Chatterjee, P. 2008. Democracy and Economic Transformation in India. *Economic and Political Weekly*, 43(16): 53–62.

Chattopadhyay, M. 1982. Role of Female Labour in Indian Agriculture. *Social Scientist*, 10(7): 43–54.

Chaudhry, P. 1994. *The Veiled Woman: Shifting Gender Equations in Rural Haryana, 1880–1990*, New Delhi: Oxford University Press.

Chavan, P. 2010. How 'Rural' Is India's Agricultural Credit? *The Hindu*, August 12, 2010. Accessed from (www.thehindu.com/opinion/op-ed/how-rural-is-indias-agricultural-credit/article566888.ece) on 17.09.2016.

Chavan, P. 2015. Rural Credit Cooperatives in Maharashtra: A Tale of Growing Divides. *Economic and Political Weekly*, 50(38): 52–60.

Chavan, P. and Bedamatta, R. 2006. Trends in Agricultural Wages in India 1964 to 1999–2000. *Economic and Political Weekly*, 41(38): 4041–4051.

Chithelen, I. 1980. Sugar Cooperative in Maharashtra. *Social Scientist*, 9(5/6): 55–61.

Chithelen, I. 1985. Origins of Co-Operative Sugar Industry in Maharashtra. *Economic and Political Weekly*, 20(14): 604–612.

Choksey, R. D. 1955. *Economic Life in the Bombay Deccan (1818–1939)*, Mumbai: Asia Publishing House.

Clark, C. 1940. *The Conditions of Economic Progress*, London: Palgrave Macmillan.

Colburn, F. D. 1982. Current Studies of Peasants and Rural Development: Applications of the Political Economy Approach. *World Politics*, 34(3): 437–449.

Connell, J. *et al.* 1976. *Migration from Rural Areas: The Evidence from Village Studies*, Oxford: Oxford University Press.

Corbridge, S. 2010. The Political Economy of Development in India. In P. R. Brass (ed.) *Routledge Handbook of South Asian Politics: India, Pakistan, Bangladesh, Sri Lanka and Nepal*, London: Routledge, pp. 305–318.

Cornwall, A. *et al.* 2008. Introduction: Reclaiming Feminism: Gender and Neoliberalism. *IDS Bulletin*, 39(6): 1–9.

da Corta, L. and Venkateshwarlu, D. 1999. Unfree Relations and the Feminisation of Agricultural Labour in Andhra Pradesh, 1970–95. *The Journal of Peasant Studies*, 26(2–3): 71–139.

Dandekar, H. 1978. *Rural Development: Lessons from a Village in Deccan Maharashtra*, PhD dissertation (unpublished) submitted to the University of California.

Dandekar, V. M. 1973. Introduction. In S. H. Deshpande (ed.) *Economy of Maharashtra*, Pune: Samaj Prabodhan Sanstha.

Dandekar, V. M. and Khudanpur, G. J. 1957. *Working of Bombay Tenancy Act 1948: Report of Investigation*, Pune: Gokhale Institute of Politics and Economics.

Dantwala, M. L. 1970. From Stagnation to Growth. *Indian Economic Journal*, 18(2): 165–192.

Das, P.V. 2015. *Colonialism, Development, and the Environment: Railways and Deforestation in British India, 1860–1884*, New York: Palgrave Macmillan.

Dasgupta, B. 1980. *The New Agrarian Technology and India*, New Delhi: Palgrave Macmillan.

Dash, A.P. *et al.* 1998. *Failure of Cotton Crop and Its Impact on Farmers*, Pune: Vaikunth Mehta National Institute of Cooperative Management.

Datar, C. 2007. Failure of National Rural Employment Guarantee Scheme in Maharashtra. *Economic and Political Weekly*, 42(34): 3454–3457.

Deaton, A. and Dreze, J. 2002. Poverty and Inequality in India: A Re-Examination. *Economic and Political Weekly*, 37(36): 3729–3748.

Deere, C. D. 2009. The Feminization of Agriculture? The Impact of Economic Restructuring in Rural Latin America. In S. Razavi (ed.) *The Gendered Impacts of Liberalization: Towards 'Embedded Liberalism'?* New York: Routledge.

De Haan, A. 2002. Migration and Livelihoods in Historical Perspectives: A Case Study of Bihar, India. *Journal of Development Studies*, 38(5): 115–142.

De Haan, A. 2011. *Inclusive Growth? Labour Migration and Poverty in India*, Working Paper 513, The Hague: International Institute of Social Sciences.

Deininger, K. 1999. *Making Negotiated Land Reform Work: Initial Experience from Brazil, Colombia, and South Africa (English)*. Policy, Research working paper; no. WPS 2040. Washington, DC: World Bank.

Deininger, K. and Feder, G. 1998. *Land Institutions and Land Markets*, Policy Research Working Paper, No. 2014, The World Bank Development Research Group.

Desai, G.M. *et al*. 1978. *Cultivators' Experience of High Yielding Varieties of Cotton (A Macro Study in Gujarat)*, Ahmedabad: Centre for Management in Agriculture, IIM.

Desai, S. *et al*. 2010. *Human Development in India: Challenges for a Society in Transition*, New York: Oxford University Press.

Deshingkar, P. 2006. *Internal Migration, Poverty and Development in Asia*, Briefing Paper, London: Overseas Development Institute.

Deshingkar, P. and Farrington, J. (eds.) 2009. *Circular Migration and Multi Locational Livelihoods Strategies in Rural India*, New Delhi: Oxford University Press.

Deshingkar, P. *et al*. 2006. Changing Livelihood Contexts in the Study Locations. In J. Farrington *et al*. (eds.) *Policy Windows and Livelihoods Futures*, New Delhi: Oxford University Press.

Deshingkar, P. and Start, D. 2003. *Seasonal Migration for Livelihoods in India: Coping, Accumulation and Exclusion*, Working Paper 220, London: Overseas Development Institute.

Deshingkar, P. *et al*. 2008. Circular Migration in Madhya Pradesh: Changing Patterns and Social Protection Needs. *The European Journal of Development Research*, 20(4): 612–628.

Deshpande, R. S. 1998. Land Reforms and Agrarian Structure in Maharashtra. *Journal of Indian School of Political Economy*, 10(1): 1–24.

Deshpande, R. S. 2002. Suicide by Farmers in Karnataka: Agrarian Distress and Possible Alleviatory Steps. *Economic and Political Weekly*, 37(26): 2601–2610.

Deshpande, R. S. and Arora, S. 2010. Editor's Introduction. In R. S. Deshpande and S. Arora (eds.) *Agrarian Crisis and Farmer Suicides*, New Delhi: Sage Publications.

Deshpande, R. S. *et al*. 1992. *State Cooperative Interface: A Study of Sugar Cooperatives of Maharashtra*, Pune: Gokhale Institute of Politics and Economics.

Dev, S. M. 1995. Alleviating Poverty: Maharashtra Employment Guarantee Scheme. *Economic and Political Weekly*, 30(41/42): 2663–2676.

Dev, S. M. and Mungekar, B. L. 1996. Maharashtra Agricultural Development: A Blueprint. *Economic and Political Weekly*, 31(13): A38–A48.

Dev, S. M. and Ranade, A. 2001. Employment Guarantee Scheme and Employment Security. In S. M. Dev *et al.* (eds.) *Social and Economic Security in India*, New Delhi: Institute for Human Development.

Dev, S. M. and Ravi, C. 2007. Poverty and Inequality: All India and States, 1983–2005. *Economic and Political Weekly*, 42(6): 509–521.

Dhanagare, D. N. 1975. Prosperity and Debt in Rural Punjab. *Sociological Bulletin*, 28(1–2): 120–130.

Dhanagare, D. N. 1983. *Peasant Movements in India 1920–1950*, New Delhi: Oxford University Press.

Dhanagare, D. N. 1987. Green Revolution and Social Inequalities in Rural India. *Economic and Political Weekly*, 22(19, 20, and 21): 137–144.

Dhanagare, D. N. 1990. Shetkari Sanghatana: The Farmers' Movement in Maharashtra-Background and Ideology. *Social Action*, 40(4): 347–369.

Dhanagare, D. N. 1994. The Class Character of Politics of Farmers Movement in Maharashtra. *The Journal of Peasant Studies*, 21(3/4): 72–94.

Dhanagare, D. N. 2014. Negative Returns of Ambivalence: Electoral Politics of the Farmers' Movement, 1980–2014. *Economic and Political Weekly*, 49(49): 40–50.

Dhanagare, D. N. 2016a. *Populism and Power: Farmers' Movement in Western India (1980–2014)*, New Delhi: Routledge.

Dhanagare, D. N. 2016b. Declining Credibility of the Neoliberal State and Agrarian Crisis in India: Some Observations. In B. B. Mohanty (ed.) *Critical Perspectives on Agrarian Transition: India in the Global Debate*, New Delhi: Routledge, pp. 164–196.

Diao, X. and Pratt, A. N. 2007. Growth Options and Poverty Reduction in Ethiopia: An Economy-Wide Model Analysis. *Food Policy*, 32(2): 205–228.

Divekar, V. D. 1983. Western India. In D. Kumar (ed.) *The Cambridge Economic History of India, Volume 2: c. 1757–1970*, New York: Cambridge University Press.

Dollar, D. and Kraay, A. 2002. Spreading the Wealth. *Foreign Affairs*, January/February, 120–133.

Downes, A. S. 2012. The Global Economic Crisis and Labour Markets in the Small States of the Caribbean. In P. Utting *et al.* (eds.) *The Global Crisis and Transformative Social Change*, New York: Palgrave Macmillan, pp. 161–176.

Dreze, J. and Khera, R. 2009. The Battle for Employment Guarantee. *Frontline*, 26(1): 3–16.

Dreze, J. and Sen, A. 1989. A Case Study: The Maharashtra Drought of 1970–73. In J. Dreze and A. Sen, *Hunger and Public Action*, Oxford: Clarendon Press, pp. 126–132.

Dubey, A. *et al.* 2006. Surplus Labour, Social Structure and Rural to Urban Migration: Evidence from Indian Data. *The European Journal of Development Research*, 18(1): 86–104.

Durkheim, E. 1933. *The Division of Labour in Society* (translated by G. Simpson), New York: Palgrave Macmillan.

Durkheim, E. 1952. *Suicide: A Study in Sociology*, London: Routledge and Kegan Paul.

Durkheim, E. 1984. *The Division of Labour in Society* (an introduction by L. Coser, translated by W. D. Hall), London: The Macmillan Press Ltd.

Dutt, G. and Ravallion, M. 1992. *Behavioural Responses to Work Fares: Evidence for Rural India*, Washington, DC: The World Bank.

Dutt, R. C. 1904. *The Economic History of India in the Victorian Age, Vol. II: From the Accession of Queen Victoria in 1837 to the Commencement of the Twentieth Century*, London: Kegan Paul.

Duvvury, N. 1989. Women in Agriculture: A Review of Indian Literature. *Economic and Political Weekly*, 24(43): WS96–WS112.

Dyson, T. and Maharatna, A. 1992. Bihar Famine, 1966–67 and Maharashtra Drought, 1970–73: The Demographic Consequences. *Economic and Political Weekly*, 27(26): 1325–1332.

Echeverri-Gent, J. 1988. Guaranteed Employment in an Indian State: The Maharashtra Experience. *Asian Survey*, 28(12): 1294–1310.

Edelman, M. 2003. Transnational Peasant and Farmer Movements and Networks. In H. Anheier *et al.* (eds.) *Global Civil Society*, London: Oxford University Press, pp. 185–220.

Elson, D.2002. Gender Justice, Human Rights and Neoliberal Economic Policie. In M. Molyneux and S. Razavi (eds.) *Gendered Dimensions of Development in Gender Justice, Development and Rights*, London: UNRISD and Oxford University Press.

Engels, F. 1950. The Peasant Question in France and Germany. In *K. Marx and F. Engels: Selected Works*, Vol. 2, London: Lawrence and Wishart.

Epstein, T. S. 1973. *South India: Yesterday, Today and Tomorrow: Mysore Villages Revisited*, London: Palgrave Macmillan.

Erlich, A. 1960. *The Soviet Industrialization Debate, 1924–1928*, Cambridge: Harvard University Press.

Evans, G. 1987. Sources of Peasant Consciousness in South-East Asia: A Survey. *Social History*, 12(2): 193–211.

Farmer, B. H. 1986. Perspectives on the Green Revolution in South Asia. *Modern Asian Studies*, 20(1): 175–199.

Feeny, D. 1983. The Moral or the Rational Peasant? Competing Hypotheses of Collective Action. *Journal of Asian Studies*, 42(4): 769–789.

Frankel, R. F. 1971. *India's Green Revolution: Economic Gains and Political Costs*, Princeton, NJ: Princeton University Press.

Franzini, M. 2006. Social Costs, Social Rights and the Limits of Free Market Capitalism: A Re-Reading of Kapp. In W. Elsner *et al.* (eds.) *Social*

Costs and Public Action in Modern Capitalism: Essays Inspired by Karl William Kapp's Theory of Social Costs, London: Routledge, pp. 56–71.

Friedmann, H. 2006. Focusing on Agriculture: A Comment on Henry Bernstein's Is There an Agrarian Question in the 21st Century?. *Canadian Journal of Development Studies*, 27(4): 461–465.

Fuchs, S. 1972. Land Scarcity and Land Hunger among Some Aboriginal Tribes of Western Central India. In K. S. Singh (ed.) *Tribal Situation In India*, Simla: Indian Institute of Advanced Study, pp. 367–373.

Fukazawa, H. 1972. Rural Servants in the 18th Century Maharashtrian Village: Demiurgic or Jajmani System? *Hitotsubashi Journal of Economics*, 12(2): 14–40.

Fukazawa, H. 1976. Maharashtrian Village Community in the Deccan Riots of 1875. *Hitotsubashi Journal of Economics*, 16(2): 17–28.

Gadre, N. A. and Mahalle, Y. P. 1985. Participation of Female Farm Labour Under Changing Agriculture in Vidarbha. *Indian Journal of Agricultural Economics*, 40(3).

Gaiha, R. 1996. How Dependent Are the Rural Poor on the Employment Guarantee Scheme in India? *The Journal of Development Studies*, 32(5): 669–694.

Ganesh-Kumar, A. *et al.* 2004. Employment Guarantee for Rural India. *Economic and Political Weekly*, 39(51): 5359–5361.

Garcia, Z. *et al.* 2006. *Agriculture, Trade Negotiations and Gender*, Food and Agricultural Organisation (FAO) of United Nations, Rome: Gender and Population Division.

Gare, G. M. 1973. *Socio-Economic Study of the Scheduled Castes in Rural Maharashtra*, Pune: Gokhale Institute of Politics and Economics.

Garikipati, S. 2006. *Feminization of Agricultural Labour and Women's Domestic Status: Evidence from Labour Households in India*, No. 30, Research Paper Series, Great Britain: University of Liverpool Management School.

Garikipati, S. 2009. Landless But Not Assetless: Female Agricultural Labour on the Road to Better Status, Evidence from India. *The Journal of Peasant Studies*, 36(3): 517–545.

Garikipati, S. and Pfaffenzeller, S. 2012. The Gendered Burden of Liberalisation: The Impact of India's Economic Reforms on its Female Agricultural Labour. *Journal of International Development*, 24(7): 841–864.

Ghosh, J. 2013. Women's Work in the India in the Early 21st Century, Conference Paper, India Today: Looking back, looking forward. Accessed from (www.sundarayya.org/sites/default/files/papers/jayati.pdf) on 05.11.2016.

Gibbon, P. *et al.* 1993. *A Blighted Harvest: The World Bank and African Agriculture in the Eighties*, London: James Currey.

Giddens, A. 1966. A Typology of Suicide. *European Journal of Sociology*, 11(2): 276–295.

Giddens, A. 1971. *Capitalism and Modern Social Theory: An Analysis of the Writings of Marx, Durkheim and Max Weber*, Cambridge: Cambridge University Press.

Gill, S. S. 1994. Framers' Movement and Agrarian Change in the Green Revolution Belt of North-West India. *The Journal of Peasant Studies*, 21(3/4): 195–211.

Gilmartin, D. 2003. Water and Waste: Nature, Productivity and Colonialism in the Indus Basin. *Economic and Political Weekly*, 38(48): 5057–5065.

GO 1993. Farmers Movement Fighting for Liberalisation. *Economic and Political Weekly*, 28(50): 2708–2710.

GoM 2013. *Report of the High Level Committee on Balanced Regional Development Issues in Maharashtra, 2013, Planning Department*, Mumbai: Government of Maharashtra.

GoM. 2016. Maharashtra State Seed Scenario. Accessed from ⟨http://www.mahaagri.gov.in/level3PdfDisp.aspx?Id=2&subid=1&sub2id=1&FileName=Seed_G_info.pdf⟩ on 15.09.2016.

Gough, K. 1968. Peasant Resistance and Revolt in South India. *Pacific Affairs*, 14(4): 526–544.

Gough, K. 1981. *Rural Society in Southeast India*, Cambridge: Cambridge University Press.

Government of British India. 1911. *Central Provinces District Gazetteers – Amraoti District Volume A. Descriptive* (edited by S.V. Fitzgerald and A.E. Nelson), Bombay: Caxton Works.

Government of British India. 1927a. *Central Provinces District Gazetteers – Amraoti District (Statistical Tables 1891–1926)*, Nagpur: Government Press.

Government of British India. 1927b. *Central Provinces District Gazetteers – Yeotmal District (Statistical Tables 1891–1926)*, Nagpur: Government Press.

Government of India. 1975. *Census of India 1971: Series- ii Maharashtra Part II-C (i), Social and Cultural Tables*, Mumbai: Directorate of Census Operations.

Government of India. 1983a. *Census of India 1981: Series-12 Maharashtra Scheduled Castes and Scheduled Tribes, Primary Census Abstract*, Mumbai: Directorate of Census Operations.

Government of India. 1991. *Cost of Cultivation of Principal Crops in India, New Delhi: Directorate of Economics and Statistics*, Department of Agriculture and Cooperation.

Government of India. 1993. *Census of India 1991: Series-14 Maharashtra Part II-B (ii) Primary Census Abstract, Scheduled Castes and Scheduled Tribes*, Mumbai: Directorate of Census Operations.

Government of India. 1998. *Reports of the Commission for Agricultural Costs and Prices for the Crops Sown during 1998–99 Season*, New Delhi: Department of Agriculture and Cooperation.

Greenough, P. 1983. Indulgence and Abundance in Asian Peasant Values: A Bengali Case in Point. *Journal of Asian Studies*, 42(4): 831–850.

Griffin, K. 1974. *The Political Economy of Agrarian Change: An Essay on the Green Revolution*, London: Palgrave Macmillan.

Guérin, I. *et al.* 2009. Neobondage, Seasonal Migration and Job Brokers: Cane Cutters in Tamil Nadu. In J. Breman *et al.* (eds.) *India's Unfree Workforce: Old and New Practices of Labour Bondage*, New Delhi: Oxford University Press, pp. 233–258.

Guha, A. 1972. Raw Cotton of Western India: Output, Transportation and Marketing 1750–1850. *The Indian Economic and Social History Review*, 9(1): 1–42.

Guha, R. 1983. *Elementary Aspects of Peasant Insurgency in Colonial India*, New Delhi: Oxford University Press.

Guha, R. 2008. *India after Gandhi: The History of the World's Largest Democracy*, London: Pan Macmillan.

Guha, S. 1985. *The Agrarian Economy of the Bombay Deccan, 1818–1941*, New Delhi: Oxford University Press.

Guha, S. 1987. The Land Market in Upland Maharashtra C 1820–1960–1. *The Indian Economic and Social History Review*, 24(2): 118–144.

Gulati, A. 1998. Indian Agriculture in an Open Economy, Will It Prosper? In J. A. Isher and I. M. D. Little (eds.) *India's Economic Reforms and Development: Essays for Manmohan Singh*, New Delhi: Oxford University Press, pp. 122–146.

Gulati, A. and Kelly, T. 1999. *Trade Liberalisation and Indian Agriculture*, New Delhi: Oxford University Press.

Gulati, L. 1975. Female Work Participation: A Study of Inter-State Differences. *Economic and Political Weekly*, 10(1/2): 35–37, 39–42.

Gupta, A. 2006. Peasants and Global Environmentalism. In N. Haenn and R. Wilk (eds.) *The Environment in Anthropology: A Reader in Ecology, Culture, and Sustainable Living*, New York: NYU Press, pp. 302–324.

Gupta, D. 2005. Whither the Indian Village: Culture and Agriculture in 'Rural' India. *Economic and Political Weekly*, 40(8): 751–758.

Habib, I. 1995. *Essays in Indian History: Towards a Marxist Perception*, New Delhi: Tulika.

Hamilton, C. H. 1939. The Social Effects of Recent Trends in the Mechanization of Agriculture. *Rural Sociology*, 4(1): 3–19.

Hardiman, D. (ed.) 1992. *Peasant Resistance in India, 1858–1914*, New Delhi: Oxford University Press.

Harnetty, P. 1972. *Imperialism and Free Trade: Lancashire and India in the Mid-Nineteenth Century*, Manchester: Manchester University Press.

Harriss, J. 1982. *Capitalism and Peasant Farming: Agrarian Structure and Ideology in Northern Tamil Nadu*, Mumbai: Oxford University Press.

Harriss, J. 1991. The Green Revolution in NortArcot: Economic Trends, Household Mobility and the Politics of an 'Awkward Class'. In P. B. R. Hazell *et al.* (eds.) *The Green Revolution Reconsidered: The Impact of High-Yielding Rice Varieties in South India*, New Delhi: Oxford University Press.

Harvey, D. 2007a. *A Brief History of Neoliberalism*, New York: Oxford University Press.

Harvey, D. 2007b. Neoliberalism as Creative Destruction. *The Annals of the American Academy*, 610 (March): 22–44.

Hatekar, N. 1996. Information and Incentives: Pringle's Ricardian Experiment in the Nineteenth Century Deccan Countryside. *The Indian Economic and Social History Review*, 33(4): 438–453.

Hatekar, N. 2003. *A Study of Sugar Factories in Western Maharashtra*, Nashik: Centre of Development Research and Documentation.

Hayami, Y. and Ruttan, V. W. 1985. *Agricultural Development: An International Perspective*, Baltimore: Johns Hopkins University Press.

Hazell, P. B. R. and Ramasamy, C. (eds.) 1991. *The Green Revolution Reconsidered: The Impact of High-Yielding Rice Varieties in South India*, Baltimore and London: Johns Hopkins University Press.

Hazell, P. B. R. *et al.* 1991. Economic Changes among Village Households. In P. B. R. Hazell *et al.* (eds.) *The Green Revolution Reconsidered: The Impact of High-Yielding Rice Varieties in South India*, New Delhi: Oxford University Press.

Hefner, R. 1990. *The Political Economy of Mountain Java: An Interpretive History*, Berkeley, CA and Los Angeles: University of California Press.

Hensman, R. 2015. Alternatives to Neoliberalism in India. In L. Pradella and T. Marois (eds.) *Polarising Development: Alternatives to Neoliberalism and the Crisis*, London: Pluto Press, pp. 203–213.

Heston, A. W. 1973. Official Yields Per Acre in India, 1886–1947: Some Questions of Interpretation. *Indian Economic & Social History Review*, 10(4): 303–332.

Heyer, J. 2016. Loosening Ties of Patriarchy in Agrarian Transition in Tamil Nadu. In B. B. Mohanty (ed.) *Critical Perspectives on Agrarian Transition: India in the Global Debate*, New York: Routledge.

Ho, P. *et al.* (eds.) 2003. Rural Development in Transitional China: The New Agriculture. *A Special Issue of the Journal of Peasant Studies*, 30(3 and 4).

Hobsbawm, E. 1969. *Bandits*, London: Wiedenfeld and Nicholson.

Hobsbawm, E. 1994. *Age of Extremes: The Short Twentieth Century 1914–1991*, London: Michael Joseph.

Hopcroft, R. L. 1994. The Social Origin of Agrarian Change in Late Medieval England. *American Journal of Sociology*, 99(6): 1559–1595.

Huang, P.C. C. 1985. *The Peasant Economy and Social Change in North China*, Stanford, CA: Stanford University Press.

Huang, Y. 2015. Can Capitalist Farms Defeat Family Farms? The Dynamics of Capitalist Accumulation in Shrimp Aquaculture in South China. *Journal of Agrarian Change*, 15(3)(July): 392–412.

Hurd, J. 1975. Railways and the Expansion of Markets in India, 1861–1921. *Explorations in Economic History*, 12(3): 263–288.

Irz, X. *et al.* 2001. Agricultural Productivity Growth and Poverty Alleviation. *Development Policy Review*, 19(4): 449–466.

Iversen, V. *et al.* 2013. On the Colonial Origins of Agricultural Development in India: A Re-Examination of Banerjee and Iyer, 'History, Institutions and Economic Performance'. *The Journal of Development Studies*, 49(12): 1631–1646.

Iyer, K.G. and Manick, M. S. 2000. *Indebtedness, Impoverishment and Suicides in Rural Punjab*, New Delhi: India Publishers and Distributors.

Jackson, C. 2003. Gender Analysis of Land: Beyond Land Rights for Women? *Journal of Agrarian Change*, 3(4): 453–480.

Jackson, C. and Rao, N. 2009. Gender Inequality and Agrarian Change in Liberalising India. In S. Razavi (ed.) *The Gendered Impacts of Liberalization: Towards 'Embedded Liberalism'?*, London: Routledge, pp. 63–98.

Jadhav, V. 2006. Elite Politics and Maharashtra's Employment Guarantee Scheme. *Economic and Political Weekly*, 41(50): 5157–5162.

Jaggar, A. 2009. The Philosophical Challenges of Global Gender Justice. *Philosophical Topics*, 37(2): 1–15.

Jayadev, A. *et al.* 2007. Patterns of Wealth Disparities in India during the Liberalisation Era. *Economic and Political Weekly*, 42(38): 3853–3863.

Jenkins, R. 1999. *Democratic Politics and Economic Reform in India*, New York: Cambridge University Press.

Jesim, P. 2006. *Some Features of Migration and Labour Mobility in the Leather Accessories Manufacture In India: A Study of the Informal Sector Industry in Dharavi, Mumbai*, Working Paper 2006/06, New Delhi: Institute for Studies in Industrial Development.

Jodhka, S. S. 2000. Prejudice' without 'Pollution'? Scheduled Castes in Contemporary Punjab. *Journal of Indian School of Political Economy*, 12(3/4): 381–404.

Joekes, S. 1999. A Gender-Analytical Perspective on Trade and Sustainable Development. In *Trade, Sustainable Development and Gender*, New York and Geneva: United Nations conference on Trade and Development (UNCTAD), pp. 33–59 (No. 47).

Johnson, H. 2004. Subsistence and Control: The Persistence of the Peasantry in the Developing World. *Undercurrent*, 1(1): 54.

Johnson, K. 2005. Globalisation at the Crossroads of Tradition and Modernity in Rural India. *Sociological Bulletin*, 54(1): 40–58.

Johnston, B. F. and Mellor, J. W. 1961. The Role of Agriculture in Economic Development. *American Economic Review*, 51(4): 566–593.

Jose, A. V. 1978. Real Wages, Employment and Income of Agricultural Labourers. *Economic and Political Weekly*, 13(12): 16–20.

Joseph, S. 2006. Power of the People: Political Mobilisation and Guaranteed Employment. *Economic and Political Weekly*, 41(50): 5149–5156.

Joshi, C. K. and Alshi, M. R. 1985. Impact of High-Yielding Varieties on Employment Potential of Female Labour: A Study in Akola District (Maharashtra). *Indian Journal of Agricultural Economics*, 40(3): 230–234.

Joshi, S. 2010. *Down to Earth*, New Delhi: Academic Foundation.

Kajale, J. and Shroff, S. 2011. *Impact of NREGA on Wage Rates, Food Security and Rural Urban Migration in Maharashtra*, Unpublished Report, Pune: Agro-Economic Research Centre, Gokhale Institute of Politics and Economics.

Kajale, J. and Shroff, S. 2012. Employment and Asset Creation under NREGA in Maharashtra: Realities and Lessons. *Journal of Agricultural Development & Policy*, 22(1): 1–10.

Kalamkar, S. S. 2011. Patterns and Determinants of Agricultural Growth in Maharashtra. *Artha Vijayana*, 50(2): 156–181.

Kale, S. S. 2014. *Electrifying India: Regional Political Economies of Development*, Stanford, CA: Stanford University Press.

Kamat, A. R. 1980. Politico-Economic Developments in Maharashtra: A Review of the Post-Independence Period. *Economic and Political Weekly*, 15(39 and 40): 1627–1630, 1669–1678.

Kapp, K. W. 1950/1971. *Social Cost of Private Enterprise*, New York: Schocken Books.

Kar, S. and Saktivel, S. 2007. Reforms and Regional Inequality in India. *Economic and Political Weekly*, 42(7): 69–77.

Karan, A. 2003. Changing Patterns of Migration from Rural Bihar. In G. Iyer (ed.) *Migrant Labour and Human Rights in India*, New Delhi: Kanishka Publishers, pp. 102–139.

Kaviraj, S. 2010. *The Trajectories of the Indian State: Politics and Ideas*, New Delhi: Permanent Black.

Kay, C. 2002. Chile's Neoliberal Agrarian Transformation and the Peasantry. *Journal of Agrarian Change*, 2(4): 464–501.

Kay, C. 2004. Rural Livelihoods and Peasant Futures. In R. N. Gwynne and C. Kay (eds.) *Latin America Transformed: Globalization and Modernity*, 2nd Edition, London: Arnold.

Keatinge, G. 1912. *Rural Economy in the Bombay Deccan*, Mumbai: Longmans, Green & Co.

Keshri, K. and Bhagat, R. 2010. Temporary and Seasonal Migration in India. *Genus*, 66(3): 25–45.

Keshri, K. and Bhagat, R. 2012. Temporary and Seasonal Migration: Regional Pattern, Characteristics and Associated Factors. *Economic and Political Weekly*, 47(4): 81–88.

Keyes, C. F. 1983. Peasant Strategies in Asian Societies: Moral and Rational Economic Approaches: A Symposium. *The Journal of Asian Studies*, 42(4): 753–768.

Klassen, J. 2015. Hegemony in Question: US Primacy, Multi-Polarity and Global Resistance. In L. Pradella and T. Marois (eds.) *Polarising Development: Alternatives to Neoliberalism and the Crisis*, London: Pluto Press, pp. 74–85.

Krishnaraj, M. *et al.* 2004. Does EGS Require Restructuring for Poverty Alleviation and Gender Equality? II: Gender Concerns and Issues for Restructuring. *Economic and Political Weekly*, 39(17): 1741–1747.

Kulkarni, S. D.1974. Alienation of Adivasi Lands, Government Not Serious. *Economic and Political Weekly*, 9(35): 1469–1471.

Kulkarni, S. D. *et al.* 2008. *Women and Land Rights in Maharashtra: Exploring the Facilitating and Constraining Factors in Achieving Resource Rights*, Pune: Society for Promoting Participative Ecosystem Management (SOPPECOM).

Kumar, D. 1965. *Land and Caste in South India*, New York: Cambridge University Press.

Kumar, N. 2016. *Unraveling Farmer Suicides in India: Egoism and Masculinity in Peasant Life*, London: Oxford University Press.

Kumar, P. and Chakraborty, D. 2016. *MGNREGA: Employment, Wages and Migration in Rural India*, New Delhi: Routledge.

Kumar, R. 1965. The Deccan Riot of 1875. *The Journal of Asian Studies*, 24(4): 613–635.

Kumar, R. 1968. *Western India in the Nineteenth Century: A Study in the Social History of Maharashtra*, London: Routledge and Kegan Paul.

Kumari, R. 1989. *Brides Are Not for Burning: Dowry Victims in India*, New Delhi: Radiant Publishers.

Kundu, A. and Sarangi, N. 2007. Migration, Employment Status and Poverty: An Analysis across Urban Centres. *Economic and Political Weekly*, 24(4): 299–306.

Kundu, A. and Saraswati, L. R. 2012. Migration and Exclusionary Urbanisation in India. *Economic and Political Weekly*, 47(26–27): 219–227.

Kurtz, M. J. 2000. Understanding Peasant Revolution: From Concept to Theory and Case. *Theory and Society*, 29(1): 93–124.

Kurtz, M. J. 2004. *Free Market Democracy and the Chilean and Mexican Countryside*, New York: Cambridge University Press.

Ladejinsky, W. 1973. Drought in Maharashtra (Not in a Hundred Years). *Economic and Political Weekly*, 8(7): 383–396.

Lal, D. 1976. Agricultural Growth, Real Wages and the Rural Poor in India. *Economic and Political Weekly*, 12(26): 47–61.

Lastarria-Cornhiel, S. 2006. *Feminization of Agriculture: Trends and Driving Forces*, Background Paper for the World Development Report 2008. Accessed from ⟨http://siteresources.worldbank.org/INTWDRS/

Resources/477365-1327599046334/8394679-1327599874257/
LastarriaCornhiel_FeminizationOfAgri.pdf) on 11. 10. 2016.

Lele, J. 1981. *Elite Pluralism and Class Rule: Political Development in Maharashtra, India*, Toronto: University of Toronto Press.

Lele, J. 1990. Caste, Class and Dominance: Political Mobilization in Maharashtra. In F. R. Frankel and M. S. Rao (eds.) *Dominance and State Power in Modern India: Decline of a Social Order*, Vol. 2, New Delhi: Oxford University Press, pp. 115–211.

Lenin, V. I. 1964. *The Development of Capitalism in Russia*, Moscow: Progress Publisher.

Lenneberg, C. 1988. Sharad Joshi and the Farmers: The Middle Peasant Lives! *Pacific Affairs*, 61(3): 446–464.

Levien, M. 2015. From Primitive Accumulation to Regimes of Dispossession: Six Theses on India's Land Question. *Economic and Political Weekly*, 50(22): 146–156.

Lewis, A. 1954. Economic Development with Unlimited Supplies of Labour, originally published by The Manchester School (reprint in A.N. Agarwala and S.P. Singh (eds.) 1990. *The Economics of Underdevelopment: A Series of Articles and Papers*, New Delhi: Oxford University Press, pp. 400–449).

Li, T. M. 2009. To Make Live or Let Die? Rural Dispossession and Protection of Surplus Populations. *Antipode*, 41(S1): 66–93.

Li, T. M. 2014. *Land's End: Capitalist Relations on an Indigenous Frontier*, London: Duke University Press.

Lieberson, J. 1981. The Silent Majority. *The New York Review of Books*, October 22.

Lindberg, S. 1994. New Farmers' Movement in India as Structural Response and Collective Identity Formation: The Cases of the Shetkari Sanghatana and the BKU. *The Journal of Peasant Studies*, 21(3/4): 95–125.

Little, D. 1989. *Understanding Peasant China: Case Studies in the Philosophy of Social Science*, New Haven, CT and London: Yale University Press.

Lockwood, B. *et al.* 1971. *The High Yielding Varieties Programme in India, Part 1, Programme Evaluation Organization, Planning Commission of India and Department of Economics*, Australian National University.

Lockwood, D. 1982. Fatalism: Durkheim's Hidden Theory of Order. In A. Giddens and G. Mackenzie (eds.) *Social Class and the Division of Labour: Essays in Honour of Ilya Neustadt*, Cambridge: Cambridge University Press.

Longford, E.1971. *Wellington: The Years of the Sword*, London: World Books.

Ludden, D. 1999. *The New Cambridge History of India (IV. 4): An Agrarian History of South Asia*, Cambridge: Cambridge University Press.

MacIntyre, A. 1971. *Against the Self-Images of the Age: Essays on Ideology and Philosophy*, London: Gerald Duckworth.

Malinowski, B.1926. *Crime and Custom in Savage Society*, New York: Harcourt Brace.

Mann, H. H. 1917. *Land and Labour in a Deccan Village*, London: Oxford University Press.

Mann, H. H. 1967. A Deccan Village Under the Peshwas. In H. H. Mann, *The Social Framework of Agriculture: India Middle East England* (Edited by Daniel Thorner), Bombay: Vora and Co., Publishers Private Ltd

Marius-Gnanou, K. 2008. Debt Bondage, Seasonal Migration and Alternative Issues: Lessons from Tamil Nadu (India). *Autrepart*, (46): 127–142.

Marks, S. R. 1975. Durkheim's Theory of Anomie. *American Journal of Sociology*, 80(2): 329–363.

Marois, T. and Pradella, L. 2015. Polarising Development: Introducing Alternatives to Neoliberalism and the Crisis. In L. Pradella and T. Marois (eds.) *Polarising Development: Alternatives to Neoliberalism and the Crisis*, London: Pluto Press, pp. 1–12.

Marx, K. 1867/2007. *Capital: A Critique of Political Economy*, Vol. 1 and 3, New York: Cosimo Classics.

McAlpin, M. B. 1974. Railroads, Prices, and Peasant Rationality: India 1860–1900. *The Journal of Economic History*, 34(3): 662–684.

McGinn, P. 2009. *Capital, 'Development' and Canal Irrigation in Colonial India*, Working Paper 209, Bangalore: Institute for Social and Economic Change.

McMillan, R. T. 1949. *Social Aspects of Farm Mechanization in Oklahoma: Stillwater*, Oklahoma: Oklahoma Agricultural Experiment Station, Bulletin No. B-339.

Mellor, J. W. 1976. *The New Economics of Growth: A Strategy for India and the Developing World*, New York: Cornell University Press.

Mencher, J. P. 1978. *Agriculture and Social Structure in Tamil Nadu: Past Origins, Present Transformations and Future Prospects*, New Delhi: Allied Publishers Private Ltd.

Mencher, J. P. 1980. The Lessons and Non-Lessons of Kerala: Agricultural Labourers and Poverty. *Economic and Political Weekly*, 15(41/43): 1781–1802.

Mencher, J. P. and Saradamoni, K. 1982. Muddy Feet, Dirty Hands: Rice Production and Female Agricultural Labour. *Economic and Political Weekly*, 17(52): A149–A153, A155–A167.

Mies, M. 1976a. A Peasants' Movement in Maharashtra Its Development and Its Perspectives. *Journal of Contemporary Asia*, 6(2): 172–185.

Mies, M. 1976b. The Shahada Movement: A Peasant Movement in Maharashtra (India): Its Development and Its Perspectives. *The Journal of Peasant Studies*, 3(4): 472–482.

Mies, M. 1986. *Indian Women in Subsistence and Agricultural Labour*, New Delhi: Vistaar Publications.

Migdal, J. S. 1974. *Peasants, Politics and Revolution: Pressures toward Political and Social Change in the Third World*, Princeton, NJ: Princeton University Press.

Migdal, J. S. 2001. *State in Society Studying How States and Societies Transform and Constitute One Another*, Cambridge: Cambridge University Press.

Miley, J. D. 1972. Structural Change and the Durkheimian Legacy: A Macro-Social Analysis of Suicide Rates. *American Journal of Sociology*, 78(3): 657–673.

Mishra, D. K. (ed.) 2016. *Internal Migration in Contemporary India*, New Delhi: Sage Publications.

Mishra, S. 2006a. *Suicide of Farmers in Maharashtra*, Mumbai: Indira Gandhi Institute of Development Research.

Mishra, S. 2006b. Farmers' Suicides in Maharashtra. *Economic and Political Weekly*, 41(16): 1538–1545.

Mishra, S. 2009. Agrarian Distress and Farmers' Suicides in Maharashtra. In D. N. Reddy and S. Mishra (eds.) *Agrarian Crisis in India*, New Delhi: Oxford University Press.

Mishra, S. and Panda, M. 2006. *Growth and Poverty in Maharashtra*, Working Paper, Mumbai: Indira Gandhi Institute of Development Research.

Misra, S. B. 2014. Growth and Structure of Rural Non-Farm Employment in Maharashtra: Reflections from NSS Data in the Post Reform Period. *Procedia Economics and Finance*, 11: 137–151.

Mitra, S. K. 1992. *Power, Protest and Participation: Local Elites and the Politics of Development in India*, London: Routledge.

Mitra, S. K. and Shroff, S. 2007. Farmers' Suicides in Maharashtra. *Economic and Political Weekly*, 42(49): 73–77.

Moghadam, V. 2005. *Globalizing Women: Transnational Feminist Networks*, Baltimore: Johns Hopkins University Press.

Mohan, R. 2004. Agricultural Credit in India: Status, Issues and Future Agenda. *Reserve Bank of India Bulletin*, (November): 993–1007.

Mohanty, B. B. 1997. State and Tribal Relationship in Orissa. *Indian Anthropologist*, 27(1): 1–17.

Mohanty, B. B. 1999. Agricultural Modernisation and Social Inequality, Case Study of Satara District. *Economic and Political Weekly*, 34(26): A50–A61.

Mohanty, B. B. 2000a. *Agricultural Modernisation and the Trend of Social Inequality in Rural Maharashtra*, Gokhale Institute Mimeograph Series No. 51, Pune: Gokhale Institute of Politics and Economics.

Mohanty, B. B. 2000b. Agricultural Modernisation in Rural Orissa: Land Transfer and Ownership Pattern. *Sociological Bulletin*, 49(1/2): 63–90.

Mohanty, B. B. 2001a. Suicides of Farmers in Maharashtra: A Socio-Economic Analysis. *Review of Development and Change*, 6(2): 146–189.

Mohanty, B. B. 2001b. Land Distribution among Scheduled Castes and Tribes. *Economic and Political Weekly*, 36(40): 3857–3868.

Mohanty, B. B. 2001c. *Land Holding and Use Pattern among Scheduled Castes and Scheduled Tribes in Maharashtra*, Pune: Gokhale Institute of Politics and Economics.

Mohanty, B. B. 2005. 'We Are Like the Living Dead': Farmer Suicides in Maharashtra, Western India. *Journal of Peasant Studies*, 32(2): 243–276.

Mohanty, B. B. 2009. Regional Disparity in Agricultural Development of Maharashtra. *Economic and Political Weekly*, 44(6): 63–69.

Mohanty, B. B. 2012. Agrarian Studies in Indian Sociology. In B. B. Mohanty (ed.) *Agrarian Change and Mobilisation*, New Delhi: Sage Publications, pp. xxv–lxiv.

Mohanty, B. B. 2013. Farmers Suicide in India: Durkheim's Types. *Economic and Political Weekly*, 48(21): 45–54.

Mohanty, B. B. 2016. Introduction: Agrarian Transition: From Classic to Current Debates. In B. B. Mohanty (ed.) *Critical Perspectives on Agrarian Transition: India in the Global Debate*, New Delhi: Routledge, pp. 1–40.

Mohanty, B. B. (ed.) 2016. *Critical Perspectives on Agrarian Transition: India in the Global Debate*, New York: Routledge.

Mohanty, B. B. and Lenka, P. K. 2016. Neoliberal Reforms, Agrarian Capitalism and the Peasantry. In B. B. Mohanty (ed.) *Critical Perspectives on Agrarian Transition: India in the Global Debate*, New Delhi: Routledge, pp. 164–196.

Mohanty, B. B. and Shroff, S. 2003. *Market Imperfections and Farmers' Distress in Maharashtra*, Pune: Gokhale Institute of Politics and Economics.

Mohanty, B. B. and Shroff, S. 2004. Farmers' Suicides in Maharashtra. *Economic and Political Weekly*, 39(52): 5599–5606.

Moise, E. 1982. The Moral Economy Dispute. *Bulletin of Concerned Asian Scholars*, 14(1): 72–77.

Moore, B., Jr. 1966. *Social Origins of Dictatorship and Democracy: Lord and Peasant in the Making of the Modern World*, Boston: Beacon Press.

Morrison, K. 1995. *Marx, Durkheim, Weber, and Foundations of Modern Social Thought*, London: Sage Publications.

Mosse, D. *et al.* 1997. *Seasonal Labour Migration in Tribal (Bhil) Western India, Report to DFID-India, New Delhi*, KRIBP Working Paper, Swansea: Centre for Development Studies, University of Wales.

Mukherji, P. N. 1979. Naxalite Movement and the Peasant Revolt in North. In M.S.A. Rao (ed.) *Social Movements in India*, New Delhi: Manohar Publication.

Munck, G. L. and Snyder, R. 2007. *Passion, Craft, and Method in Comparative Politics*, Baltimore: Johns Hopkins University Press.

Murdia, R. 1975. Land Allotment and Land Alienation Policies and Programmes for Scheduled Castes and Tribes. *Economic and Political Weekly*, 10(32): 1201–1214.

Nadkarni, A. S. 1988. *Impact of Agricultural Developmental Programmes on Scheduled Castes and Scheduled Tribes Population in Maharashtra*, Pune: Gokhale Institute of Politics and Economics.

Nadkarni, M. V. 1987. *Farmers' Movement in India*, New Delhi: Allied Publishers Private Ltd.

Nagaraj, K. 2008. *Farmers' Suicides in India: Magnitudes, Trends and Spatial Patterns*, Madras: Madras Institute of Development Studies.

Nand, B. 1980. *The Deccan Peasant Uprising-1875*, Unpublished M. Phil Thesis, Jawaharlal University, New Delhi.

Nandy, D. 2013. *Child Rights Situation Analysis: Children of Families Engaged in Sugarcane Farming in Maharashtra*, Pune: Save the Children.

Nanekar, K. R. 1968. *Land Reforms in Vidarbha Region: An Inquiry into the Implementation of Land Reforms in the Vidarbha Region*, Calcutta: Oxford and IBH Publishing Company.

Narain, D. 1977. Growth and Productivity in Indian Agriculture. *Indian Journal of Agricultural Economics*, 32(1): 1.

Narayanamoorthy, A. and Deshpande, R. S. 2005. *Where Water Seeps! Towards a New Phase in India's Irrigation Reforms*, New Delhi: Academic Foundation.

Narula, S. and Macwan, M. 2001. 'Untouchability': The Economic Exclusion of the Dalits in India, The International Council on Human Rights Policy International Seminar on Racism: Economic Roots of Discrimination, Geneva, 24–25 January.

Navale, J. *et al.* 2014. Stone Quarry Workers of Wagholi: A Study Report Prepared for 'Beyond the Resource Curse Charting a Path to Sustainable Livelihood for Mineral-Dependent Communities', ARC Discovery Project led by Kuntala Lahiri-Dutt, Canberra: Australian National University. Accessed from (http://asmasiapacific.org/wp-content/uploads/2015/02/Santulan-Study-Report.compressed.pdf) on 18. 08. 2016.

Nayar, B. R. 1989. *India's Mixed Economy: The Role of Ideology and Interest in Is Development*, Mumbai: Popular Prakashan.

Nayyar, D. 1978. Industrial Development in India: Some Reflections on Growth and Stagnation. *Economic and Political Weekly*, 13(31–33): 1265–1278.

Nayyar, R. 1987. Female Participation Rates in Rural India. *Economic and Political Weekly*, 22(51): 2207–2209+2211–2216.

Neelamma, H. 2010. *Seasonal Migration of Labourers in Maharashtra*, Pune: Gokhale Institute of Politics and Economics.

Nelson-Richards, M. 1988. *Beyond the Sociology of Agrarian Transformation: Economy and Sociology in Zambia, Nepal and Zanzibar*, Vol. 20, Brill.

Nirmala, A. K. 2003. *Market Imperfections and Farmers' Distress in Andhra Pradesh*, Visakhapatnam: Andhra University.

Nurkse, R. 1953. *Problems of Capital Formation in Underdeveloped Countries*, New York: Oxford University Press.

O'Hanlon, R. 1985. *Caste, Conflict and Ideology: Mahatma Jotirao Phule and Low Caste Protest in Nineteenth Century Western India*, Cambridge: Cambridge University Press.

Omvedt, G. 1975. Rural Origins of Women's Liberation in India, *Social Scientist*, 4(4/5), Special Number on Women (November–December): 40–54.

Omvedt, G. 1976. *Cultural Revolt in a Colonial Society the Non-Brahmin Movement in Western India: 1873 to 1930*, Mumbai: Scientific Socialist Education Trust.

Omvedt, G. 1980. Cane Farmers' Movement. *Economic and Political Weekly*, 15(49): 2041–2042.

Omvedt, G. 1993. *Reinventing Revolution: New Social Movements and the Socialist Tradition in India*, New York: M. E. Sharpe.

Omvedt, G. 1994a. *Dalits and the Democratic Revolution, Dr. Ambedkar and the Dalit Movement in Colonial India*, New Delhi: Sage Publications.

Omvedt, G.1994b. 'We Want the Return for Our Sweat': The New Peasant Movement in India and the Formation of a National Agricultural Policy. *The Journal of Peasant Studies*, 21(3–4): 126–164.

Omvedt, G. 1999. Dalit Suicides? *The Hindu*, April 24.

Oommen, T. K.1975. Agrarian Legislations and Movements as Sources of Change. *Economic and Political Weekly*, 10(40): 1581–1584.

Oughton, E. 1982. The Maharashtra Droughts of 1970–73: An Analysis of Scarcity. *Oxford Bulletin of Economics and Statistics*, 44(3): 169–197.

Overton, M. 1996. *Agricultural Revolution in England: The Transformation of the Agrarian Economy 1500–1850*, Cambridge: Cambridge University Press.

Oya, C. 2005. Stick and Carrots for Farmers in Developing Countries: Agrarian Neoliberalism in Theory and Practice. In A. Saad-Filho and D. Johnston (eds.) *Neoliberalism: A Critical Reader*, London: Pluto-Press, pp. 127–134.

Oya, C. 2010. Agro-Pessimism, Capitalism and Agrarian Change: Trajectories and Contradictions in Sub-Saharan Africa. In V. Padayachee (ed.) *Political Economy of Africa*, London: Routledge, pp. 85–109.

Palshikar, S. and Deshpande, R. 2003. Maharashtra: Challenges before the Congress System'. *Journal of the Indian School of Political Economy*, Special Number (January–June): 97–122.

Panagariya, A. 2005. Agricultural Liberalisation and the Least Developed Countries: Six Fallacies. *The World Economy*, 28(9): 1277–1299.

Panagariya, A. 2008. *India: The Emerging Giant*, New York: Oxford University Press.

Parthasarathy, G. 1970. *Green Revolution and Weaker Sections*, Mumbai: Thacker and Co.

Parthasarathy, G. and Shameem 1998. Suicides of Cotton Farmers in Andhra Pradesh: An Exploratory Study. *Economic and Political Weekly*, 33(13): 720–726.

Patel, M. L. 1974. *Changing Land Problems of Tribal India*, Bhopal: Progress Publishers.

Patnaik, U. 1996. Export-Oriented Agriculture and Food Security in Developing Countries and India. *Economic and Political Weekly*, 31(35/37): 2429–2449.

Patnaik, U. 2007. Neoliberalism and Rural Poverty in India. *Economic and Political Weekly*, 42(30): 3132–3150.

Patnaik, U. 2013. Poverty Trends in India 2004–05 to 2009–10: Updating Poverty Estimates and Comparing Official Figures. *Economic and Political Weekly*, 48(40): 43–58.

Patnaik, U. 2016. Capitalist Trajectories of Global Interdependence and Welfare Outcomes: The Lessons of History for the Present. In B. B. Mohanty (ed.) *Critical Perspectives on Agrarian Transition: India in the Global Debate*, London: Routledge.

Patnaik, U. and Moyo, S. 2011. *The Agrarian Question in the Neoliberal Era: Primitive Accumulation and the Peasantry*, Oxford: Pambazuka Press.

Pauw, K. and Thurlow, J. 2011. Agricultural Growth, Poverty, and Nutrition in Tanzania. *Food Policy*, 36(6): 795–804.

Peers, D. M. 2013. *India Under Colonial Rule: 1700–1885*, London: Routledge.

Peletz, M.G. 1983. Moral and Political Economies in South-East Asia. *Comparative Studies in Society and History*, 25(4): 731–739.

Perkins, J. H. 1997. *Geopolitics and the Green Revolution: Wheat, Genes, and the Cold War*, New York: Oxford University Press.

Petras, J. and Veltmeyer, H. 2011. *Social Movements in Latin America: Neoliberalism and Popular Resistance*, New York: Palgrave Macmillan.

Phansalkar, S. J. 2005. Political Economy of Irrigation Development in Vidarbha. *Journal of Indian School of Political Economy*, 17(4): 66–96.

Polyani, K. 1944. *The Great Transformation: The Political and Economic Origins of Our Time*, New York: Rinehart.

Popkin, S. L. 1979. *The Rational Peasant: The Political Economy of Rural Society in Vietnam*, Berkeley, CA: University of California Press.

Powell, E. H. 1958. Occupation, Status and Suicide: Toward a Redefinition of Anomie. *American Sociological Review*, 23(2): 131–139.

Prasad, C. S. 1999. Suicide Deaths and Quality of Indian Cotton: Perspectives from History of Technology and Khadi Movement. *Economic and Political Weekly*, 34(5): PE12–PE21.

Prügl, E. *et al.* 2013. *Gender and Agriculture after Neoliberalism?* Report of a workshop organized by the United Nations, Research Institute for Social Development (UNRISD) and the Graduate Institute of International and Development Studies (IHEID), 19–20 July 2012, Villa Rigot, Geneva.

Punalekar, S.P. 1998. Growth, Inequities and Tensions: A Case Study of Sangli District, Maharashtra. In K.L. Sharma (ed.) *Caste and Class in India*, New Delhi: Rawat Publications.

Radhakrishnan, P. 1989. *Peasant Struggles: Land Reforms and Social Change in Malabar 1836–1982*, New Delhi: Sage Publications.

Raikes, P. 2000. Modernization and Adjustment in African Peasant Agriculture. In D. Bryceson *et al.* (eds.) *Disappearing Peasantries? Rural Labour in Africa, Asia and Latin America*, London: ITDG Publishing, pp. 64–80.

Rajan, S. I. (ed.) 2011. *India Migration Report 2011: Migration, Identities and Conflict*, London: Routledge.

Rajasekar, D. 1988. *Land Transfers and Family Partitioning: A Historical Study of an Andhra Village*, Oxford: IBTT Publishing Co, New Delhi: Centre for Development Studies, Thiruvananthapuram.

Rajasekaran, N. 1996. *Trends in Operational Holding in Maharashtra: An Analysis of Determinants*, Pune: Gokhale Institute of Politics and Economics.

Rajasekaran, N. 1998. Land Reforms in Maharashtra: A Regional Analysis, *Review of Development and Change*, 3(2): 238–263.

Rakowski, C. A. 2000. Obstacles and Opportunities to Women's Empowerment Under Neoliberal Reform. In R. L. Harris and M. J. Seid (eds.) *Critical Perspectives on Globalization and Neoliberalism in the Developing Countries*, Leiden, Netherlands: Brill.

Randhawa, M. S. 1983. *A History of Agriculture in India, Volume III, 1757–1947*, New Delhi: Indian Council of Agricultural Research.

Rao, B. J. 1987. *Land Alienation in Tribal Areas of Andhra Pradesh*, Warangal: Kakatiya University.

Rao, B. J. 1988. Struggling for Production Conditions and Producing Conditions for Emancipations: Women and Water in Rural Maharashtra. *Capitalism Nature Socialism*, 1(2): 65–82.

Rao, C. H. H. 1980. *Technological Change and Distribution of Gains in Indian Agriculture*, New Delhi: Palgrave Macmillan.

Rao, C. H. H. 1985. Changes in Rural Poverty in India: Implications for Agricultural Growth. *IASSI Quarterly Newsletter*, 4(3–4): 7–14.

Rao, N. 2005. Women's Rights to Land and Assets: Experience of Mainstreaming Gender in Development Projects. *Economic and Political Weekly*, 40(44/45): 4701–4708.

Rao, N. 2008. *'Good Women Do Not Inherit Land': Politics of Land and Gender in India*, New Delhi: Social Science Press and Orient Blackswan.

Rao, P. V. 2009. New Insights into the Debates on Rural Indebtedness in 19th Century Deccan. *Economic and Political Weekly*, 44(4): 55–61.

Rao, S. 2011. Work and Empowerment: Women and Agriculture in South India. *The Journal of Development Studies*, 47(2): 294–315.

Rao, V.M. 1972. Land Transfers in Rural Communities: Some Findings in a Ryotwari Region. *Economic and Political Weekly*, 7(40): A133–A144.

Rao, V.M. and Gopalappa, D. V. 2004. Agricultural Growth and Farmer Distress: Tentative Perspectives from Karnataka. *Economic and Political Weekly*, 39(2): 5591–5598.

Raper, A. 1946. The Role of Agricultural Technology in Southern Social Change. *Social Forces*, 25(1): 21–30.

Rath, N. and Mitra, A. K. 1989. Economics of Irrigation in Water Scarce Regions: A Study of Maharashtra. *ArthaVijnana*, 31(1).

Rawal, V. 2013. Changes in the Distribution of Operational Holdings in Rural India: A Study of National Sample Survey Data. *Review of Agrarian Studies*, 3(2): 73–104.

Razavi, S. 2002. Introduction. In S. Razavi (ed.) *Shifting Burdens: Gender and Agrarian Change under Neoliberalism*, USA: Kumarian Press.

Razavi, S. 2003. Introduction: Agrarian Change, Gender and Land Rights. *Journal of Agrarian Change*, 3(1–2): 2–32.

RDRC 1876. *Deccan Riots Commission: Appendix B: Action of the Law and the Civil Courts on the Agricultural Debtors: Preliminary Digest of Evidence*, Mumbai: Government Central Press.

RDRC 1878. *Report of the Commission Appointed in India to Inquire into the Causes of the Riots Which Took Place in the Year 1875, in the Poona and Ahmadnagar Districts of the Bombay Presidency*, London: George Edward Eyre and William Spottiswoode.

Reddy, D. N. 1990. *Rural Migrant Labour in Andhra Pradesh*, Report Submitted to the National Commission on Rural Labour, Government of India.

Reddy, D. N. and Mishra, S. 2009. Agriculture in the Reform Regime. In D. N. Reddy and S. Mishra (eds.) *Agrarian Crisis in India*, New Delhi: Oxford University Press.

Reddy, D. N. and Mishra, S. (eds.) 2009. *Agrarian Crisis in India*, New Delhi: Oxford University Press.

Redfield, R. 1960. *The Little Community and Peasant Society and Culture*, Chicago: Chicago University Press.

Riesebrodt, M. 1986. From Patriarchalism to Capitalism: The Theoretical Context of Max Weber's Agrarian Studies (1892–93). *Economy and Society*, 15(4): 476–502.

Rodgers, G. B. 1983. *Poverty Ten Years On: Incomes and Work among the Poor of Rural Bihar*, Working Paper 130, Geneva: International Labour Organization, Population and Labour Policies Programme.

Rodgers, G. B. and Rodgers, J. 2010. *Inclusion or Exclusion on the Periphery? Rural Bihar in India's Economic Growth*, Working Paper 3, New Delhi: Institute for Human Development.

Rodrigues, L. 1998. *Rural Political Protest in Western India*, New Delhi: Oxford University Press.

Rogaly, B. *et al.* 2001. Seasonal Migration, Social Change and Migrants' Rights: Lessons from West Bengal. *Economic and Political Weekly*, 36(49): 4547–4559.

Rosenthal, D. B. 1977. *The Expansive Elite: District Politics and State Policymaking in India*, Berkeley, CA: University of California Press.

Russell, R.V. 1916. *The Tribes and Castes of the Central Provinces of India*, Vol. 4, London: Palgrave Macmillan.

Rutten, M. 1995. *Farms and Factories: Social Profile of Large Farmers and Rural Industrialists in West India*, New Delhi: Oxford University Press.

Sahu, G. B. *et al.* 2004. Credit Constraints and Distress Sales in Rural India: Evidence from Kalahandi District, Orissa. *The Journal of Peasant Studies*, 31(2): 210–241.

Sainath, P. 2009. Neo-Liberal Terrorism in India: The Largest Wave of Suicides in History. *Countepunch*, February 12. Accessed from ⟨http://agrariancrisis.in/2009/02/13/neo-liberal-terrorism-in-india-the-largest-wave-of-suicides-in-history/⟩ on 01. 11. 2016.

Sala-i-Martin, X. 2002. *The World Distribution of Income (Estimated from Individual Country Distributions)*, NBER Working Paper 8933, Cambridge: National Bureau of Economic Research.

Sanyal, K. 2007. *Rethinking Capitalist Development: Primitive Accumulation, Golernmentality and Post-Colonial Capitalism*, New Delhi: Routledge.

Sarap, K. 1991. *Interlinked Agrarian Markets in Rural India*, New Delhi: Sage Publications.

Sarkar, S. 1983. *Modern India 1885–1947*, New York: Palgrave Macmillan.

Sarma, E.A.S. 2004. Is the Rural Economy Breaking Down? Farmers' Suicides in Andhra Pradesh. *Economic and Political Weekly*, 39(28): 3087–3089.

Sawant, S. D. *et al.* 1999. *Agricultural Development in Maharashtra Problems and Prospects*, Mumbai: National Bank for Agriculture and Rural Development.

Saxena, K. B. 2011. Land Reforms: Unfinished Agenda or Reversal of Policy. In M. Mohanty (ed.) *India-Social Development Report 2010: The Land Question and the Marginalised*, New Delhi: Oxford University Press.

Schmidt, K. J. 2015. *An Atlas and Survey of South Asian History*, New York: Routledge.

Schultz, T. W. 1964. *Transforming Traditional Agriculture*, New Haven, CT: Yale University Press.

Scott, J. C. 1976. *The Moral Economy of the Peasant: Rebellion and Subsistence in Southeast Asia*, New Haven, CT: Yale University Press.

Scott, J. C. 2012. *Decoding Subaltern Politics: Ideology, Disguise, and Resistance in Agrarian Politics*, New York: Routledge.

Sekhar, T. V. 2012. Rural Demography in India. In L. J. Kulcsár (ed.) *International Handbook of Rural Demography*, New York: Springer.

Sen, A. and Himanshu 2004. Poverty and Inequality in India. *Economic and Political Weekly*, 39(38/39): 4247–4263 and 4361–4375.

Sen, A. and Himanshu 2005. Poverty and Inequality in India: Getting Closer to the Truth. Accessed from (www.networkideas.org). Reprinted in Deaton, A. and Kozel, V. (eds.) 2005. *Data and Dogma: The Great Indian Poverty Debate*, New Delhi: Palgrave Macmillan, pp. 306–370.

Shah, C. H. and Sawant, S. D. 1973. Implementation of Land Tenancy Law in Western Maharashtra. In S. H. Deshpande (ed.) *Economy of Maharashtra*, Pune: Samaj Prabodhan Sanstha.

Shah, D. 2012. *Implementation of NREGA in Maharashtra: Experiences, Challenges and Ways Forward*, MPRA Paper No. 39270. Accessed from (https://mpra.ub.uni-muenchen.de/39270/) on 10.10.2016.

Shah, M. *et al.* 2007. Rural Credit in 20th Century India: Overview of History and Perspectives. *Economic and Political Weekly*, 42(15): 1351–1364.

Sharma, S. D. 2009. *China and India in the Age of Globalization*, New York: Cambridge University Press.

Shiva, V. and Jafri, A. H. 1998. *Seeds of Suicide: The Ecological and Human Costs of Globalisation of Agriculture*, New Delhi: Research Foundation for Science, Technology and Ecology.

Shroff, S. 2003. *Building Up of an Efficient Marketing System to Obviate the Need for a Large Scale State Intervention in Maharashtra*, Agro Economic Research Centre, Pune: Gokhale Institute of Politics and Economics.

Singh, L. *et al.* 2016. *Agrarian Distress and Farmer Suicides in North India*, New Delhi: Routledge.

Sirsikar, V. M. 1995. *Politics of Modern Maharashtra*, Mumbai: Orient Longman Limited.

Smita 2008. *Distress Seasonal Migration and Its Impact on Children's Education*, Research Monograph No. 28, CREATE, Centre for International Education, University of Sussex.

Sparr, P. (ed.) 1995. *Mortgaging Women's Lives: Feminist Critiques of Structural Adjustment*, London: Zed Books.

Srivastava, R. S. 2005. *Bonded Labour in India: Its Incidence and Pattern*, Geneva: International Labour Office.

Srivastava, R. S. 2011. Labour Migration in India: Recent Trends, Patterns and Policy Issues. *The Indian Journal of Labour Economics*, 54(3): 411–456.

Srivastava, R.S. and Singh, R. 2006. Rural Wages during the 1990s: A Re-Estimation. *Economic and Political Weekly*, 41(38): 4053–4062.

Stiglitz, J.E. 2006. *Making Globalization Work*, New York: W. W. Norton & Company.

Still, C. 2014. *Dalits in Neoliberal India: Mobility or Marginalisation?* New Delhi: Routledge.

Stone, I. 1984. *Canal Irrigation in British India: Perspectives on Technological Change in a Peasant Economy*, Cambridge: Cambridge University Press.

Ganguly, A. and Talwar, S. 2003. Feminization of India's Agricultural Workforce. *Labour Education*, 2–3(131–132): 29–33.

Tambe, S. 2004. Collective Subjectivity, Democracy and Domination: The MJVA in Marathwada, India. *Current Sociology*, 52(4): 671–691.

Teerink, R. 1995. Migration and Its Impact in Khandeshi Women in Sugarcane Harvest. In L. S. Sandbergen (ed.) *Women and Seasonal Labour Migration*, New Delhi: Sage Publications, pp. 210–300.

Teltumbde, A. 1996. Impact of Economic Reforms on Dalits in India, Paper presented at the Seminar on 'Economic Reforms and Dalits in India' organised by the University of Oxford, Oxford, UK.

Thirtle, C. *et al.* 2003. The Impact of Research-Led Agricultural Productivity Growth on Poverty Reduction in Africa, Asia and Latin America. *World Development*, 31(12): 1959–1975.

Thompson, E. P. 1971. The Moral Economy of the English Crowd in the Eighteenth Century. *Past and Present*, 50(1): 76–136.

Thorner, D. 1966. Chayanov's Concept of Peasant Economy. In D. Thorner *et al.* (eds.) *A. V. Chayanov on the Theory of Peasant Economy*, Homewood, IL: Richard. D. Irwin, Inc., pp. xi–xxiii.

Thorner, D. and Thorner, A. 1962/2005. *Land and Labour in India*, Mumbai: Asia Publishing House.

Timmer, C.P. 1969. The Turnip, the New Husbandry, and the English Agricultural Revolution. *The Quarterly Journal of Economics*, 83(3): 375–395.

Timmer, P. 1988. The Agricultural Transformation. In H. Chenery and T. N. Srinivasan (eds.) *Handbook of Development Economics*, Vol. 1. Amsterdam: North-Holland, pp. 275–331.

Todaro, M. P. 1969. A Model of Labour Migration and Urban Unemployment in Less Developed Countries. *The American Economic Review*, 29(1): 138–148.

Tomaskovic-Devey, D. and Roscigno, V. J. 1997. Uneven Development and Local Inequality in the US. South: The Role of Outside Investment, Landed Elites, and Racial Dynamics. *Sociological Forum*, 12(4): 565–597.

Tomlinson, B. R. 1993. *The New Cambridge History of India (III): The Economy of Modern India, 1860–1970*, Cambridge: Cambridge University Press.

Touraine, A. 2001. *Beyond Neoliberalism*, Cambridge: Polity Press.

Tripathy, A. K. 2014. *Agricultural Prices and Production in Post-Reform India*, New Delhi: Routledge.

Tucker, V. 1999. The Myth of Development: A Critique of a Eurocentric Discourse. In R. Munk and D. O'Hearn (eds.) *Critical Development Theory Contributions to a New Paradigm*, London: Zed Books.

Usami, Y. 2011. A Note on Recent Trends in Wage Rates in Rural India. *Review of Agrarian Studies*, 1(1): 149–182.

Van der Ploeg, J.D. 2010. The Peasantries of the Twenty-First Century: The Commoditization Debate Revisited. *The Journal of Peasant Studies*, 37(1): 1–30.

Vanhaute, E. 2012. Peasants, Peasantries and (De)peasantisation in the Capitalist World-System. In S. J. Babones and C. Chase-Dunn (eds.) *Routledge Hand Book of World-Systems Analysis*, New York: Routledge.

Varshney, A. 1994. *Democracy, Development and the Countryside: Urban-Rural Struggles in India*, New York: Cambridge University Press.

Vasavi, A. R. 1999. Agrarian Distress in Bidar: Market, State and Suicides. *Economic and Political Weekly*, 34(32): 2263–2268.

Vasavi, A. R. 2012. *Shadow Space: Suicides and the Predicament of Rural India*, Harayana: Three Essays Collective.

Venkatesan, V. 2002. *Institutionalising Panchayati Raj in India*, New Delhi: Concept Publishing Company.

Venugopal, V. 2015. Neoliberalism as Concept. *Economy and Society*, 44(2): 165–187. DOI: 10.1080/03085147.2015.1013356.

Vepa, S. 2005. Feminisation of Agriculture and Marginalisation of Their Economic Stake. *Economic and Political Weekly*, 40(25): 2563–2568.

Vertika and Rodrigues, V. 2015. The Indian State and the Unfinished Task of Land Reforms: A Critical Analysis. In H. Ha (ed.) *Land and Disaster Management Strategies in Asia*, Singapore: Springer, pp. 29–48.

Vicol, M. 2015. Corporatisation of Rural Spaces: Contract Farming as Local Scale Land Grabs in Maharashtra, India, Conference Paper No. 38, An international academic conference on 'Land grabbing, conflict and agrarian-environmental transformations: Perspectives from East and Southeast Asia', 5-6 June 2015, Chiang Mai University.

Walker, K. L. M. 2008. Neoliberalism on the Ground in Rural India: Predatory Growth: Agrarian Crisis: Internal Colonisation, and the Intensification of Class Struggle. *Journal of Peasant Studies*, 35(4): 557–620.

Walker, T. S. and Ryan, J. G. 1990. *Village and Household Economies in India's Semi-Arid Tropics*, Baltimore: Johns Hopkins University Press.

Washbrook, D. 1994. The Commercialization of Agriculture in Colonial India. *Modern Asian Studies*, 28(1): 129–164.

Whitcombe, E. 1972. *Agrarian Conditions in Northern India: The United Provinces Under British Rule, 1860–1900*, Vol. 1, Berkeley, CA: University of California Press.

Whitehead, A. and Lockwood, M. 1999. Gendering Poverty: A Review of Six World Bank African Poverty Assessments. *Development and Change*, 30(3): 525–555.

Whitehead, A. and Tsikata, D. 2001. *Policy Discourses on Women's Land Rights in Sub Saharan Africa*, Geneva: UNRISD, Mimeo.

Wolf, E. R. 1969. *Peasant Wars of the Twentieth Century*, New York: Harper and Row.

World Bank 2002. *Global Economic Prospects and the Developing Countries 2002: Making Trade Work for the World's Poor*, Washington, DC: The World Bank.

World Development Report 2008. *Agriculture for Development*, Washington, DC: The World Bank.

World Development Report 2009. *Reshaping Economic Geography*, Washington, DC: The World Bank.

World Development Report 2012. *Gender Equality and Development*, Washington, DC: The World Bank.

Wright, A. and Wolford, W. 2003. *To Inherit the Earth: The Landless Movement and the Struggle for a New Brazil*, Oakland, CA: Food First Books.

Wrigley, E. A. 1988. *Continuity, Chance and Change: The Character of the Industrial Revolution in England*, Cambridge: Cambridge University Press.

Xaxa, V. 2001. Protective Discrimination: Why Scheduled Tribes Lag behind Scheduled Castes. *Economic and Political Weekly*, 36(29).

Young, B. and Hoppe, H. 2003. *The Doha Development Round, Gender and Social Reproduction*, Dialogue on Globalization, Occasional Paper No.7, Berlin: Freidrich-Ebert-Stiftung.

Young, I. M. 2011. *Responsibility for Justice*, Oxford: Oxford University Press.

Index